Reproduction and Sexuality in Marine Fishes

Reproduction and Sexuality in Marine Fishes

Patterns and Processes

Edited by

Kathleen S. Cole

UNIVERSITY OF CALIFORNIA PRESS

Berkeley   Los Angeles   London

University of California Press, one of the most distinguished university presses in the United States, enriches lives around the world by advancing scholarship in the humanities, social sciences, and natural sciences. Its activities are supported by the UC Press Foundation and by philanthropic contributions from individuals and institutions. For more information, visit www.ucpress.edu.

For a digital version of this book, see the press website.

University of California Press
Berkeley and Los Angeles, California

University of California Press, Ltd.
London, England

Library of Congress Cataloging-in-Publication Data

Reproduction and sexuality in marine fishes : patterns and processes / Kathleen S. Cole, editor.
    p. cm.
  Includes bibliographical references and index.
  ISBN 978-0-520-26433-5 (cloth : alk. paper) 1. Marine fishes—Reproduction. 2. Marine fishes—Sexual behavior.
I. Cole, Kathleen S. (Kathleen Sabina), 1950–
  QL620.R47 2010
  597.177—dc22

                2010027053

16   15   14   13   12   11   10
10 9 8 7 6 5 4 3 2 1

The paper used in this publication meets the minimum requirements of ANSI/NISO Z39.48-1992 (R 1997)(*Permanence of Paper*).

Cover illustration: Detail from "Sauvez la mer, sauvez les poissons" (Save the Sea, Save the Fish), a poster by David Lance Goines, 1998.

*This book is dedicated to the oceans and to the fishes
that live in them.*

CONTENTS

CONTRIBUTORS

KATHLEEN S. COLE
Department of Zoology
University of Hawaii at Mānoa
Honolulu, Hawaii, United States
colek@hawaii.edu

BRUCE B. COLLETTE
National Marine Fisheries Service
    Systematics Laboratory
Smithsonian Institution
Washington DC, United States
collettb@si.edu

JOHN GODWIN
Department of Biology and W. M. Keck
    Center for Behavioral Biology
North Carolina State University
Raleigh, North Carolina, United States
John_Godwin@ncsu.edu

PHILIP A. HASTINGS
Marine Biology Research Division
Scripps Institution of Oceanography
University of California, San Diego
La Jolla, California, United States
phastings@ucsd.edu

INGRID M. KAATZ
Department of Environmental and
    Forest Biology
State University of New York
College of Environmental Science and
    Forestry
Syracuse, New York, United States
ingridmkaatz1@yahoo.com

FREDERIEKE J. KROON
CSIRO Sustainable Ecosystems
Atherton, Queensland, Australia
frederieke.kroon@csiro.au

TETSUO KUWAMURA
School of International Liberal
    Studies
Chukyo University
Yagoto-honmachi, Nagoya, Japan
kuwamura@lets.chukyo-u.ac.jp

PHILLIP S. LOBEL
Department of Biology
Boston University
Boston, Massachusetts, United States
plobel@bu.edu

CARLOTTA MAZZOLDI
Department of Biology
University of Padova
Padova, Italy
carlotta.mazzoldi@unipd.it

PHILIP L. MUNDAY
ARC Centre of Excellence for Coral Reef
  Studies and School of Marine and Tropical
  Biology
James Cook University
Townsville, Queensland, Australia
philip.munday@jcu.edu.au

MARTA MUÑOZ
Àrea de Zoologia
Departament Ciències Ambientals
Universitat de Girona, Campus de Montilivi
Girona, Spain
marta.munyoz@udg.edu

JOHN A. MUSICK
Virginia Institute of Marine Science
Gloucester Point, Virginia, United
  States
jmusick@vims.edu

CHRISTOPHER W. PETERSEN
College of the Atlantic
Bar Harbor, Maine, United States
chrisp@coa.edu

AARON N. RICE
Bioacoustics Research Program
Cornell Laboratory of Ornithology
Cornell University
Ithaca, New York, United States
arice@cornell.edu

PREFACE

When Rachel Carson's *Silent Spring* was released in 1962, it raised an alarm regarding the disappearance of North American songbirds due to pesticide toxins. Everyone could identify with the loss of birds from our immediate environments and recognize the consequences of species extinction happening in their own back yards. Unfortunately, most of us are not so readily attuned to the sounds and rhythms of the oceans. Yet we are facing no less of a crisis in the current decline of marine fish populations worldwide. Ever-intensifying fishing practices, increasing climate and temperature oscillations, and significant losses of coastal habitats are all contributing to the decline, and possible disappearance, of marine fish species worldwide. But unlike Rachel Carson's songbirds, the fishes of the world's oceans have limited public appeal. While other warm-blooded animals are perceived as kindred spirits, a more muted response is generated when it comes to fish. This no doubt is due in part to the very different appearance of fishes, with their finned bodies and fixed facial expressions. In addition, most marine fish species live in an effectively hidden universe that the average individual rarely, if ever, visits or otherwise experiences. Unfortunately, for fishes occupying this invisible world, the adage "out of sight, out of mind" is particularly fitting.

This volume, however, is less about the challenges to survival that currently face marine fishes worldwide and more about the celebration of their continued existence. The chapters included here focus on biological patterns and processes that are most critical for species survival: those of reproduction and sexuality. The included authors have labored to reveal the amazing diversity and complexity that is to be found among many marine fish taxa. The chapter themes range from reproductive and sexual patterns in several major fish taxa, including pelagics, to processes associated with acoustic behavior, neuroendocrinology, fertilization, and reproductive ontogeny. Topic coverage includes the evolution of reproduction in cartilaginous fishes, the amazing reproductive diversity in scorpaeniforms, and mating systems and hermaphroditism in blennioids and gobiids, respectively. Both blennies and gobies represent some of the smallest-sized and least conspicuous of marine fishes, and these two taxa make up the majority of the cryptobenthic fish communities of coastal and reef habitats. They also constitute a critical link between bottom-end and higher trophic levels that is essential for energy cycling in these environments. Yet very little is known regarding the most basic aspects of their biology. At the other extreme, epipelagic fishes, many of which are large-sized or occur in schools, are primarily open water species and in many cases are targets for commercial exploitation. As a consequence of their offshore lifestyle, much of their biology, especially their reproductive biology, remains unknown. Nowhere is this information vacuum more critically felt than within the context of current efforts to develop effective management and conservation policies for the continued existence of these highly vulnerable fishes.

The following chapters outline what is known, and what remains to be discovered, with regard to patterns and processes of reproductive and sexual biology among a variety of marine fishes. As will become evident, our knowledge deficit is much greater than our existing understanding of the biology of coastal and oceanic fishes. This book highlights the remarkable biological diversity to be found just beyond our shores, if only we would look. Hopefully, it will encourage an increase in research efforts directed toward achieving a better understanding, both of the challenges to survival that face many marine fish species, and of our mutual dependency on meeting those challenges.

Kathleen S. Cole, editor

INTRODUCTION

Kathleen S. Cole

The study of evolutionary biology is like a detective story, complete with the classic three word mantra of "what, how, and why?" Shared biological patterns of new characters or modifications of existing characters suggest the possibility, if not probability, of a shared evolutionary history. Consequently, the "what" referred to above is an initial observation of an apparent shared character trait that leads to a hypothesis of homology-based evolutionary relatedness. The question of "how" leads naturally into investigations of formative biological processes underlying observed patterns. This frequently involves the study of molecules, patterns of gene expression and regulation, endocrine production and function, and/or ontogenetic events involved in trait or pattern expression. Lastly, new characters or modifications of existing characters do not persist in a vacuum. "Why" investigates hypotheses regarding historical biotic and abiotic conditions and consequent selection pressures that may have favored the retention of a particular trait. The combination of such what, how, and why investigations inevitably leads to increased insight into the evolutionary history of taxa and a deeper understanding of evolutionary processes. Therefore, identifying both patterns and underlying processes is central to an understanding of how taxa have evolved and why they may have evolved in diverse directions.

The contents of this volume represent a selection of studies of patterns and processes associated with reproductive biology and sexuality in marine fishes. For such a broad topic, the number of possible subjects is vast, and a comprehensive coverage would fill several encyclopedias. This book, however, represents a more modest undertaking. The collection of complementary themes related to the reproductive and sexual biology of marine fishes that is presented here is viewed by the authors as a first step, hopefully of many, toward developing a more integrated understanding of the reproductive and sexual diversity that exists among marine fishes. The selection of issues addressed herein was based on two criteria: What issues have eluded comprehensive reportage in the past? And what topics are of increasing concern in light of both documented and possible future changes to our ocean environments?

The first part of this volume is entitled Patterns. Identifying patterns is the first step in any biological investigation that seeks an increased understanding of taxon and trait evolution. Hence, the examination of reproductive and sexual patterns within and among taxa becomes increasingly informative when viewed in the context of a shared environmental or evolutionary history. The five chapters in this first section examine diverse aspects of reproductive biology, sexual biology, or both within a taxon-specific or habitat-specific context.

Reproduction in marine fishes exhibits an astonishing level of biological diversity and sophistication. Nowhere is this more evident than in the chondrichthyans, which share the character of an entirely cartilaginous skeleton and constitute one of the two major groups of extant fishes. Chondrichthyans diverged from bony fishes early in the evolutionary history of jawed vertebrates. Since this divergence, their reproductive biology has evolved along several highly specialized and diverse pathways. Chapter 1, by John Musick, examines reproductive diversity in this group from a phylogenetic perspective. In doing so, he provides a wealth of detail regarding reproductive mode and morphology and offers strong support for hypothesized ancestral reproductive states and more recently derived reproductive traits within the taxon. Here, phylogenetic patterns both inform our understanding of the origins of diverse reproductive modes in chondrichthyan fishes and provide insights into the evolutionary potential for reproductive processes among vertebrates as a whole.

Chapter 2, by Bruce Collette, switches from a taxon-centric to a habitat-centric viewpoint. Oceanic epipelagic fishes live most or all of their lives in blue water environments, far from continental coasts. Many oceanic epipelagic species are of major commercial importance and provide a significant proportion of harvested fish biomass. However, the inaccessibility of their environment has made studies of their biology—and in particular their reproductive biology—extremely difficult. Consequently, we know next to nothing regarding basic biological processes, the understanding of which is critical to successful management and conservation efforts. In this chapter, Collette has gathered together all of the available information on reproduction and development of oceanic epipelagic fishes. In doing so, he has been able to highlight the limits of our knowledge, which are considerable, and identify the necessary directions for research in the immediate future. As his chapter demonstrates, the gaps in our knowledge of the reproductive biology of oceanic epipelagic fishes are extensive. Should we wish to have a better understanding of how to proceed in our efforts to conserve both commercially important epipelagic species and, by association, the oceanic epipelagic ichthyofauna community (upon which the effects of commercially important species removal are unknown), this chapter provides essential reading.

Chapter 3, by Marta Muñoz, returns to a taxon-based perspective and offers a phylogenetic treatment for reproductive traits among the Scorpaeniformes. This taxon, as indicated in the chapter introduction, is one of the most morphologically diverse of all teleostean orders and, perhaps not surprisingly, is also one of the most reproductively diverse. Scorpaenids as a group exhibit both internal and external fertilization modes. Among internal fertilizers, offspring may be released from the maternal body at varying stages of development. Among externally spawning species in which ova are released directly into the environment prior to fertilization, the resulting embryos may be anchored, drift free, or raft with other embryos. As this chapter reveals, an impressive array of reproductive variations is exhibited within this taxon. With such a variety, the Scorpaeniformes offer a unique opportunity to investigate the evolution of diverse reproductive modes, an opportunity of which the author takes full advantage by applying a phylogenetic framework to her coverage and analysis.

Blennioid fishes comprise almost 900 described species, most of which are associated with coastal environments. Yet, as Philip Hastings and Christopher Petersen point out in chapter 4, essentially nothing has been published regarding the reproductive biology of more than 90% of blennioid species. The authors seek to redress this omission by presenting a review of spawning, mate choice, and parental care within an evolutionary context. The ecological importance of this taxon cannot be overstated. Blennioids, along with gobioids, constitute the dominant component of cryptic reef and coastal fishes and as such play a key role in coastal and reef community ecology, trophic cycling, and within-habitat energy flow. This chapter provides an informative review of their reproductive biology and in so doing focuses much-needed attention on this under-recognized group.

Gobiid fishes constitute the second largest family of vertebrates, and with more than 235 genera and over 1,400 species, they are by a considerable margin the largest marine family of fishes. As such, they comprise a dominant taxon of tropical coastal and coral reef fish communities. However, because of their small size and cryptic lifestyles, they are significantly underrepresented in both ichthyological surveys and in studies of coral reef community ecology. As a result, much remains unknown regarding their biodiversity and basic biology. Among gobiids the reproductive mode of external fertilization, oviposition of demersal eggs, and parental care until hatching is relatively simple. What makes this group unusual is the capability of some taxa to produce both ova and sperm during adult life. While this ability among marine fish taxa is not limited to gobiids, the diversity of patterns of sexual expression in this group reaches a level unmatched in other hermaphroditic fish taxa. Chapter 5, by Kathleen Cole, examines this aspect of reproductive biology among hermaphroditic goby taxa from a phylogenetic viewpoint. Patterns of gonad morphology are examined as possible clade traits in order to investigate whether hermaphroditism has multiple origins within the Gobiidae and to what extent shared patterns of gonad morphology may be predictive of phylogenetic relatedness. By investigating the origins of differing sexual patterns in this group, we can hopefully begin to understand why sexual lability, which is so rare in all other vertebrate groups, is so common among marine fishes.

The second part of this volume is entitled Processes. Reproductive processes reflect the drivers of adaptive innovations. How developmental processes may have become modified to generate variable sexual expression is the topic of Chapter 6, also by Kathleen Cole. Here, events associated with reproductive morphogenesis in teleosts are examined to determine whether diverse patterns of reproductive morphology among hermaphroditic gobiids can be explained in terms of ontogenetic modifications. This chapter reviews the chain of ontogenetic events, starting with the early formation of the germ cell line and carrying through to the formation of a sexually differentiated gonad, to evaluate various possibilities regarding the ontogenetic origins of labile sexual expression among hermaphroditic gobiid fishes. The hypotheses put forward in this chapter remain to be tested. However, they do offer a blueprint for the further exploration of regulatory mechanisms and morphogenic processes associated with sexual determination and sexual expression in gobiids and other hermaphroditic teleosts.

Living in an aquatic environment is not unique to marine fishes. However, the physical and chemical nature of marine systems differs substantially from freshwater environments, and many aspects of reproduction in marine fishes reflect this. Nowhere is this more evident than in the topic of Chapter 7. This chapter by Christopher Petersen and Carlotta Mazzoldi deals with fertilization success in marine fishes. As the authors query in their introduction, what could be more straightforward? But as their chapter reveals, processes associated with fertilization success are anything but. Whether or not sperm meets egg and becomes a zygote reflects the outcome of a delicate balance of numerous factors, including spawning mode, mating behavior, sperm quality, and environmental conditions. We live in a world where ocean environments are clearly changing in terms of temperature, acidity, CO_2, and calcium carbonate levels. At the same time, protein dependence on fishery resources is growing at an unprecedented rate, and the documentation of fishery depletions is ever increasing. Consequently, our understanding of fertilization processes in marine fishes is central to understanding the factors that influence reproductive success. Current efforts for the effective management of commercially important fish species and the conservation of noncommercial species make this a timely chapter in any discussion of reproduction in marine fishes.

Among marine fishes, the topic of reproduction would not be complete without the inclusion of a discussion of ecological and evolutionary patterns of sexuality. Fishes exhibit the greatest variety of sexual expression of all vertebrate taxa. Gonochorism, which refers to the expression of constant (i.e., fixed) sex among individuals, is nearly universal among most vertebrate taxa. Among marine fishes, however, reportage of labile sexual patterns is becoming increasingly common. Early research in this area focused on documentation of the phenomenon (e.g., Chapter 5) and the development of evolutionary models based on individual costs and benefits. Subsequently, attention turned to the internal and external regulators of sexual expression (see Chapter 6). The drivers of this research include not only scientific interest but, more recently, the documented effects of environmental factors on sexual development in fishes as well. In light of recent findings on the effects of warming waters on adult reproductive output and on the masculinization of genetic females among some fish species, a fresh look at processes associated with teleost sexuality is timely.

Chapter 8, by Philip Munday, Tetsuo Kuwamura, and Frederieke Kroon, explores the unusual phenomenon of bidirectional sex change, consisting of the alternating expression of male and female function. The initial discovery of this phenomenon in a fish species was instrumental in challenging our fundamental assumptions regarding the nature of vertebrate sexuality. The concept that behavior, brain function, endogenous endocrine levels, and reproductive system morphology could all shift not once, but numerous times, opened up new avenues of exploration for improving our understanding of vertebrate sexuality. In this chapter, taxon-specific features of reproductive biology, social systems, and environmental factors are all explored in an effort to answer the question as to why such a complex sexual pattern has evolved in some marine fish taxa.

The early findings of sequential sexual function (i.e., unidirectional sex change) among fishes, followed by the discovery of serial sexuality as described in Chapter 8, made it clear that the vertebrate model of sexual genotype determining sexual phenotype was in need of a reexamination as to how, precisely, sexuality is regulated and how external social conditions can influence that regulation. In Chapter 9, John Godwin investigates how physiological and neurobiological mechanisms in concert with social environmental cues can generate discrete sexual phenotypes within a common genotype. At the heart of this investigation is the question as to how external environments can generate differing physiological and neural responses that are responsible for altering biological processes as complex as sexual expression and function. Here, the sexual phenotype of an individual is not a singular condition. Rather, sexual phenotype is independent of any specific genotype and can have multiple expressions within a single individual. As stated in the author's introduction, identifying related regulatory mechanisms and how they respond to external cues is a hopeful approach to better our understanding of evolutionary patterns of sexual expression.

The three chapters examining various aspects of labile sexual expression that are included in Part Two collectively engage in the what, how, and why of teleost sexuality. Our increased understanding of this topic is a matter of growing urgency. Changing environmental conditions associated with direct anthropogenic disturbance and climate change are becoming increasingly associated with abnormalities in sex ratios among both freshwater and marine fishes. In a number of cases, genotypic females are masculinized by environmental conditions and irreversibly develop into phenotypic males. The possibility of increasingly unbalanced, male-biased sex ratios among numerous fish species, including commercially important ones, clearly has serious implications for both coastal ecology and human welfare.

The sensory environment of the world oceans is also changing. The anthropogenic addition of increased noise levels has as yet unknown impacts on fish reproduction. Sound travels farther and faster in water than in air. Therefore, the introduction of artificial noise can interfere substantially with reproductive behaviors that include the production and reception of auditory signals. As discussed in Chapter 10 by Phillip Lobel, Ingrid Kaatz, and Aaron Rice, sound production associated with reproductive behavior among marine fishes serves to synchronize the behavior of potential mates, thereby leading to successful fertilization. In oceanic environments, coastal waters receive the most exposure to introduced artificial noise. However, increases in shipping activities, commercial fishing, and off-shore drilling are also steadily increasing artificial noise levels in offshore waters. Acoustic interference in marine waters will only grow with time, and the effects on fish reproduction are virtually unknown. This final chapter demonstrates how sound production and acoustical communication are integral to the reproductive biology of many marine fish species, yet are frequently overlooked in reproductive studies.

The overall goal of this volume is to raise awareness of how little we know about the reproductive and sexual biology of most marine fishes, and how critical this lack of knowledge may be. We are facing a future that holds the promise of progressively declining resources within our oceans, decreasing biodiversity in marine floral and faunal communities, and the potentially irreversible depletion of much of the world's primary source of protein. For many third world countries with tropical coasts, fish collecting and fish rearing for the aquarium trade and local consumption provide important sources of sustenance and income. Yet the ways in which both commercial and noncommercial fishes constituting a major component of coastal and reef communities contribute to ecological stability in those environments are poorly understood. Our best hope for the protection, effective management, and ultimate conservation of the fishes of marine environments lies in having a comprehensive knowledge of their basic biology, and especially of their reproductive biology. The authors hope that this volume moves us closer toward that goal.

PART ONE

Patterns

Chondrichthyan Reproduction

John A. Musick

The chondrichthyan fishes have evolved separately from the Osteichthyes (Euteleostomi) since the dawn of gnathostomy more than 450 million years ago (Miller 2003; Kikugawa et al. 2004). Indeed, the chondrichthyans may be the oldest gnathostome group, perhaps having evolved from some thelodont agnathan ancestor in the Silurian (Marss et al. 2002). Whatever their origins, the Chondrichthyes and Osteichthyes underwent rapid divergent radiations during the Devonian (Miles 1967). This early divergence resulted in quite different reproductive trajectories in the two clades, probably initiated by high egg and larval predation from the newly evolved gnathostomes themselves (Musick & Ellis 2005). Osteichthyan reproductive evolution has been based on oviparity, with vulnerable ova lacking a maternally derived protective shell or case. Consequently, several adaptations have evolved multiple times to decrease egg predation (nest building, parental care, viviparity, etc.), or to maintain fitness despite predation (production of huge numbers of pelagic eggs). In contrast, chondrichthyan reproductive evolution has been based on lecithotrophic viviparity (i.e., yolk provides sole source of nutrients during development), matrotrophy (i.e., nutrients include maternally derived supplements), and in a small number of clades, oviparity with protective leathery egg cases. This chapter will review the evidence for this conclusion recently proposed by Musick & Ellis (2005).

THE CHONDRICHTHYAN REPRODUCTIVE SYSTEM

The elasmobranch reproductive system is predicated exclusively on internal fertilization. All male chondrichthyans have intromittent organs called claspers

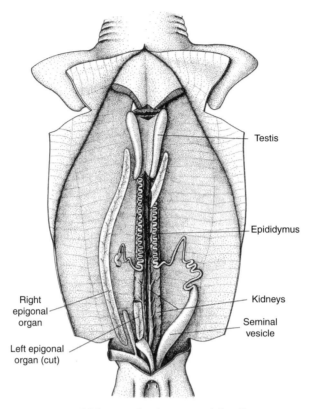

FIGURE 1.1. Male reproductive system (after Castro, 1973, Texas A&M Press by permission).

(myxopterygia), which are paired, grooved extensions of the posterior base of the pelvic fin, supported by cartilaginous endoskeletal elements (Compagno 1999).

Sperm is produced by the paired testes, then discharged into the ductus efferens and passed onto the convoluted epididymis, from which it passes on to the vas deferens and seminal vesicle (Conrath 2005)(Figure 1.1). The vas deferens and seminal vesicle function as storage areas for the semen, which may be packaged into spermatozeugmata or spermatophores (Wourms 1977). The paired seminal vesicles empty into a single urogenital sinus, which leads into a common cloaca. From there the semen enters the clasper grooves. Most male Chondrichthyes also posses siphon sacs, paired subcutaneous muscular bladders located just anterior to the base of the claspers. Each sac opens posteriorly through the apopyle and ends blindly anteriorly (Gilbert & Heath 1972). During copulation, the male usually inserts a single clasper into the female's urogenital opening and the siphon

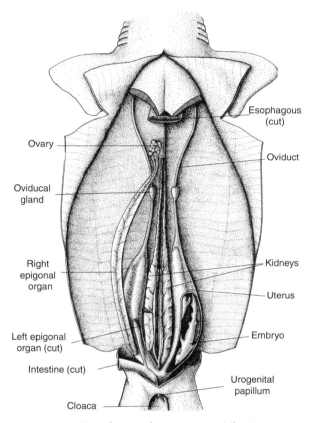

FIGURE 1.2. Female reproductive system (after Castro, 1973, Texas A&M Press by permission).

sac, which fills with seawater, functions under pressure to squirt semen into the female oviduct (Conrath 2005). The evolution of claspers has involved not only the coordinated development of the muscular siphon sac but also the muscles required to maneuver the clasper during copulation (Musick & Ellis 2005). The presence of claspers and prismatic skeletal calcification (i.e., a mineralized form of cartilage) are the two principal synapomorphies that define the Chondrichthyes (Grogan & Lund 2004). Thus, claspers and internal fertilization probably have been defining features of the group since its origin (Musick & Ellis 2005).

Prominent in both Figures 1.1 and 1.2 are the epigonal organs—long, white, strap-like bodies closely associated with the reproductive system in both sexes—however, epigonal organs are composed of myeloid tissue and are part of the immune system (Luer et al. 2004).

The female reproductive system (Figure 1.2) in Chondrichthyes is comprised of one or two ovaries, and paired oviducts into which the eggs enter through funnel-shaped ostia (Hamlett & Koob 1999). The oviducts pass into paired oviducal (= shell, = nidmental) glands. These are complex structures and histologically differentiated into four zones (Hamlett et al. 1998). Fertilization takes place in the anterior part of the oviducal glands, or just forward of them, in the oviduct. After an egg is fertilized, the oviducal gland surrounds the egg with a jelly coat and other egg investments and a tertiary egg envelope to form an egg capsule (Hamlett et al. 2005b). In many species, the oviducal gland may also store sperm in the posterior section from a few weeks to more than a year, leading to delayed fertilization after mating (Conrath 2005). The largest and most complex oviducal glands occur in oviparous species (Hamlett et al. 1998; Musick & Ellis 2005). Egg capsules pass out of the oviducal gland through the isthmus to the paired uteri. Function of the uteri varies depending on the reproductive mode of the species at hand. In yolk-sac viviparous and oophagous species, the uterus regulates the uterine environment; supplies oxygen, water, and minerals; and regulates wastes (Hamlett & Koob 1999). In other matrotrophic viviparous species, the uterus provides the above services, but it also produces nutritious mucous in limited histotrophs, and copious uterine "milk" in lipid histotrophs, and is the site of embryonic placentation in placental species (Hamlett & Hysell 1998). The uterus in oviparous species contributes to polymerization and scleratization of the egg capsule, which may be retained for several days before oviposition (Hamlett & Hysell 1998).

The uteri unite posteriorly to form a cervical and urogenital sinus (Hamlett & Koob 1999), which empties into the cloaca. In many species of viviparous sharks, only the right ovary develops, but the rest of the reproductive system is paired (Conrath 2005). Conversely, in many myliobatid rays, the entire reproductive system on the right side may be reduced.

CHONDRICHTHYAN MODES OF REPRODUCTION

Although the extant chondrichthyans are a relatively small class of vertebrates, including about 1100 species of elasmobranchs (sharks and rays) and 30+ species of holocephalans (chimaeras), they exhibit a surprising diversity of reproductive modes (Hamlett et al. 2005b).

These modes may be classified into lecithotrophic and matrotrophic based on whether fetal nutrition is supported solely by the yolk in the egg or augmented by additional maternal input of nutrients during development (Wourms 1981)(Table 1.1). Lecithotrophy includes two forms of oviparity (single and multiple) and one form of viviparity (yolk sack viviparity). Matrotrophy includes five different forms of viviparity (Wourms 1981; Hamlett et al. 2005b; Musick & Ellis 2005)(Table 1.1).

TABLE 1.1 Chondrichthyan reproductive modes

Reproductive strategies	Lecithotrophic	Matrotrophic
Oviparity		
Single	+	
Multiple	+	
Viviparity		
Yolk-sac	+	
Limited histotrophy		+
Lipid histotrophy		+
Carcharhinid oophagy		+
Lamnid oophagy		+
Placental		+

Oviparity

Oviparous chondrichthyans all deposit benthic eggs with leathery, structurally complex shells (Hamlett & Koob 1999). Chondrichthyan oviparity is limited to clades that are benthic in habit. Single oviparity, in which eggs are usually deposited on the sea floor in pairs, one from each uterus, is the only form of reproduction in the extant holocephalans. However, this group is but a small relic of a once diverse group of Mississippian chondrichthyans within which viviparity has been well documented in different taxa (Lund 1980, 1990; Grogan 1993, 2000, 2009, unpublished data). Evidence of oviparity is sparse in the Bear Gulch, the most intensively studied Mississippian fossil deposit, despite the high quality of preservation there (Grogan & Lund 2004). Within the elasmobranchs, oviparity occurs in only a small number of clades (some speciose).

Single oviparity is the sole form of reproduction in the horn sharks (Heterodontiformes), the batoid family Rajidae (skates), and in most cat sharks (Scyliorhinidae) and occurs along with various forms of viviparity in the carpet sharks (Orectolobiformes). In single oviparity, eggs are usually deposited every few days over a period of months. This results in an annual fecundity of 20 to 100 or more eggs per year in most species (Musick & Ellis 2005), an order of magnitude higher than that of viviparous elasmobranchs of similar size. Oviparity in elasmobranchs has evolved as an adaptation to increase fecundity in groups in which most members have small body size (<100cm TL) and thus limited uterine capacity (Musick & Ellis 2005). In addition, oviparity in small elasmobranchs may represent a form of "bet hedging" (Stearns 1992). Small individuals suffer a proportionally higher predation rate than do large individuals (Cortés 2004). If a pregnant viviparous shark is eaten, her immediate fitness is zero, whereas if an oviparous species is predated, her most recently produced offspring may still survive (Frisk et al. 2002; Musick & Ellis 2005). Multiple oviparity (Table 1.1) occurs in a small number of Scyliorhinidae and represents an evolutionary reversal. In this reproductive mode, females retain developing eggs in the uterus for most of the developmental period, then deposit them before they hatch (Nakaya 1975). This obviously limits the fecundity and probably has evolved in response to very high egg predation rates (Musick & Ellis 2005). The same may be said about a small number of scyliorhinids in the terminal sub-tribe Galeini, which have reverted to yolk-sac viviparity (Musick & Ellis 2005).

Lecithotrophic Viviparity

Yolk-sac viviparity is the simplest form of viviparity, wherein the developing eggs are retained within the uterus until parturition and fetal nutrition is supplied solely by the yolk and thus is lecithotrophic (Hamlett et al. 2005b). This form of reproduction is basal and most widespread in elasmobranchs and is present in all extant orders except the Heterodontiformes, which are oviparous, and the Lamniformes, which have a more advanced form of viviparity (oophagy)(Musick & Ellis 2005)(Figure 1.3). Yolk-sac viviparity occurs in many species formally classified as "ovoviviparous." The term ovoviviparous was abandoned because some of the species so classified actually exhibited a limited form of matrotrophy (Ranzi 1934; Budker 1958; Hoar 1969). Subsequently, "ovoviviparity" was replaced by the term "aplacental viviparity," which included three major modes of elasmobranch reproduction (yolk-sac viviparity, histotrophy, and oophagy), thus obscuring the true reproductive diversity in the group. In addition to being based on a negative attribute (lack of a placenta), by inference, the term elevated the relative importance of placental viviparity, a mode of reproduction restricted to a small number of terminal nodes within the Carcharhiniformes (Musick & Ellis 2005). The term "aplacental viviparity" should be abandoned and replaced with "yolk-sac viviparity," "histotrophy," or "oophagy" as appropriate.

Matrotrophic Viviparity

Mucoid (Limited) Histotrophy.   Mucoid histotrophy is the simplest form of matrotrophic viviparity wherein developing embryos receive additional nutrients above those supplied in the yolk (Hamlett et al. 2005b; Musick & Ellis 2005) by ingesting mucus produced by the uterus. This form of matrotrophy may be insidious and difficult to detect without obtaining ash-free dry weights from newly fertilized ova to compare with those of full -term embryos (Ranzi 1934; Needham 1942, Hamlett et al. 2005b). During embryogenesis, nutrients are expended to support the metabolic requirements for embryonic maintenance, growth, and development. Thus, in truly lecithotrophic species, more than a 20 percent reduction of ash-free dry weight should occur during development from egg to term embryo (Hamlett et al. 2005b). Ranzi (1932, 1934) noted early on that although some lecithotrophic species of Torpediniformes and Squaliformes lost 23 to 46 percent organic content during development, other squaliforms and some Triakidae supposed to be lecithotrophic actually gained 1 to 369 percent in organic content. Evidence for mucoid histotrophy may also be provided by histological examination of the uterine walls, which should exhibit high mucus secretory activity at least during early and midterm development (Hamlett et al. 2005b). Mucoid histotrophy appears to be widespread among viviparous groups, and further research is needed to determine the frequency of this reproductive mode (Hamlett et al. 2005b).

Lipid Histotrophy.   Lipid histrophy is restricted to the myliobatiform stingrays. This reproductive mode involves the secretion of a lipid-rich histotroph from highly developed secretory structures called trophonemata located in the uterine lining. Embryos supported by lipid histotrophy may undergo an increase in organic content of 1680 to 4900 percent (Needham 1942).

Oophagy.   Oophagy is a form of matrotrophic viviparity where embryonic development is supplemented by the mother's production of unfertilized eggs, which are ingested by the embryo. (Musick & Ellis 2005). This nominal mode of reproduction has evolved twice among elasmobranchs: in the lamniforms, and in the small carcharhiniform family, Pseudotriakidae. The mechanics of oophagy in these two groups differ and are not homologous (Musick & Ellis 2005). Oophagy is the only mode of reproduction known in the lamniforms, where large numbers of unfertilized eggs are produced by the mother and ingested by the embryos during most of the pregnancy (Gilmore et al. 2005). Adelphophagy (intrauterine cannibalism), where the first embryo that develops in each uterus attacks and eats its developing siblings, is an extension of oophagy and is known to occur in only one species, the sand tiger (*Carcharias taurus*). After the embryos have eaten their siblings, subsequent development in this species is supported through oophagy, as in all other Lamniformes (Gilmore et al. 2005). Adelphophagy results in the birth of only two large (>1m TL) neonates, one from each uterus.

In the carcharhiniform Pseudotriakidae, a number of unfertilized eggs are included within the egg envelope with the embryo, which then ingests the eggs during development. No further unfertilized eggs are produced to support the developing embryos above those included in the egg envelope, but the Pseudotriakidae may also be limited histotrophs (Yano 1992, 1993).

Placental Viviparity.   Placental viviparity is present in five higher families within the Carcharhiniformes. The "placenta" in elasmobranchs is analogous, but not homologous, to that in mammals and has been termed a yolk-sac placenta (Hamlett et al. 2005b). In elasmobranchs, the yolk sac forms the attachment with the uterine epithelium and the yolk stalk elongates to form an umbilical cord. The developing embryos are maintained in separate uterine compartments. All placental sharks utilize yolk stores from the egg for initial development, and then mucoid histotrophy before, and for some species, even during placentation (Hamlett 1989; Hamlett et al. 2005b).

ELASMOBRANCH PHYLOGENY

The Neoselachii are a monophyletic sub-class that includes all living elasmobranchs as well as some extinct Mesozoic forms and possibly, a small number of Paleozoic fossils (Maisey et al. 2007). Historical classifications of modern elasmobranchs have recognized two major clades: the Batoidei and the Selachii (Bigelow & Schroder 1948, 1953). This classification was radically changed in the 1990s following morphological cladistic analyses that placed the Batoidei as a terminal group within the squalomorph sharks in a new clade, the Hypnosqualea (Shirai 1992, 1996; de Carvalho 1996), an arrangement that was in conflict with the paleontological data. The earliest known batoids had separated from and were concurrent with the earliest heterodontiform, hexanchiform, and orectolobiform sharks by the Jurassic if not earlier (Thies 1983; Capetta 1987; Maisey et al. 2007), contradicting the batoid terminal position in the cladistic analysis. This contradiction was resolved by more recent molecular and paleontological analyses (Douady et al. 2003; Naylor et al. 2005; Maisey et al. 2007), which clearly showed the Batoidei to be the sister group to the Selachii. Cladistic misclassification of the Batoidei based on morphology may have been mitigated by homoplasies shared by the benthic, dorsoventrally flattened batoids and the squalean Squantiniformes (Nelson, 2006).

PHYLOGENETIC REPRODUCTIVE PATTERNS

Six major modes of elasmobranch reproduction have been mapped on the most recently accepted cladogram for neoselachians (Figure 1.3)(Musick & Ellis 2005; Naylor et al. 2005; Nelson 2006). This figure provides considerable insight into the evolutionary polarity of the various modes of reproduction. Among the 10 major clades within the Batoidei, oviparity has evolved in only one family, Rajidae, which is clearly terminal and derived from yolk-sac viviparous ancestors (McEachran & Aschliman 2004). Lipid histotrophy has evolved in the myliobatid stingrays. The mode of reproduction found in all the batoid basal clades is yolk-sac viviparity (Musick & Ellis 2005).

Within the Selachii two superorders, the Squalomorphii and the Galeomorphii, have been recognized. Oviparity is absent in the Squalomorphii and yolk-sac viviparity is the plesiomorphic mode of reproduction in all five orders (Figure 1.3).

Limited histotrophy is a derived state present in some squalomorphs, and more research is required to determine the extent of this cryptic, derived mode of fetal nutrition (Hamlett et al 2005b; Musick & Ellis 2005).

Reproductive patterns among the Galeomorphii are more diverse and complex (Musick & Ellis 2005). This superorder has been divided into two major, distantly-related clades, the Heterodontoidea and Galeoidea (Musick & Ellis 2005; Naylor et al. 2005; Maisey et al. 2007)(Figure 1.3). The Heterodontoidea includes only one relict order (Heterodontiformes) limited to one family with a handful of species each restricted to a specific coastal region of the world. The extant heterodontoids are all oviparous with complex, unique corkscrew-shaped egg capsules and specialized oviducal glands (Hamlett et al. 1998).

Among the Galeoidea, the Orectolobiformes are basal and have been divided into two suborders: the Parascylloidei, and Orectoloboidei (Goto 2001; Maisey et al. 2007). Whereas the Parascylloidei is comprised of one family of small, benthic, oviparous sharks (Compagno 2001), Orectoloboidei includes two superfamilies which contain six families, four of which are yolk-sac viviparous, and two of which are oviparous (the small benthic speciose Hemiscyliidae and the larger monotypic Stegostomatidae). The two oldest families, Brachaeluridae (lower Jurassic, 180mya) and Orectolobidae (middle Jurassic, 160mya)(Capetta et al. 1993) are both in the superfamily Orectoloboideia, and both have yolk-sac viviparity (Musick & Ellis 2005).

The order Lamniformes, as far as known, is uniformly oophagous (including adelphophagy in *C. taurus*)(Figure 1.3). The earliest stages of development in lamniformes are supported by the yolk sac, and yolk-sac viviparity appears to have been a necessary ancestral precursor to oophagy.

The Carcharhiniformes have usually been subdivided into two suborders: the Scyliorhinoidei and the Carcharhinoidei (Compagno 1988, 1999) (Figure 1.4). The Scyliorhinoidei, as presently recognized (Musick & Ellis 2005; Maisey et al. 2007), includes two families: the Scyliorhinidae and the Proscylliidae. The Scyliorhinidae, a speciose, benthic group of small, bathyal sharks, was considered to be the most primitive group of Carcharhiniformes (White 1937). This conclusion was based on their attenuate body and caudal fin, the posteriorly placed dorsals and poor vertebral calcifications. However, an attenuate body and tail, with posteriorly placed dorsals is typical of many benthic sharks (including the Orectolobiformes)(Compagno 1988, 1999), and reduced vertebral calcification is found in most bathyal sharks including many of the squalomorphs. Further, Compagno (1988) suggested that given that lamnoids and carcharhinoids are sister groups, the proscylliid body form with a high, forward-placed first dorsal, may be primitive for the Carcharhiniformes, with the scyliorhinids derived.

The proscylliids are currently comprised of three genera: *Ctenacis* and *Eridacnis,* which have yolk-sac viviparity, and *Proscyllium,* which is oviparous. *Ctenacis*

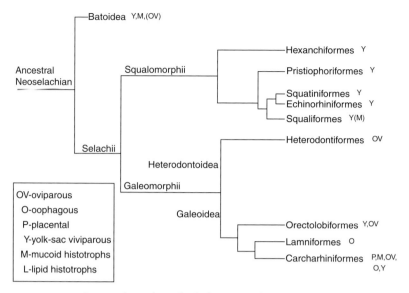

FIGURE 1.3. Recent elasmobranch phylogeny, with reproductive modes mapped for major clades.

and *Eridacnis* are more closely related to each other than either is to *Proscyllium*, and the latter is closest to the Scyliorhinidae, sharing characteristics with *Schroederichthys* (Compagno 1988). Thus, the position of *Proscyllium* is equivocal as it may be the primitive sister group to the Scyliorhinidae (Figure 1.4). Regardless, *Ctenacis* and *Eridacnis* comprise a yolk-sac viviparous clade that is most likely the primitive sister group to the rest of the Carcharhiniformes (Musick & Ellis 2005). In addition, the sister group relationship between Lamniformes (which has no oviparous clades) and Carcharhiniformes would dictate a viviparous group to be basal in the latter. Other than the scyliorhinids, which are mostly oviparous, with a reversal to yolk-sac viviparity in a small number of species in the subtribe Galeini, all the other families of Carcharhiniformes are viviparous: the Pseudotriakidae are oophagous and probably have mucoid histotrophy (Yano 1992, 1993); the monotypic Leptochariidae has a form of placental viviparity; the Triakidae have placental viviparity and/or mucoid histotrophy, the Hemigaleidae are placental, and all of the Carcharhinidae save one, the tiger shark (*Galeocerdo cuvieri*), are placental. *Galeocerdo*, considered to be the most primitive member of the family, is yolk-sac viviparous.

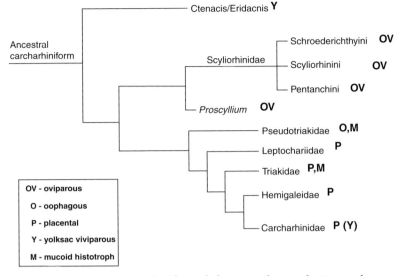

FIGURE 1.4. Recent carcharhiniform phylogeny with reproductive modes mapped. Note equivocal position of *Proscylium*.

YOLK-SAC VIVIPARITY: THE PLESIOMORPHIC STATE

The hypothesis that yolk-sac viviparity is unequivocally the plesiomorphic reproductive state in neoselachians, and plausibly in the Chondrichthyes, is supported by multiple sources of evidence:

Female reproductive system. The simplest, least specialized oviducal glands and uteri are found in species with yolk-sac viviparity (Hamlett et al. 2005a; Musick & Ellis 2005).

Male reproductive system. All male Chondrichthyes, both recent and fossil (so far as known) possess claspers (Grogan & Lund 2004; Musick & Ellis 2005). These have no other use than internal fertilization during copulation. Internal fertilization along with the presence of well-developed uteri in females provides the potential, if not the probability, for viviparity.

Urea retention. Chondrichthyans retain urea in marine environments as the principal mechanism to maintain osmotic equilibrium (Smith 1953). Urea is produced through the ornothine-urea cycle, which is found in all gnathostome classes except the Aves, in which it has been lost (Griffith 1991). The ornothine-urea cycle probably evolved in early gnathostomes as a means of detoxifying ammonia produced by catabolism in internally developing embryos (Griffith 1991). Thus, viviparity must have evolved quite early in the evolution of chondrichthyans or their ancestors. Conversion of ammonia to urea would not be necessary in oviparous marine forms because ammonia is rapidly diluted and carried away in seawater. (In recent oviparous chondrichthyans with horny egg capsules, pores in the leathery shell and embryonic movements produce seawater flushing through the egg case.) In early chondrichthyans, post-embryonic urea retention probably evolved through paedomorphosis, and allowed more efficient osmoregulation and greater euryhalinity than in their isosmotic, stenohaline ancestors (Musick & Ellis 2005).

Parsimony. Earlier studies that have been based on oviparity as the plesiomorphic reproductive state in modern elasmobranchs are less parsimonious than one based on plesiomorphous yolk-sac viviparity. In the oviparity "camp," Wourms & Lombardi (1992) estimated that viviparity evolved from oviparity 18 to 20 times. Dulvy & Reynolds (1997) estimated there were 9 to 10 transitions from oviparity to viviparity. Blackburn (2005) invoked 15 transitions from oviparity to viviparity among recent elasmobranchs. These three and related studies were based in part on older elasmobranch classifications and either assumed that oviparity was plesiomorphous, or used an oviparous out-group, the holocephalans, in their cladistic analyses. (Recent paleontological evidence suggests that the Paleozoic basal group from which both the Holocephalii and Elasmobranchii evolved was viviparous [Grogan & Lund unpublished data]). Thus the possibility that viviparity might be plesiomorphous was never considered or adequately tested. In contrast, hypothesizing yolk-sac viviparity as the plesiomorphic state, and analyzing the most recently accepted elasmobranch phylogeny (Naylor et al. 2005; Nelson 2006; Maisey et al. 2007), Musick & Ellis (2005) found six transitions from viviparity to oviparity (Table 1.2). Thus, with yolk-sac viviparity as ancestral there are 3 to14 fewer transitions.

Phylogeny. Yolk sac viviparity is unequivocally the plesiomorphous reproductive state in the Batoidea, and in the selachien Squalomorphii. The latter is

TABLE 1.2 Parsimony and the plesiomorphic reproductive state

State			
Plesiomorphic →	*Apomorphic*	Transitions	Source
Oviparity	Viviparity	18–20	Wourms 1992
Oviparity	Viviparity	9–10	Dulvy & Reynolds 1997
Oviparity	Viviparity	15	Blackburn 2005
Viviparity	Oviparity	6	Musick & Ellis 2005

the sister group to the Galeomorphii, in which reproductive patterns in the most primitive clades are not so well defined; Heterodontiformes are oviparous, and Orectolobiformes includes both oviparous and viviparous clades. However, in the latter order the oldest families (Brachaeluridae, Orectolobidae) are yolk sac viviparous. Thus the preponderance of evidence supports the hypothesis that yolk sac viviparity is plesiomorphous in neoselachians.

Paleontology. Viviparity has been well documented in Paleozoic chondrichthyans, whereas evidence of oviparity is sparse (Lund 1980, 1990; Grogan 1993; Grogan & Lund 2000, 2009, submitted). In addition, viviparity has recently been documented among the Devonian Placodermi, the Paleozoic sister group to the Chondrichthyes, and perhaps all the other gnathostome groups (Long et al. 2008, 2009).

Plesiomorphic yolk-sac viviparity simplifies the pattern of reproductive evolution in living elasmobranchs and forms the unspecialized basis for all other modes of elasmobranch reproduction. Early chondrichthyans possessed intromittent organs, internal fertilization, viviparity, and precocial young. Thus, their eggs and developing embryos had a large measure of protection from newly evolving gnathostome egg predators. In contrast, the Actinopterygii with vulnerable unprotected eggs and small larvae, evolved adaptations to decrease egg and larval predation (egg hiding, nest building, parental protection), or to increase fitness by producing huge numbers of small pelagic eggs (Balon 1975). Such adaptations have been superfluous for the chondrichthyans.

ACKNOWLEDGMENTS

This is VIMS Contribution 3014. Thanks are due to José Castro and the Texas A&M Press for use of Figures 1.1 and 1.2 (modified). Eileen Grogan, Dick Lund, and Nick Dulvy contributed to stimulating conversations leading to the ideas presented in this paper. "Ain't science fun?" (S. P. Applegate pers. comm.)

REFERENCES

Balon EK (1975). Reproductive guilds of fishes: A proposal and definition. Journal of Fisheries Research Board of Canada 32: 821–864.

Bigelow HB, Schroeder WC (1948). Sharks. In: Tee-Van J, Breder CM, Parr AE, Schroeder WC, and Schultz LP (eds.), *Fishes of the Western North Atlantic.* Memoirs of the Sears Foundation for Marine Research, Yale University, New Haven, Connecticut, pp. 59–546.

Bigelow HB, Schroeder WC (1953). *Fishes of the Western North Atlantic, Part II, Sawfishes, Guitarfishes, Skates and Rays; Chimaeroids.* Memoirs of the Sears Foundation for Marine Research, Yale University, New Haven, Connecticut.

Blackburn DG (2005). Evolutionary origins of viviparity in fishes. In: Grier HJ, Uribe MC (eds.), *Viviparous Fishes*. New Life Publications, Homestead, Florida, pp. 287–301.

Budker P (1958). La viviparité chez les sélaciens. In: Grasse P (ed.), *Traité de Zoologie*. Masson, Paris, pp. 1755–1790,

Capetta H (1987). Chondrichthyes II: Mesozoic and Cenozoic Elasmobranchii, vol. 3, In: Schultze HP (ed.), *Handbook of Paleoichthyology*. Verlag Friedrich Pfeil, Munich, Germany.

Capetta H, Duffin CJ, Zidek J (1993). Chondrichthyes. In: Benton MJ (ed.), *The Fossil Record*. Chapman and Hall, London, pp. 539–609.

Castro J (1973). *Sharks of North American Waters*. Texas A&M Press, College Station, Texas.

Compagno LJV (1988). *Sharks of the Order Carcharhiniformes*. Princeton University Press, Princeton, New Jersey.

Compagno LJV (1999). Systematics and body form. In: Hamlett WC (ed.), *Sharks, Skates, and Rays: The Biology of Elasmobranch Fishes*. Johns Hopkins University Press, Baltimore and London, pp. 1–42.

Compagno LJV (2001). Sharks of the World: An Annotated and Illustrated Catalogue of Shark Species Known to Date, Vol. 2. Bullhead, Mackerel and Carpet Sharks (Heterodontiformes, Lamniformes and Orectolobiformes). FAO Species Catalogue for Fishery Purposes 1(2), FAO, Rome.

Conrath CL (2005). Reproductive biology. In: Musick JA and Bonfil R (eds.), Management Techniques for Elasmobranch Fisheries. FAO Fisheries Technical Paper 474, pp. 103–127.

Cortés E (2004). Life-history patterns, demography, and population dynamics. In: Carrier J, Musick J and Heithaus M (eds.), *The Biology of Sharks and Their Relatives*. CRC Press, Boca Raton, Florida, pp. 449–470.

de Carvalho MR (1996). Higher level elasmobranch phylogeny, basal squaleans, and paraphyly. In: Stassny MLJ, Parenti LR, Johnson DG (eds.), *Interrelationships in Fishes*. Academic Press, London, pp. 35–62.

Douady CJ, Dosay M., Shivji MS, Stanhope MJ (2003). Molecular phylogenetic evidence refuting the hypothesis of Batoidea (rays and skates) as derived sharks. Molecular Phylogenetics and Evolution 26: 215–221.

Dulvy NK, Reynolds JD (1997). Evolutionary transitions among egg-laying, live-bearing, and maternal inputs in sharks and rays. Proceedings of the Royal Society B. 264(1386): 1309–1315.

Frisk MG, Miller TJ, Fogarty MJ (2002). The population dynamics of little skate *Leucoraja erinacea*, winter skate *Leucoraja ocellata*, and barndoor skate *Dipturus laevis*: predicting exploitation limits using matrix analyses. ICES Journal of Marine Science 59: 576–586.

Gilbert PW, Heath GW (1972). The clasper-siphon sac mechanism in *Squalus acanthias* and *Mustelus canis*. Comparative Biochemistry and Physiology 42a: 97–119.

Gilmore RG, Putz O, Dodrill JW (2005). Oophagy, intrauterine cannibalism and reproductive strategy in Lamnoid sharks. In: Hamlett WC (ed.), *Reproductive Biology and Phylogeny of Chondrichthyes*. Science Publishers, Inc., Enfield, New Hampshire, pp. 435–462.

Goto T (2001). Comparative anatomy, phylogeny and cladistic classification of the order Orectolobiformes (Chondrichthyes, Elasmobranchii). Memoirs of the Graduate School of Fisheries Sciences, Hokkaido University 48(1): 1–100.

Griffith RW (1991). Guppies, toadfish, lungfish, coelacanths and frogs: A scenario for the evolution of urea retention in fishes. In: Musick JA, Bruton MN, Balon EK (eds.), The biology of *Latimeria chalumnae* and the evolution of Coelacanths. Environmental Biology of Fishes 32: 1–4, The Hague, pp. 199–218.

Grogan ED (1993). The structure of the holocephalan head and the relationships of the Chondrichthyes. Ph.D. dissertation, School of Marine Science, College of William and Mary, Williamsburg, Virginia.

Grogan ED, Lund R (2000). *Debeerius ellefseni* (Fam. Nov., Gen. Nov., Spec. Nov.), an autodiastylic chondrichthyan from the Mississippian Bear Gulch Limestone of Montana (USA), the relationships of the Chondrichthyes, and comments on gnathostome evolution. Journal of Morphology 243(3): 219–245.

Grogan ED, Lund R (2004). Origin and relationships of early Chondrichthyes. In: Carrier JC, Musick JA, Heithaus MR (eds.), *Biology of Sharks and Their Relatives*. CRC Press, Boca Raton, Florida, pp. 3–31.

Grogan ED, Lund R (2009). Live birth and superfetation in a 318 million year old Carboniferous chondrichthyan. Joint Meeting of Ichthyologists and Herpetologists, Portland, Oregon, 22–27 July 2009 (Abst.)

Hamlett WC (1989). Evolution and morphogenesis of the placenta in sharks. In: Hamlett WC, Tota B (eds.), Eighth International Symposium on Morphological Sciences, Rome, Italy. Journal of Experimental Zoology, Suppl. 2: 35–52.

Hamlett WC, Hysell MK (1998). Uterine specializations in elasmobranchs. Journal of Experimental Zoology 282(4–5): 438–459.

Hamlett WC, Koob T (1999). Female reproductive system. In: Hamlett WC (ed.), *Sharks, Skates, and Rays: The Biology of Elasmobranch Fishes*. Johns Hopkins University Press, Baltimore and London, pp. 398–443.

Hamlett WC, Knight DP, Koob TJ, Jezior M, Loung T, Rozycki T, Brunette N, Hysell MK (1998). Survey of oviducal gland structure and function in elasmobranchs. Journal of Experimental Zoology 282: 399–420.

Hamlett WC, Knight DP, Pereira FTV, Steele J, Sever DM (2005a). Oviducal glands in Chondrichthyans. In: Hamlett WC (ed.), *Reproductive Biology and Phylogeny of Chondrichthyes*. Science Publishers, Inc., Enfield, New Hampshire, pp. 301–336.

Hamlett WC, Kormarik CG, Storrie M, Serevy B, Walker TI (2005b). Chondrichthyan parity, lecithotrophy and matrotrophy. In: Hamlett WC (ed.), *Reproductive Biology and Phylogeny of Chondrichthyes*. Science Publishers Inc., Enfield, New Hampshire, pp. 395–434.

Hoar WS (1969). Reproduction. In: Hoar WS, Randall DJ (eds.), *Fish Physiology, Volume III, Reproduction and Growth; Bioluminescence, Pigments and Poisons*. Academic Press, New York and London, pp. 1–72.

Kikugawa K, Katoh K, Kuraku S, Sakurai H, Ishida O, Iwabe N, Miyata T (2004). Basal jawed vertebrate phylogeny inferred from multiple nuclear SNA-coded genes. BioMed Central Biology 2: 3.

Long JA, Trinajstic K, Young GC, Senden T (2008). Live birth in the Devonian period. Nature 453: 650–652.

Long JA, Trinajstic K, Johanson Z (2009). Devonian arthrodire embryos and the origin of internal fertilization in the vertebrates. Nature 457: 1124–1127

Luer C, Walsh CJ, Bodine AB (2004). The immune system in sharks and rays. In: Carrier JC, Musick JA, Heithaus MR (eds.), *Biology of Sharks and Their Relatives*. CRC Press, Boca Raton, Florida, pp. 369–398.

Lund R (1980). Viviparity and intrauterine feeding in a new holocephalan fish from the Lower Carboniferous of Montana. Science 209: 697–699.

Lund R (1990). Chondrichthyan life history styles as revealed by the 320 million year old Mississippian of Montana. Environmental Biology of Fishes 27(1): 1–19.

Maisey JG, Naylor GJP, Ward D (2007). Mesozoic elasmobranchs, neoselachian phylogeny, and the rise of modern neoselachian diversity, In: Arratria G, Tintori A (eds.), *Mesozoic Fishes III. Systematics, Paleoenvironments and Biodiversity*. Verlag Pfeil, Munich, Germany, pp. 17–56.

Marss T, Wilson MVH, Thorsteinsson R (2002). New thelodont (Agnatha) and possible chondrichthyan (Gnathostomata) taxa established in the Silurian and lower Devonian of the Canadian Arctic Archipelago. Proceedings of the Estonian Academy of Science Geology 51(2): 88–120.

McEachran JD, Aschliman N (2004). Phylogeny of Batoidea. In: Carrier JC, Musick JA, Heithaus MR (eds.), *Biology of Sharks and Their Relatives*. CRC Press, Boca Raton, Florida, pp. 79–114.

Miles RS (1967). Observations on the ptyctodont fish, *Rhamphodopsis* Watson. Journal of the Linnean Society of London, Zoology 47: 99–120.

Miller RF (2003). The oldest articulated chondrichthyan from the Early Devonian period. Nature 425: 501–504.

Musick JA, Ellis J (2005). Reproductive evolution of chondrichthyans. In: Hamlett WC (ed.), *Reproductive Biology and Phylogeny of Chondrichthyans*. Science Publishers Inc., Enfield, New Hampshire, pp. 45–79.

Nakaya K (1975). Taxonomy, comparative anatomy and phylogeny of Japanese catsharks, Scyliorhinidae. Memoirs of the Faculty of Fisheries, Hokkaido University 23(1): 1–94.

Naylor GJP, Ryburn JA, Fedrigo O, Lopez JA (2005). Phylogenetic relationships among the major lineages of modern elasmobranchs. In: Hamlett WC (ed.), *Reproductive Biology and Phylogeny of Chondrichthyans*. Science Publishers Inc., Enfield, New Hampshire, pp. 1–44.

Needham J (1942). *Biochemistry and Morphogenesis*. Cambridge University Press, Cambridge.

Nelson J (2006). *Fishes of the World*. 4th ed. John Wiley & Sons, Hoboken, New Jersey.

Ranzi S (1932). Le basi fisio-morfologische dello sviluppo embrionale dei Selaci—Parti I. Pubblicazioni Della Stazione Zoologíca di Napoli 13: 209–240.

Ranzi S (1934). Le basi fisio-morfologische dello sviluppo embrionale dei Selaci—Parti II and III. Pubblicazioni Della Stazione Zoologíca di Napoli 13: 331–437.

Shirai S (1992). *Squalean Phylogeny. A New Framework of "Squaloid" Sharks and Related Taxa*. Hokkaido University Press, Sapporo.

Shirai S (1996). Phylogenetic interrelationships of neoselachians (Chondrichthyes, Euselachii). In: Stiassny MLG, Parenti LR, Johnson GD (eds.), *Interrelationships of Fishes*. Academic Press, San Diego, London, pp. 9–34.

Smith HW (1953). *From Fish to Philosopher*. Little, Brown and Co., Boston.

Stearns SC (1992). *The Evolution of Life Histories*. Oxford University Press, Oxford.

Thies, D (1983). Jurazeitlicher Neoselachier aus Deutschland und S. England. Courier Forshungsinstitut Senckenberg 58: 1–116.

White EG (1937). Interrelationships of the elasmobranchs with a key to the order Galea. Bulletin of the American Museum of Natural History 74: 25–138.

Wourms JP (1977). Reproduction and development in chondrichthyan fishes. American Zoologist 17: 379–410.

Wourms JP (1981). Viviparity: The maternal-fetal relationship in fishes. American Zoologist 21(2): 473–515

Wourms JP, Lombardi J (1992). Reflections on the evolution of piscine viviparity. American Zoologist 32: 276–293.

Wourms JP, Grove BD, Lombardi J (1988). The maternal-embryonic relationship in viviparous fishes. In: Hoar WS, Randall DJ (eds.), *Fish Physiology*, vol. 2. Academic Press, San Diego, pp. 1–134.

Yano K (1992). Comments on the reproductive mode of the false cat shark *Pseudotriakis microdon*. Copeia 1992(2): 460–468.

Yano K (1993). Reproductive biology of the slender smoothhound, *Gollum attenuata*, collected from New Zealand waters. Environmental Biology of Fishes 38: 59–71.

2

Reproduction and Development in Epipelagic Fishes

Bruce B. Collette

The marine pelagic environment is the largest realm on Planet Ocean, constituting 99% of the biosphere volume and supplying about 80% of the fish consumed by humans (Angel 1993; Game et al. 2009). The epipelagic or holoepipelagic region of Parin (1970) is a thin upper fraction of the pelagic realm. Epipelagic fishes spend all or almost all of their lives in the open ocean, mostly above the thermocline, usually in the upper 20 to 30 meters, although many species move deeper to feed. This treatment of the reproductive biology and development of epipelagic fishes excludes sharks and deeper-dwelling mesopelagic fishes. It also excludes Sargassum-associated species such as *Histrio histrio* (Antennariidae), *Syngnathus pelagicus* (Syngnathidae), and *Xanthichthys ringens* (Balistidae), which are representatives of benthic or inshore families that secondarily become members of the Sargassum complex. I have selected representatives of true oceanic epipelagic fishes from groups of fishes with which I am most familiar. Those treated in this chapter belong to four orders: (1) Lampriformes (Lampridae, opahs); (2) Beloniformes (Scomberesocidae, sauries; Belonidae, needlefishes; Hemiramphidae, halfbeaks; and Exocoetidae, flyingfishes); (3) three suborders of the Perciformes: Xiphioidei (Xiphiidae, swordfish; and Istiophoridae, billfishes), Percoidei (Echeneidae, remoras), and Scombroidei (Scombridae, tunas and mackerels); and (4) Tetraodontiformes (Molidae, ocean sunfishes). Other groups of epipelagic fishes, such as the Rainbow Runner (*Elagatis bipinnulata)* of the Carangidae and many species of the suborder Stromateoidei, are omitted due to lack of time and space.

Relatively little is known about the reproduction of most oceanic epipelagic fishes, particularly on aspects of reproduction that rely on direct field observations of courtship and mating (Iversen et al. 1970), although there are observations on captive Pacific Bonito and Yellowfin Tuna (Magnuson & Prescott 1966; Margulies et al. 2007). Selected species accounts will summarize available information on (see Table 2.1): distribution; maximum size; all-tackle game fish record (International Game Fish Association [IGFA] 2009); longevity; sexual dimorphism; size at first maturity; size at 50% maturity (when available); spawning location, season, and temperature; migrations; breeding habits; fecundity; egg characteristics; and sources of larval illustrations followed by comments on fishery importance and threat status of the species using the Red List categories of the International Union for the Conservation of Nature (IUCN 2009). Problems facing epipelagic fishes include finding a mate in the right place and at the right time of year for successful reproduction and having mechanisms for keeping eggs near the surface where young fish can find adequate food supplies.

Several species of tunas and billfishes are highly migratory, tolerating a wide thermal range while feeding but returning to warmer waters for spawning (Boyce et al. 2008). This is particularly true of the three species of bluefin tunas. Most epipelagic fishes show little external differentiation between the sexes, although female billfishes are usually larger than males, and male dolphinfishes and tunas are larger than females. Epipelagic fishes practice external fertilization, and functional hermaphroditism is typically absent (Sadovy & Liu 2008). This chapter will concentrate on a sample of oceanic epipelagic fishes and will not treat coastal pelagic species such as most jacks (Carangidae), bonitos (*Sarda*), mackerels (*Scomber* and *Rastrelliger*), Spanish mackerels (*Scomberomorus*), or inshore needlefishes (*Strongylura*) and halfbeaks (*Hyporhamphus*).

ORDER LAMPRIFORMES

The Lampriformes contains seven families and about 21 species of highly modified epipelagic fishes characterized by a unique type of protrusible upper jaw (Olney 1984).

Family Lampridae

Lampris guttatus (Brünnich 1788), Opah.   Found worldwide in tropical and temperate waters and replaced in the Southern Hemisphere by *L. immaculata* Gilchrist 1904. Maximum size at least 185 cm TL and 220 to 275 kg, commonly to 120 cm. IGFA all-tackle record 73.9 kg. Spawning behavior unknown. Fecundity of a 963-mm FL female from Puerto Rico estimated as 7.2 to 9.7 million eggs. A 119-cm FL running ripe female caught in February in the South Pacific contained large eggs, 2.3 mm in diameter with no oil globules. Larvae illustrated by Olney (1984, 2005). Opah are taken incidentally by tuna longliners and are an excellent food fish. See Bane 1965; Klawe 1966; Olney 1984, 2005.

ORDER BELONIFORMES

The Beloniformes (or Synentognathi) is an order of atherinomorph fishes containing two suborders, six families, 37 genera, and at least 230 species (Rosen & Parenti 1981; Collette et al. 1984a; Collette 2004). Species of one suborder, Adrianichthyoidei, inhabit Asian fresh and/or brackish waters. Most species of the five families of Exocoetoidei are tropical epipelagic marine fishes, but the internally fertilizing Zenarchopteridae and most genera of Belonidae and Hemiramphidae are restricted to coastal marine waters or freshwater.

Development has long been of interest in beloniform fishes (Schlesinger 1909; Nichols & Breder 1928; Collette et al. 1984a; Lovejoy 2000; Collette 2003; Lovejoy et al. 2004). Most beloniform fishes produce large spherical eggs with attaching filaments, characters they share with other atherinomorph fishes (Rosen & Parenti 1981). Freshwater and estuarine genera of halfbeaks in the Asian family Zenarchopteridae practice internal fertilization and three genera are viviparous (Meisner & Collette 1999). Adrianichthyid eggs are the smallest (1.0 to 1.5 mm in diameter), followed by exocoetids (generally 1.5 to 2.0 mm), Hemiramphidae (typically 1.5 to 2.5 mm), Scomberesocidae (slightly elliptical, 1.5 to 2.5 mm), and belonid eggs, which are generally the largest (most 3.4 mm). The eggs typically have a homogeneous yolk and a relatively small perivitelline space. The incubation period is relatively long in exocoetoids (Kovalevskaya 1982). Belonids hatch at the largest sizes (6.8 to 14.4 mm), followed by halfbeaks (4.8 to 11 mm), sauries (at least as small as 6.0 to 8.5 mm), flyingfishes (3.5 to 6.1 mm), and adrianichthyids (3.5 to 4.5 mm)(Collette et al. 1984a). Fin formation generally begins during the embryonic stage or soon after hatching. Caudal, dorsal, and anal fins generally form first, followed by the pectorals, and lastly, the pelvics (except in exocoetids, where the pectorals form last). During post-embryonic development, exocoetoids undergo complex changes in barbel development, beak length, melanistic dorsal fin lobe, body bars, and pelvic fin pigmentation, and these features are important both in phylogeny and identification of species.

Family Scomberesocidae (Sauries)

The Scomberesocidae is the sister group of the Belonidae (discussed next) and is defined by the series of 4 to 7 finlets behind the dorsal and anal fins. This family and the Belonidae form the superfamily Scomberesocoidea based on two derived characters: presence of a premaxillary canal and upper jaw at least slightly elongate (Collette et al. 1984a). Maximum size of the two large species, *Cololabis saira* and *Scomberesox saurus*, 400 to 762 mm SL respectively; maximum size of the two dwarf species, *C. adocoetus*, 126 mm and *S. simulans*, 68 mm.

The four species of sauries are either placed in four monotypic genera (Hubbs & Wisner 1980): *Scomberesox* and its dwarf derivative *Nanichthys*, and *Cololabis*

TABLE 2.1 Summary of some reproductive information for selected epipelagic fishes

Family/Species	Max. size[a]	Longevity (years)	Sex. dimorph.[b]	Size at first maturity	Size at 50% maturity	Spawning time/ season	Spawning temp. (°C)
Lampridae							
Lampris guttatus	105 cm FL	?	?	?	?	?	?
Scomberesocidae							
Cololabis saira	400 mm SL	2	no	?	?	all	?
C. adocoetus	126 mm SL	<2?	no	?	?	?	?
Scomberesox saurus	762 mm SL	2?	?	?	?	?	?
Belonidae							
Ablennes hians	70 cm SL	?	no	?	?	?	?
Platybelone argalus	38 cm SL	?	no	?	?	May–June	?
Tylosurus acus	129 cm SL	?	no	?	?	?	?
T. crocodilus	105 cm SL	?	no	?	?	May–June	?
Hemiramphidae							
Euleptorhamphus velox	310 mm SL	?	no	?	?	March	?
Hemiramphus balao	280 mm SL	?	no	?	?	spring–summer	?
H. brasiliensis	350 mm SL	?	no	?	?	spring–summer	?
Oxyporhamphus micropterus	185 mm SL	?	no	120 mm SL	?	?	?
Exocoetidae							
Exocoetus obtusirostris	196 mm SL	1	no	140–150 mm SL	?	?	?
E. volitans	189 mm SL	1+	no	140 mm SL	?	all	?
Coryphaenidae							
Coryphaena hippurus	200 cm FL	4	male> female		458 mm FL	all	?
C. equiselis	75 cm FL	?	male> female		?	?	>21

Spawning migration	Fecundity[c]	Batch spawning	Egg diameter (mm)	Oil globule	Filaments	Fishery importance[d]	Threat status[e]
?	7.2–4.7M	?	23	no	no	*	LC
?	?	?	1.5–1.8	no	yes	**	LC
?	?	?	2	no	no	no	LC
?	?	?	2.3–2.5	no	no	*	LC
?	660	?	3.0–3.5	no	yes	+	LC
?	2100	?	>1.5–1.8	?	yes	+	LC
?	?	?	3.2–4.0	?	yes	+	LC
?	25–31T	?	4.0–4.6	?	yes	+	LC
?	?	?	?	?	yes?	no	LC
?	3700	yes	1.6	?	yes	+	LC
?	1200	yes	2.4	?	yes	+	LC
?	?	?	1.8–2.1	no	short	no	LC
?	10,300	yes	?	no	no		LC
?	?	yes	1.7–3.0	no?	no	no	LC
?	58T–1.5M	yes	1.3	one	no	***	LC
?	?	yes?	1.35	?	no	+	LC

(continued)

TABLE 2.1 (*continued*)

Family/Species	Max. size[a]	Longevity (years)	Sex. dimorph.[b]	Size at first maturity	Size at 50% maturity	Spawning time/ season	Spawning temp. (°C)
Xiphiidae							
Xiphias gladius	445 cm TL	9+	female > male	70–100 cm	?	?	?
Istiophoridae							
Istiophorus platypterus	340 cm TL	11+	female > male	162 cm	?	all	
Istiompax indica	448 cm TL	?	female > male	140–160 LJFL	?	all	27–28
Kajikia albida	280 cm TL	12+	female > male	130–140 LJFL	?	spring	20–29
K. audax	350 cm TL	12+	female > male	140–160 LJFL	?	spring– summer	27.5– 31.9
Makaira nigricans	440 cm TL	17	female > male	130–140 LJFL	?	Mar– Nov	20–29
Tetrapturus angustirostris	200 cm TL	?	female > male	?	?	winter	25
T. belone	240 cm BL	?	?	?	?	winter– spring	?
T. georgii	160 cm BL	?	?	?	?	?	?
T. pfluegeri	200 cm BL	4	no	?	?	?	?
Scombridae							
Acanthocybium solandri	210 cm FL	9	female > male	?	93–102 cm FL	May– Oct	?
Allothunnus fallai	96 cm FL	?	?	71.5 cm FL	?	summer	?
Auxis rochei	53 cm FL	5	no	35–37 cm FL	24 cm	all	24+
A. thazard	58 cm FL	4	no?	29–35 cm FL	30– 34 cm	all	24+
Gasterochisma melampus	195 cm FL	?	?	?	?	?	?
Katsuwonus pelamis	108 cm FL	8	?	40–45 cm FL	?	Apr– Nov	24+
Thunnus alalunga	127 cm FL	13	no	85– 97 cm	?	all	24+
T. albacares	>200 cm FL	6+	no	50–60 cm FL	62–92 cm FL	all	24+

Spawning migration	Fecundity[c]	Batch spawning	Egg diameter (mm)	Oil globule	Filaments	Fishery importance[d]	Threat status[e]
yes	2–5M	?	1.6–1.8	one	no	***	NT
	1–19.5M	yes	1.3	one	no	**	VU
yes	~40M	?	?	?	no?	**	NT
yes	?	no?	?	?	no?	**	VU
?	11–29M	?	~1.0	?	no?	**	VU
yes	?	?	~1.0	?	no?	**	VU
?	?	?	1.3–1.6	?	no?	*	
?	?	?	1.48	?	no?	*	LC
?	?	?	?	?	no?	*	LC
?	?	?	?	?	no?	*	LC
?	6M	yes	0.8	?	no?	**	LC
?	?	?	?	?	no?	*	LC
?	31–148T	yes	0.82–0.88	one	no?	*	LC
?	1.37M	?	0.84–0.92	one	no?	*	LC
?	?	?	?	?	no?	*	LC
no	80T–2M	yes	0.80–1.17	one	no?	***	LC
yes	2–3M	yes	0.84–0.94	one	no	***	VU
little	163T–8M	yes	0.90–1.04	one	no	***	LC

(continued)

TABLE 2.1 (*continued*)

Family/Species	Max. size[a]	Longevity (years)	Sex. dimorph.[b]	Size at first maturity	Size at 50% maturity	Spawning time/ season	Spawning temp. (°C)
Scombridae (continued)							
T. atlanticus	110 cm FL	8	male> female	48–52 cm FL	50–52 cm FL	Apr– Nov	24+?
T. maccoyii	225 cm FL	20+	no	120 cm FL	152 cm	summer	24+
T. obesus	>200 cm FL	16+	no	80–102 cm FL	102–135 cm	all	24–30
T. orientalis	300 cm FL	16	no	150 cm FL	?	Apr– July	24+
T. thynnus	>300 cm FL	35+	?	115–200 cm FL	?	?	?
T. tonggol	130 cm FL	5+	?	?	39.6 mm	Jan–Apr Aug–Sept	?
Luvaridae							
Luvarus imperialis	200 cm	?	?	?	?	spring– summer	?
Molidae							
Masturus lanceolatus	259 cm TL	?	?	?	?	?	?
Mola mola	368 cm TL	?	?	?	?	Aug–Oct	?
Ranzania laevis	200 cm TL	?	?	?	?	?	?

NOTE: A solitary question mark indicates unknown; information followed by a question mark indicates that the information is a good inference based on data for closely related species but there is no direct support for it.

[a] Size: FL = fork length; SL = standard length; LJFL = lower jaw fork length (for billfishes); BL = body length, end of opercle to base of tail.

[b] Sex. dimorphism: "No?" and "yes?" indicate the author's opinions based on phylogeny.

[c] Fecundity: T = thousand; M = million.

[d] Fishery importance: + indicates slight, artisanal fishing; increasing numbers of * indicates increasing importance of the fishery.

[e] Threat status: Red List categories LC = least concern; NT = near threatened; VU = vulnerable; EN = endangered; CR = critically endangered.

Spawning migration	Fecundity[c]	Batch spawning	Egg diameter (mm)	Oil globule	Filaments	Fishery importance[d]	Threat status[e]
?	?	?	?	?	no?	*	LC
yes	14–15M	yes	0.66–1.05	?	no	***	EN
yes	2.9–6.3M	yes	1.03–1.08	one	no?	***	NT
yes	5–25M	?	?	?	?	***	NT
?	?	?	1.00–1.12	one	no	***	EN
?	1.2–1.9M	?	?	one	no?	*	LC
?	47.5M	?	?	?	no?	*	LC
?	?	?	1.8	40	no	?	LC
?	300M	yes	?	multiple?	?	?	LC
?	?	?	1.42–1.65	20–30	no	?	LC

and its dwarf derivative *Elassichthys*; or in two genera, considering *Nanichthys* a synonym of *Scomberesox* and *Elassichthys* a synonym of *Cololabis* (Collette et al. 1984a). Both species of *Scomberesox* develop an elongate beak; the snout increases in length in *S. simulans* throughout its life span and in *S. saurus* until a length of about 200 mm SL. The two dwarf species, *Cololabis adocoetus* and *Scomberesox simulans*, differ convergently from the two larger species, *C. saira* and *S. saurus*, in being much smaller, losing one ovary and the swimbladder, and in having fewer vertebrae, branchiostegal rays, pectoral fin rays, and gill rakers. *Scomberesox* inhabits all three oceans; *Cololabis* is restricted to the Pacific Ocean. As a group, sauries spawn large oval eggs that contain an unpigmented yolk.

Cololabis adocoetus (Böhlke 1951).   Eastern Pacific. Maximum size 126 mm SL. Spawning period unknown. Only one ovary. Eggs within ovary 2 mm in diameter lack filaments and sculpturing. See Orton 1964; Hubbs & Wisner 1980.

Cololabis saira (Brevoort 1856), Pacific Saury.   Restricted to the North Pacific Ocean north of about 20°N. Maximum size about 400 mm SL. Longevity not more than two years. Both ovaries present. Spawn year-round with a spring peak in California Current Region. Egg slightly ovoid, 1.5 to 1.8 mm in diameter, no oil globule. Cluster of 12 to 20 filaments at one end of the egg, plus one thicker filament at the other end. Hatching at about 5 to 7 mm, flexion before hatching. Illustrations of eggs and larvae provided by Mukhacheva (1960) and Watson (1996). Of great commercial importance, particularly in Japan. See Mukhacheva 1960; Orton 1964; Hubbs & Wisner 1980; Watson 1996.

Scomberesox saurus, Atlantic Saury.   Antitropical in temperate parts of the Atlantic, Pacific, and Indian oceans. Hubbs & Wisner (1980) recognized two subspecies with the nominal subspecies, *scomberesox saurus saurus*, in the Northern Hemisphere and *S. saurus scombroides* (Richardson 1843) in the Southern Hemisphere. This saury spends most of its life in warm homogeneous surface layers of the open sea, far from shallow continental shelf waters. Sauries are one of the most abundant epipelagic planktivores inhabiting the open part of the Atlantic Ocean, feeding mainly on siphonophores, copepods, euphausiids, amphipods, fish eggs and larvae, protozoans, algae, and larvae of polychaetes, decapods, isopods, ostracods, cirripeds, and siphonophores. Sauries serve as food for many inhabitants of the sea, such as squids, Swordfish, marlins, sharks, tunas, dolphins, whales, and birds. The great abundance of sauries and their wide distribution make them an important link in the epipelagic food chain of the ocean by transferring energy from lower to higher trophic levels. Maximum size 762 mm SL. They spawn offshore between the 26.5°C isotherm in the north and the 23.5° isotherm in the south. Both ovaries developed. Greatest diameter of eggs, 2.32 to 2.52 mm, oil globules absent. Yolk clear and without any vesicles. The long filaments characteristic of *Cololabis saira* and most other beloniform eggs are absent with only numerous, uniformly spaced, short rigid bristles remaining. Incubation 14 to 16 days depending on water temperature. Hatching at 6.8 to 8.5 mm, flexion at about 6.4 mm. Complete development of fins and finlets at about 25 mm. Sauries are valuable food fishes in some parts of the world such as the Mediterranean. See Orton 1964; Nesterov & Shiganova 1976; Hubbs & Wisner 1980; Boehlert 1984; Hardy & Collette 2005.

Family Belonidae (Needlefishes)

Belonidae, the needlefishes, is the sister group of the Scomberesocidae (sauries). The family contains 10 genera and 34 species (Collette 2003), of which two genera, the monotypic *Ablennes*, and *Tylosurus* with six species, contain oceanic epipelagic species. The other genera contain either freshwater or coastal epipelagic species.

Needlefishes are oviparous and the eggs are released to the external environment prior to fertilization. There is a tendency for the right gonad to be reduced in length or even lost in some species, particularly in females. Needlefishes deposit large eggs with well-developed chorionic filaments that attach to vegetation. The filaments are typically long, numerous, and uniformly spaced over the chorion. Needlefish eggs are generally larger (2.3 to 4.3 mm in diameter) than other beloniform eggs (Collette et al. 1984a: table 90). Correlated with large egg size, belonids hatch at the largest sizes (6.8 to 14.4 mm) among beloniforms.

During post-embryonic development, needlefishes, like other beloniform fishes, undergo a number of complex changes in beak length, melanistic dorsal fin lobe, and body bars. Most species of Belonidae pass through a "halfbeak" stage in which the lower jaw, but not the upper jaw, is greatly elongated. Juveniles of *Belone belone* remain in the halfbeak stage for a longer time than other needlefishes. *Petalichthys* and *Platybelone* also remain in the halfbeak stage for a long time. Comparative development of *Platybelone* (as *Strongylura longleyi*), *Strongylura marina*, *S. notata*, and two species of *Tylosurus* (*T. acus* and *T. crocodilus*) has been illustrated by Breder (1932: figs. 7 and 10, plates 1 and 2). *Tylosurus crocodilus* lacks a halfbeak stage, with upper and lower jaws growing at the same rate from larval to adult stages of development (Breder 1932: plate 2, fig. 2, as *T. raphidoma*).

Ablennes and *Tylosurus* share a prominent, enlarged, melanistic lobe in the posterior part of the dorsal fin. Other genera of needlefishes lack any trace of this posterior dorsal lobe. Breder (1932: plates 3–5) illustrated the development of this posterior lobe in *T. acus* and *T. crocodilus* and its absence in *Strongylura* and *Platybelone*. Juveniles of two species of *Tylosurus*, *T. gavialoides* and *T. acus* (see Collette & Parin 1970: fig. 12), and *Ablennes hians* have bars. These bars are retained in adult *Ablennes* as is the melanistic posterior dorsal fin lobe.

Ablennes hians (Valenciennes 1846), Flat Needlefish.   Worldwide in tropical and subtropical waters. Maximum size 70 cm SL, 63 cm body length. IGFA all-tackle record 4.8 kg. Spawning season unknown. Oviparous; only the right gonad developed. A 278 mm female had 660 eggs, 3.0 to 3.5 mm in diameter, oil globules absent. Uniformly spaced tufts of filaments on chorion, 1 to 6 per tuft, 37 to 59 total, filaments longer than diameter of egg. The enlarged black dorsal fin lobe characteristic of juveniles is retained in adult *Ablennes*. Illustrations of juveniles 12.3 to 187 mm are in Collette (2005a). Harvested by artisanal fisheries in some countries such as India and considered a game fish by IGFA. See Chen 1988; Watson 1996; Collette 2005a.

Platybelone argalus (LeSueur 1821), Keeltail Needlefish.   Worldwide in tropical and subtropical waters with seven subspecies recognized (Collette 2003). Particularly abundant around islands. Maximum size 38.2 cm SL, 25.6 cm body length. Ripe females taken in June in the Caribbean Sea. Oviparous, right ovary longer than left. A 266 mm BL female had 944 eggs in the left ovary, 1,136 in the right. Juveniles remain in the halfbeak stage longer than most other species of needlefishes, at least to 100 mm BL. Of minor importance in artisanal fisheries. See Erdman 1977; Chen 1988; Collette 2005a.

Tylosurus acus (Lacepède 1803), Agujon Needlefish.   Worldwide in tropical and subtropical seas within the 23.9°C isotherm, except replaced in the eastern Pacific by *T. pacificus* (Steindachner 1876). Maximum size 128.5 cm SL, 95 cm body length. Oviparous, left gonad absent or greatly reduced in size, ratio of left gonad length to right 2.3 to 15.5+. Ovarian egg counts in two females, 485 and 500 mm BL (body length), 116 (2.3 to 3.0 mm in diameter) and 196 (1.0 to 1.2) in the left ovary, 1,676 (2.5 to 2.9) and 12,017 (0.9 to 1.3) in the right for totals of 1,792 and 12,313 eggs. Diameter of fertilized eggs 3.2 to 4.0 mm. Egg with uniformly spaced tufts of 2 to 3 filaments that are longer than egg diameter. Incubation about 10 to 12 days at 25.0 to 30.4°C. Hatching at 10.2 mm, flexed at hatching. Dorsal and anal fins develop prior to hatching at 168 hours post-fertilization, pectoral fin rays by hatching, and pelvic fin rays at 14 mm TL. The enlarged melanistic posterior dorsal fin lobe forms at about 23 mm BL, reaches maximum development from 169 to 244 mm, is still evident up to 605 mm, and is then resorbed. Flesh is of good quality so harvested in many tropical countries and also considered a game fish by IGFA. See Breder & Rasquin 1954; Mito 1958; Collette 2005a.

Tylosurus crocodilus (Peron & LeSueur 1821), Hound Needlefish.   Worldwide in tropical and subtropical waters, the nominal subspecies, *Tylosurus crocodilus crocodilus,* in the Atlantic and Indo-West Pacific, *T. c. fodiator* Jordan and Gilbert in the eastern Pacific. Maximum size 101.3 cm SL, 71.5 cm body length. IGFA all-tackle record 3.74 kg. Spawning in May through June in Brazil, but ripe females found in the Caribbean in October through November. Both ovaries developed but the right is longer than the left, ratio of right to left 1.1 to 1.5. An 860 mm BL (body length) female had 7,535 eggs in left ovary (3.6 to 4.4 mm in diameter), 23,721 in the right (3.9 to 4.8 mm), for a total of 31,256. Diameter of eggs 4.0 to 4.6 mm. Egg with numerous long, fine, transparent thread-like filaments. Incubation 8 to 10 days. Hatching at 10.7 to 12.0 mm, flexed at hatching. The enlarged melanistic posterior dorsal fin lobe forms at 25 to 30 mm, reaches maximum development at 100 to 200 mm, and begins to disintegrate at 200 to 250 mm. Flesh is of good quality so harvested in many tropical countries and also considered a game fish by IGFA. See Breder & Rasquin 1952; Randall 1960; Masurekar 1968; Erdman 1977; Collette 2005a.

Family Hemiramphidae (Halfbeaks)

Hemiramphidae, the halfbeaks, is the sister group of the Exocoetidae, the flyingfishes, which together form the superfamily Exocoetoidae (Collette et al. 1984a). Most halfbeaks have an elongate lower jaw that distinguishes them from the flyingfishes, which lack an elongate lower jaw, and from the needlefishes (Belonidae) and sauries (Scomberesocidae), which have both jaws elongate.

The Hemiramphidae contains nine genera and subgenera and at least 63 species and subspecies (Collette 2004). Three genera contain epipelagic oceanic species: *Oxyporhamphus* (two species), *Euleptorhamphus* (two species), and *Hemiramphus* (10 species). The family Zenarchopteridae contains five genera and 54 sexually dimorphic Indo-West Pacific estuarine or freshwater species that were previously included in the Hemiramphidae.

Halfbeak eggs are typically 1.5 to 2.5 mm in diameter and have attaching filaments, although these are greatly reduced in length in the pelagic eggs of *Oxyporhamphus*. Halfbeaks hatch at a size (4.8 to 11 mm) smaller than needlefishes but larger than flyingfishes and sauries (Collette et al. 1984a). Larvae are well developed at hatching with partially to fully pigmented eyes, an open mouth, fully flexed notochord, developing rays in the dorsal, anal, and caudal fins, and a small to moderate sized yolk sac (Watson 1996). A preanal fin fold is typically present throughout much of the larval stage. Like other beloniform fishes, post-embryonic development of halfbeaks includes complex changes in beak length, melanistic dorsal fin lobe, body bars, and pelvic fin pigmentation. Adults of four genera lack the elongate lower jaw that characterizes most members of this family, but juveniles of all four genera have a distinct beak. Juveniles of *Hemiramphus* and *Oxyporhamphus* develop a darkened posterior lobe on the dorsal fin similar to that present in two genera of needlefishes, *Ablennes* and *Tylosurus*. The ten species of *Hemiramphus* have a series of broad vertical bars on the body during some stages of their development. Body bars are retained for different periods of time during development: all body bars are lost before 120 mm SL in *He. bermudensis* and *He.*

brasiliensis but are retained past 175 mm in *He. balao*, and all are retained throughout life in the Indo-Pacific *He. far*. Species of *Hemiramphus* also have pigmented pelvic and caudal fins as juveniles. Patterns of pelvic fin pigmentation divide the genus into two species groups, one with pigment concentrated proximally on the fin (*He. balao* group), and the other with pigment absent basally and concentrated distally (*He. brasiliensis* group, including *He. bermudensis*). As indicated below, much less information is available on reproduction compared to post-fertilization stages, among open ocean halfbeaks.

Halfbeaks are valued food fishes in many parts of the world such as Australia and New Zealand (Collette 1974). They are utilized for food in the West Indies and South America but are not presently considered an important resource in the United States. Their value to man in the western Atlantic is largely as forage and bait for a wide variety of important food and game species such as tunas, Spanish mackerels, billfishes, dolphinfishes, Bluefish, and also sea birds. The most direct use of halfbeaks in the western North Atlantic is as bait for some of the game species mentioned above (Berkeley et al. 1975; McBride et al. 1996).

Euleptorhamphus velox, Flying Halfbeak.   *Euleptorhamphus velox* Poey 1868 in the Atlantic and *E. viridis* (van Hasselt 1823) in the Pacific. Maximum size 310 mm SL. Ripe females found in the Caribbean in March. Reproductive biology of both species is poorly known. Illustrations of larvae from 8.6 to 135 mm SL are in Collette (2005b). Neither species is of any fisheries interest. See Erdman 1977; Chen 1988; Watson 1996; Collette 2005b.

Oxyporhamphus micropterus (Valenciennes 1847), Smallwing Flyingfish.   The nominal subspecies, *Oxyporhamphus micropterus micropterus,* is widespread in tropical and subtropical waters of the Indo-Pacific and is replaced in the Atlantic by *O. micropterus similis* Bruun 1935. Maximum size 185 mm SL. A second species, *O. convexus* (Weber & de Beaufort 1922) is found in the Indo-Pacific. Females are ripe at about 120 mm SL. Ripe females have been found in March, August, and November in the Caribbean Sea, suggesting year round spawning. Egg diameter 1.8 to 2.1 mm, no oil globules. The chorion is decorated with 74 to 120, very short (0.08 to 0.12 mm), filaments, unlike the condition in most other beloniforms. Hatching at 4.0 mm SL, flexion before hatching. A lower jaw beak occupies more than 20% SL at lengths of 20 to 40 mm SL, decreases rapidly to less than 5% of SL until about 50 mm SL, and disappears by about 100 mm SL. Juveniles develop a melanistic lobe in the posterior part of the dorsal fin at about 38 mm SL which remains until at least 70 mm SL. Larvae of *O. micropterus micropterus* from 3.4 to 66.0 mm SL have been illustrated by Watson (1996), of *O. micropterus similis* 4.2 to 146.0 mm by Collette (2005b). Neither species is of any fisheries interest. See Bruun 1935; Breder 1938; Khrapkova-Kovalevskaya 1963; Erdman 1977; Watson 1996; Collette 2005b.

Hemiramphus.   The most complete reproductive information for any of the 10 species of *Hemiramphus* is for two Atlantic species, *He. balao* (LeSueur 1823) Balao Halfbeak and *He. brasiliensis* Linnaeus 1758, Ballyhoo Halfbeak. Both species are widespread in tropical and subtropical waters of the Atlantic and are replaced by other species of the genus in the Indo-Pacific. Maximum size 280 to 350 mm SL and 400 mm TL, respectively. Cyclic patterns of gonadosomatic indices indicate that both species spawn during spring and summer months in south Florida. Hydration of oocytes begins in the morning and spawning occurs at dusk the same day. All mature females spawn daily, in June for *He. balao*, in April for *He. brasiliensis*. Batch fecundity in *He. balao* averages 3,734 hydrated oocytes in a 100 g female compared to 1,164 in *He. brasiliensis*. Spawning in *He. brasiliensis* occurs inshore, all along the coral reef tract of the Atlantic Ocean; spawning in *He. balao* is over deeper, more offshore waters. Illustrations of larvae of *He. balao* from 5.2 to 117 mm SL and of eggs and juveniles of *He. brasiliensis* 13.5 to 119 mm SL are in Collette (2005b). Illustrations of larvae of the eastern Pacific *He. saltator* (Gilbert & Starks 1904) from 5.2 to 14.1 mm are in Watson (1996). Valued food fishes in the West Indies and South America but more important as baitfish in Florida. Indo-West Pacific species of *Hemiramphus* are also valued food fishes. See Berkeley & Houde 1978; Watson 1996; McBride et al. 2003; McBride & Thurman 2003; Collette 2005b.

Family Exocoetidae (Flyingfishes)

Exocoetidae contains seven genera and about 50 species divided into four subfamilies (Collette et al. 1984a). Flyingfishes are a significant component of the epipelagic food chain. They feed on small zooplankton, predominantly copepods and chaetognaths. They are eaten by a large variety of predatory fishes, such as dolphinfishes, tunas, snake mackerels, and also by omastrephid squids, seabirds, and dolphins. Flyingfishes are an important fishery resource in many parts of the world (Oxenford et al. 1995). In the southeastern Caribbean, the catch of the gillnet flyingfish fishery is almost entirely the Four-wing Flyingfish *Hirundichthys affinis* (Khokiattiwong et al. 2000), so more information is available about the biology of this species than of the more oceanic *Exocoetus*.

The Exocoetinae contains the single genus *Exocoetus*, the two-wing flyingfishes, which have only the pectoral fins enlarged compared to the more advanced four-wing flyingfishes, which have both pectoral and pelvic fins enlarged. *Exocoetus* is the most oceanic genus of flyingfishes and is the only genus treated here. Five species of *Exocoetus* are recognized (Parin & Shakhovskoy 2000): circumtropical *E. volitans*, Indo-West Pacific *E. monocirrhus* Richardson, Atlantic *E. obtusirostris*, *E. gibbosus* Parin & Shakhovskoy from the southern subtropical Pacific, and *E. peruvianus* Parin & Shakhovskoy from off Peru and Ecuador in the eastern Pacific. The eggs of all five species are pelagic, without the filaments characteristic of other flyingfishes and most beloniform fishes. Juveniles either lack barbels or have a single chin barbel (*E. monocirrhus*).

Exocoetus obtusirostris (Günther 1866), Oceanic Two-Wing Flyingfish.   Widely distributed on both sides of the Atlantic, between 40°N and 40°S, including the Gulf of Mexico and Caribbean Sea in the western Atlantic and between 30°N and 30°S in the eastern Atlantic at water temperatures varying from 17.6–29.2°C. Maximum known size 196 mm SL. Females and males attain sexual maturity at about 140–150 mm SL and all females larger than 170 mm SL are in spawning condition during the year. Otoliths of mature fishes, including the largest specimens, constantly show one opaque and one hyaline zone, indicating that longevity may be about one year and all individuals die after the first reproductive season. Spawning is intermittent with up to 20 batches laid in five or more days; each batch consisting of 420 to 890 (mean 630) eggs. Estimated total fecundity averages 10,300 eggs. See Breder 1938; Cotten & Comyns 2005.

Exocoetus volitans (Linnaeus 1758), Tropical Two-wing Flyingfish.   The most abundant flying fish in offshore tropical waters of all oceans at 20.0 to 29.0°C. In the Atlantic Ocean common between 30 to 35°N and 25 to 30°S in the west and between 20 to 28°N and 20 to 25°S in the east. Maximum known size 189 mm SL. As with *E. obtusirostris*, all specimens of *E. volitans* below 140 mm SL are immature and all above 170 mm SL are ripe. The maximum age is 1+ year. Spawning is year round in the Caribbean Sea. Spawning intermittent, each batch numbering 327 to 418 (mean 370) eggs. Egg diameter 1.7 to 3.0 mm, no oil globule. Hatch at 3.5 to 4 mm, flexion before hatching. Illustrations of larvae 3.7 to 26.4 mm are in Watson (1996). See Erdman 1977; Watson 1996; Cotten & Comyns 2005.

ORDER PERCIFORMES

This is the largest order of fishes, containing 20 suborders, 160 families and more than 10,000 species (Nelson 2006). Epipelagic species from three suborders and five families are treated here.

Family Coryphaenidae (Dolphinfishes)

There are two cosmopolitan species of *Coryphaena* (Gibbs & Collette 1959): the Common Dolphinfish, *C. hippurus*; and the Pompano Dolphinfish, *C. equiselis*. Adult males develop a bony crest on front of head in both species but more dramatically in *C. hippurus*. Dolphinfishes are epipelagic, inhabiting open waters, but also approach the coast and follow ships. They show a high affinity for floating objects. Both feed mainly on fishes, but also on crustaceans and squids. Both species spawn in the open sea, probably approaching the coast as water temperatures rise.

Caught by trolling and on tuna longlines; also occasionally with purse seines. Highly appreciated food fishes.

Coryphaena equiselis (Linnaeus 1758), Pompano Dolphinfish.   Maximum size 75 cm FL, commonly to 50 cm FL. IGFA all-tackle record 3.86 kg. Much smaller and less important to fisheries than *C. hippurus*. Age at first maturity 3 to 4 months. Spawning is probably year-round at water temperatures greater than 21° C. Egg diameter 1.35 mm. Length at flexion 7.5 to 9.0 mm SL. Length at transformation 25 to 30 mm SL. Juveniles illustrated by Gibbs & Collette (1959), Ditty et al. (1994), and Ditty (2005). See Gibbs & Collette 1959; Palko et al. 1982; Ditty et al. 1994; Ditty 2005.

Coryphaena hippurus (Linnaeus 1758), Common Dolphinfish.   Maximum size 200 cm FL, commonly to 100 cm FL. IGFA all-tackle record 39.46 kg. Longevity up to four years but usually less than two years. Growth is extremely rapid, first year growth ranging from 1.43 to 4.71 mm/d. Age at first maturity three to four months in the Gulf of Mexico, six to seven months in the northeastern North Atlantic. Off North Carolina, males reach 50% maturity at 476 mm, 100% at 645 mm; females reach 50% maturity at 458 mm, 100% at 560 mm. Spawning is probably year-round in tropical regions with water temperatures greater than 21° C. In temperate areas such as North Carolina, peak spawning occurs from April through July. Batch spawner spawning at least two or three times per spawning period. Batch fecundity estimates in the west central Atlantic range from 58,000 to 1.5 million eggs and are strongly influenced by female size. Diameter of ripe eggs off Taiwan 1.0 to 1.6 mm, one oil globule present. Hatch at 3 mm TL, flexion at 7.5 to 9.0 mm SL. Juveniles illustrated by Gibbs & Collette (1959), Ditty et al. (1994), and Ditty (2005). A very highly appreciated sports fish and food fish, frequently marketed under the Hawaiian name "mahi-mahi." Caught by trolling and on tuna longlines. See Gibbs & Collette 1959; Beardsley 1967; Palko et al. 1982; Oxenford & Hunte 1986; Ditty et al. 1994; Oxenford 1999; Wu et al. 2001; Ditty 2005; Schwenke & Buckel 2008.

Family Echeneidae (Remoras)

The Echeneidae is divided into two subfamilies, four genera, and eight species (Gray et al. 2009). Six species are oceanic epipelagic; the two species of *Echeneis* are more coastal. They are perciform fishes with a transversely laminated, oval-shaped, cephalic disc on their head. This structure is derived from the spinous dorsal fin. Juveniles of some species have an elongate median caudal filament.

Remoras attach themselves to many different marine vertebrates including sharks, rays, larger teleost fishes, sea turtles, whales, and dolphins (O'Toole 1999). They may also attach to ships and various floating objects. Some species have a preference or specificity for certain hosts. *Remora australis* (Bennett 1839), the

Whalesucker, is only known from marine mammals. *Remora osteochir* (Cuvier 1829), the Marlinsucker, is usually found in the gill cavities of billfishes, particularly Sailfish and White Marlin. Frequently they occur in pairs, a male under one opercle, a female under the other (Strasburg 1964 and pers. obs.), perhaps increasing the probability of finding a mate. The preferred host of *Remora albescens* (Temminck & Schlegel 1850), the White Suckerfish, is the Manta Ray. Species of *Remora* are almost always captured on their host, where they may be found attached to the body, in the mouth, or in the gill cavity. Many species feed on parasitic copepods on their hosts (Cressey & Lachner 1970).

Breeding behavior of *Echeneis naucrates* has been described in an aquarium (Nakajima et al. 1987) but almost nothing else is known about their reproductive biology. Echeneid eggs have not been described. Young are free-swimming until about 40 to 80 mm SL. Development of the sucking disc occurs at early stages. Postflexion stage, larvae, and juveniles of some echeneids have been illustrated by Richards (2005a). See Strasburg 1964; Cressey & Lachner 1970; Nakajima et al. 1987; O'Toole 1999; Richards 2005a; Gray et al. 2009.

Family Xiphiidae (Swordfish)

Xiphias gladius (Linnaeus 1758), Swordfish.   A single cosmopolitan epi-mesopelagic, oceanic species usually found from 45°N to 45°S in surface waters warmer than 13°C. It is primarily a warm-water species that migrates toward temperate or cold waters for feeding in the summer and back to warm waters in winter for spawning and overwintering. Maximum size 445 cm TL and about 540 kg. IGFA all-tackle record 536.15 kg. Longevity 9+ years. Females are usually larger than males and most Swordfish over 140 kg or 210 cm LJFL (lower jaw fork length) are females. First spawn at five to six years of age in the Pacific. Males reach sexual maturity at about 100 cm and females at about 70 cm in the Atlantic. Ovaries contain 2 to 5 million eggs. Egg diameter 1.6 to 1.8 mm. One oil globule, about 0.4 mm in diameter. Incubation 2.5 days. Hatching size 4.2 mm NL (notochord length). Length at flexion 12 mm. Young Swordfish lack the strong pterotic and preopercular spines characteristic of juvenile Istiophoridae. Jaws start to elongate and distinct highly modified prickle-like scales form in juveniles by 7 mm TL. Although previously thought to be naked, scales persist in adults but become embedded deep in the dermis as the stratum spongiosum increases in thickness above the scale. Illustrations of juveniles are in Arata (1954), Sun (1960), Palko et al. (1981), Potthoff & Kelley (1982), Collette et al. (1984b), and Richards (2005c). A highly important food and game species. The North Atlantic stock was rated as "Endangered" in the IUCN Red List based on a 1996 assessment (IUCN 2009), but seems to be recovering. See Arata 1954; Sun 1960; Palko et al. 1981; Potthoff & Kelley 1982; Collette et al. 1984b; Nakamura 1985; Govoni et al. 2004; Richards 2005c; Wang et al. 2006.

Family Istiophoridae (Billfishes)

Billfishes include nine species of epipelagic oceanic fishes: *Istiophorus* (monotypic), *Istiompax* (monotypic), *Makaira* (monotypic), *Kajikia* (two species), and *Tetrapturus* (four species) following Collette et al. (2006). Billfishes are at or near the apex of pelagic food webs, have broad diets, grow very rapidly, have high fecundity and, in some cases show long-distance migrations (Kitchell et al. 2006). More information is available on the reproductive biology of billfishes than for most of the fishes previously discussed in this chapter. Synopses on the biology of billfish species are included in Shomura and Williams (1975). Females are usually larger than males. Oviparous, buoyant eggs, pelagic larvae. Spawning: warm months. Fecundity: 0.75 to 19 million eggs, increasing with size. Age: 9 to 12+ years. All but the smallest young billfishes are quite easily identified to family because the snout starts to elongate by 3 mm notochord length although it does not take on the adult spear shape until a length of 50 mm or longer. However, identification of larvae and juveniles to species is extremely difficult (Richards & Luthy 2005).

All are important sport fishes and many are also taken in long-lining operations and used for food. Several species are under intense fishing pressure and since the early 1980s, stock assessments have indicated that Atlantic stocks of some billfishes are overfished (Restrepo et al. 2003; Die 2006). Size limitations, encouragement of catch-and-release sport fishing, and recommendations for using circle hooks instead of J-hooks are measures designed to increase survival in catch-and-release sport fishing and may be instrumental in their successful management.

Istiophorus platypterus (Shaw & Nodder 1792), Sailfish. Widely distributed in tropical and temperate waters. Maximum size more than 340 cm TL and 100 kg. IGFA all-tackle record 100.24 kg (Pacific), 64 kg (Atlantic). Longevity 11+ years, but the sport fishery in Florida is largely dependent on fish between 6 and 18 months old. No external sexual dimorphism but females grow larger than males. Proportion of females in the catch off Taiwan increased with size as LJFL (lower jaw to fork length) increased beyond 145 cm and reached nearly 100% at sizes greater than 227 LJFL. Estimated mean LJFL at sexual maturity of females is 166 cm off Taiwan; smallest mature female, 162 cm LJFL. Spawning occurs with males and females swimming in pairs or with two or three males chasing a single female. Spawning takes place throughout the year in tropical waters. Off southeast Florida, presence of three distinct groups of maturing ovocytes in the ovaries of ripe females shows that ovocyte development is asynchronous, resulting in fractional or multiple spawning. Fecundity 1 to 19.5 million eggs, sharply increasing with female size. Egg diameter 1.3 mm, one oil globule. Length at flexion about 6 mm. Juveniles are illustrated by Gehringer (1956, 1970), Sun (1960), Ueyanagi (1964), Beardsley et al. (1975), Fritzsche (1978), Collette et al. (1984b), and Richards & Luthy (2005).

Taken as bycatch by tuna longliners and also with surface driftnets and by trolling and harpooning but more important as a valued sports fish. See Gehringer 1956, 1970; Sun 1960; Ueyanagi 1964; Beardsley et al. 1975; Fritzsche 1978; Collette et al. 1984b; Nakamura 1985; de Sylva & Breder 1997; Richards & Luthy 2005; Chiang et al. 2006; Wang et al. 2006.

Istiompax indica (Cuvier 1832), Black Marlin.   Found throughout the tropical and subtropical waters of the Indo-Pacific and extending a short distance into the South Atlantic. Maximum size more than 448 cm TL and 700 kg. IGFA all-tackle record 707.61 kg. Longevity unknown. Males and females indistinguishable externally but females attain a much larger size. Sex ratio varies with area and season. In Taiwan waters, all Black Marlin greater than 270 cm LJFL were females. Length of males at first maturity about 140 cm, 230 cm for females. Age at first maturity not known. Intensive spawning occurs in the Coral Sea, especially during October and November. Water temperatures about 27 to 28°C during spawning. A large fish (presumably a female) seen followed by several smaller fish (presumably males) off Cairns. Egg counts of ripe females totaled about 40 million. Juveniles illustrated by Ueyanagi (1960, 1964) and Nakamura (1975). Caught by tuna longliners and also a very important sports fish off Peru, Ecuador, and northeastern Australia. See Ueyanagi 1960, 1964; Nakamura 1975, 1985; Wang et al. 2006; Matsumoto & Bayliff 2008.

Kajikia albida (Poey 1860), White Marlin.   Found throughout warm waters of the Atlantic from 45°N to 45°S including the Gulf of Mexico, Caribbean Sea, and the Mediterranean. Maximum size over 280 cm TL and over 82 kg. IGFA all-tackle record 82.5 kg. Longevity 12+ years. No apparent sexual dimorphism but females attain larger sizes than males. Length at first maturity 130 cm orbit to fork length or about 20 kg in the female. Spawning areas are in deep-blue oceanic waters, generally at high surface temperatures (20 to 29°C). Migrates into subtropical waters to spawn with peak spawning in spring and early summer, March through June. Apparently spawn once a year. Eggs undescribed. Length at flexion about 6 mm. Juveniles illustrated by Gehringer (1956), Fritzsche (1978), and Richards & Luthy (2005). Over 90% of the reported landings are attributed to longline fisheries and there are also important directed recreational fisheries (Restrepo et al. 2003). Despite voluntary conservation measures, mandated minimum size limits and wide acceptance of catch-and-release, White Marlin are currently considered to be severely overfished (Restrepo et al. 2003; Jesien et al. 2006). A recent petition to declare White Marlin an endangered species in the United States was not accepted (White Marlin Review Team 2002). See Gehringer 1956; Mather et al. 1975; Fritzsche 1978; Nakamura 1985; de Sylva & Breder 1997; Restrepo et al. 2003; Richards & Luthy 2005; Jesien et al. 2006.

Kajikia audax (Philippi 1887), Striped Marlin.   Latitudinally the most widely distributed billfish occurring throughout tropical, subtropical, and temperate waters of the Pacific and Indian oceans. Maximum size exceeds 350 cm TL and 200 kg. IGFA all-tackle record 224.1 kg. Longevity 12+ years. Little size difference between males and females but proportion of females increases with size in Taiwan waters. Size at first maturity estimated to be between 140 and 160 cm eye-fork length around Taiwan and off east Africa. Ripe Striped Marlin found from May through December in the southern Gulf of California, larvae at temperatures of 27.5 to 31.5°C. Fecundity 11 to 29 million eggs. Ovarian eggs from New Zealand averaged 0.85 mm in diameter shortly before spawning so ovulated eggs should exceed 1 mm in diameter. Larvae primarily found during late spring and early summer in both hemispheres of the Pacific. Larvae in four spatially discrete regions: eastern North Pacific, eastern South Pacific, western North Pacific, and central South Pacific, suggesting spawning site fidelity. Larvae illustrated by Ueyanagi (1964), Ueyanagi & Wares (1975), and Hyde et al. (2006). An important commercial and recreational resource throughout its range, with the largest catches taken as bycatch by the pelagic longline fisheries targeting tunas. See Ueyanagi 1964; Ueyanagi & Wares 1975; Nakamura 1985; González-Armas et al. 2006; Hyde et al. 2006; Wang et al. 2006; McDowell & Graves 2008.

Makaira nigricans (Lacepède 1802), Blue Marlin.   Cosmopolitan in tropical and temperate waters. Latitudinal range varies seasonally and extends from 45°N to 35°S in the Atlantic, 48°N to 48°S in the Pacific. Maximum size over 906 kg and 420 cm TL in the Pacific, smaller in the Atlantic. IGFA all-tackle record 636 kg. Longevity 17 years. Females grow larger than males; around the Bonin Islands, all fish over 200 cm eye-fork length were females; in Taiwan waters all over 180 cm eye-fork length or 280 cm LJFL were females. Estimated size at first maturity 130 cm eye-fork length for males, 180 cm for females. Make seasonal north-south migrations. Ripe or subripe individuals found from March to October in the Caribbean, May to September in the western Pacific. Spawn between April and September in the northeast Atlantic at temperatures between 26 and 29°C. Multiple spawners, spawning every two to three days on average. Batch fecundity 2.11 to 13.5 million eggs. Eggs within ovary or intra-ovarian 1 mm in diameter. Flexion at about 6 mm. Juveniles illustrated by Gehringer (1956), Ueyanagi (1964), Fritzsche (1978), and Richards & Luthy (2005). More than 73% of reported landings are incidental to large offshore longline fisheries; other major fisheries are the directed recreational fisheries of the United States and other countries (Restrepo et al. 2003). See Gehringer 1956; Ueyanagi 1964; Rivas 1975; Erdman 1977; Fritzsche 1978; Collette et al. 1984b; Nakamura 1985; de Sylva & Breder 1997; Restrepo et al. 2003; Richards & Luthy 2005; Wang et al. 2006; Sun et al. 2009.

Tetrapturus angustirostris (Tanaka 1915), Shortbill Spearfish.   Distributed throughout the tropical and temperate waters of the Pacific and Indian Oceans. Maximum size about 2 m and 52 kg in weight. IGFA all-tackle record 36.8 kg. Females may average slightly larger than males. Spawning is believed to occur mainly during winter months, especially in warm offshore currents with surface temperatures of about 25°C. Diameters of shed eggs range from 1.3 to 1.6 mm, mean 1.442 mm, in the equatorial western Indian Ocean. Illustration of larvae provided by Sun (1960), Ueyanagi (1964), and Kikawa (1975). No special fisheries but caught incidentally by tuna longliners and rarely by trolling or sport fishing. See Sun 1960; Ueyanagi 1964; Kikawa 1975; Nakamura 1985.

Tetrapturus belone (Rafinesque 1810), Mediterranean Spearfish.   Distribution limited to the Mediterranean Sea. Maximum size exceeds 240 cm in body length and 70 kg in weight. IGFA all-tackle record 41.2 kg. Nothing is known about the biology of this species. Probably spawns in winter or spring. Pelagic eggs from the Straits of Messina averaged 1.48 mm in diameter with a single oil globule. Taken at the surface by harpoons, longlines, driftnets and set nets incidental to fishing for swordfish, bluefin tuna, and albacore. See de Sylva 1975; Nakamura 1985.

Tetrapturus georgii (Lowe 1841), Roundscale Spearfish.   Originally described from Madeira and reported from several other eastern Atlantic localities but only recently known with certainty from the western Atlantic as well (Shivji et al. 2006). Maximum size at least 160 cm body length and 21.5 kg weight. IGFA all-tackle record 31.2 kg. Little is known of the reproductive biology of this species. Taken by sports fishermen along with White Marlin. See Nakamura 1985; Shivji et al. 2006.

Tetrapturus pfluegeri (Robins & de Sylva 1963), Longbill Spearfish.   Widely distributed in Atlantic offshore waters from approximately 40°N to 35°S. Maximum size exceeds 200 cm in body length and 58 kg in weight. IGFA all-tackle record 58 kg. Longevity probably four years. No sexual dimorphism reported. First spawning probably occurs at the end of the first year and few females apparently survive beyond a second spawning. Females probably spawn once a year. Spawning takes place throughout wide areas of the tropical and subtropical Atlantic from late November to early May. Eggs undescribed. Drawing of a 368-mm juvenile included in Robins (1975). Taken as bycatch by tuna longliners and also by sports fishermen. See Robins 1975; Nakamura 1985; de Sylva & Breder 1997; Richards & Luthy 2005.

Family Scombridae (Mackerels, Tunas, and Bonitos)

The Scombridae contains 15 genera and 51 species (Collette et al. 2001) and is divided into two subfamilies: Gasterochismatinae, which contains only the peculiar southern ocean *Gasterochisma melampus*; and Scombrinae. On the basis of internal osteological characters, Collette and Chao (1975) and Collette & Russo (1985) divided the Scombrinae into two groups of tribes. The more primitive mackerels (Scombrini) and Spanish mackerels (Scomberomorini) are characterized by (i) a distinct notch in the hypural plate that supports the caudal fin rays, (ii) absence of a bony support for the median fleshy keel (when present), and (iii) preural vertebrae centra not greatly shortened as compared to those of the other vertebrae. Bonitos (tribe Sardini) are a group of four genera and seven species that are intermediate between Spanish mackerels (tribe Scomberomorini) and higher tunas (tribe Thunnini). They lack any trace of a specialized subcutaneous vascular system or dorsally projecting cartilaginous ridges on the tongue, and the bony structure underlying their median fleshy caudal peduncle keel is incompletely developed; they also lack the prominent paired frontoparietal fenestra on the dorsal surface of the skull characteristic of most Thunnini. The Thunnini contains five genera, four of which (all except *Allothunnus*) are unique among bony fishes in having countercurrent heat exchanger systems that allow them to retain metabolic heat so that the fish is warmer than the surrounding water. Three genera of this tribe (*Auxis, Euthynnus,* and *Katsuwonus*) and the yellowfin group of *Thunnus* have central and lateral heat exchangers, while the specialized bluefin group of *Thunnus* have lost the central heat exchanger and evolved very well-developed lateral heat exchangers (Graham & Dickson 2000).

Scombrids are swift, epipelagic or epi-mesopelagic predators; some species occur in coastal waters, others far from shore. Spanish mackerels, bonitos, and tunas feed on larger prey, including small fishes, crustaceans, and squids. The main predators of smaller scombrids are other predacious fishes, particularly large tunas and billfishes. Scombrids are dioecious (separate sexes) and most display little or no sexual dimorphism in structure or color pattern. Reports of courting behavior in scombrids are rare (Magnuson & Prescott 1966; Iversen et al. 1970). Males of many species attain larger sizes than females. Batch spawning of most species takes place in tropical and subtropical waters. The eggs are pelagic and hatch into planktonic larvae. Mackerels and tunas support very important commercial and recreational fisheries as well as substantial artisanal fisheries throughout the tropical and temperate waters of the world. Catches in cold and warm temperate waters predominate over tropical catches, with more than half of the world catch being taken in the northwestern Pacific, the northeastern Atlantic, and the southeastern Pacific. Many species of tunas and mackerels are the target of long-distance fisheries. The principal fishing methods used for fish schooling near the surface include purse seining, driftnetting, hook and line/bait boat fishing, and trolling; standard and deep longlining are used for (usually bigger) fish occurring at least temporarily in deeper water. Recreational fishing methods involve mostly surface trolling and pole-and-line fishing, while numerous artisanal fisheries deploy a great variety of gear including bag nets, cast nets, lift nets, gill (drift) nets, beach seines, hook-and-line, handlines, harpoons, specialized traps, and fish corrals.

Early life-history pattern: dioecious; usually no sexual dimorphism in external characters (except size); sex ratio does not deviate significantly from the expected 1:1 ratio; oviparous; asynchronous oocyte development; multiple or batch spawners; epipelagic or mesopelagic; spawning dependent on warm water temperature, usually at least 24°C (Schaefer 2001). Larvae are characterized by large heads, triangular gut, and large jaws but are very difficult to identify to species (Richards 2005b).

Acanthocybium solandri (Cuvier 1832), Wahoo.   Tropical and subtropical waters of all oceans including the Caribbean and Mediterranean seas. Nuclear and mitochondrial DNA show extensive sharing of haplotypes across the Wahoo's entire global range and analyses are unable to detect significant structure (Theisen et al. 2008). Maximum size 210 cm FL and more than 83 kg. IGFA all-tackle record 83.46 kg. Females grow larger than males. A fast growing species with a high mortality. Longevity nine years. Spawning seems to extend over a long period of time, in the western Atlantic from at least May to October. In the northern Gulf of Mexico, 50% sexual maturity in males is reached before 935 mm FL, probably at an age of one year; in females, size at 50% maturity is approximately 1,020 mm FL, at an estimated age of two years. Females are multiple batch spawners and fecundity is quite high, a mean batch fecundity of 1,146,395 ±291,210 eggs/female. An individual female might spawn every two to six days, a total of 20 to 62 times during a spawning season, resulting in a total fecundity for a five-year-old female of 30–92.8 million eggs. Mature eggs 0.8 mm in diameter. Hatching at 2.5 mm NL, flexion at about 6 mm. Drawings of larvae in Wollam (1969), Fritzsche (1978), Collette et al. (1984b) and Richards (2005b). There do not appear to be organized fisheries for Wahoo in most areas but they are targeted in the western Atlantic by both commercial and recreational fisheries and are highly appreciated as a food fish. See Wollam 1969; Erdman 1977; Fritzsche 1978; Collette & Nauen 1983; Collette et al. 1984b; Brown-Peterson et al. 2000; Oxenford et al. 2003; Richards 2005b; McBride et al. 2008; Theisen et al. 2008.

Allothunnus fallai (Serventy 1948), Slender Tuna.   Circumglobal in the Southern Ocean from 20°S to 50°S. Maximum size 96 cm FL and 12 kg. IGFA all-tackle record 11.85 kg. Size at first maturity 71.5 cm FL. Active spawning is presumed to take place during the summer months over a wide range of the temperate Indian and South Pacific oceans. Fecundity unknown. Five larvae probably referable to *Allothunnus fallai* were described and illustrated by Watanabe et al. (1966). Slender tuna are taken incidentally by tuna longliners fishing for Southern Bluefin Tuna and by purse seiners. See Watanabe et al. 1966; Collette & Nauen 1983; Graham & Dickson 2000.

Auxis rochei (Risso 1810), Bullet Tuna.   Cosmopolitan in warm waters. Maximum size 50 cm FL, common to 35 cm FL. IGFA all-tackle record 1.84 kg. Longevity five years. Males and females of equal length. Length at first maturity in the Philippines 17 cm. Length at 50% maturity of both sexes off India 24 cm, 18.8 cm in the Philippines. The spawning season varies from region to region at sea surface temperatures of 24°C or higher. In the Gulf of Mexico, peaks of batch spawning are reported from March to April; from June to August in coastal waters from Cape Hatteras to Cuba; and in the Straits of Florida, spawning begins in February. Fecundity estimates range between 31,000 and 162,800 eggs per spawning event correlated with the size of the female. Hydration occurs between 1100 and 1300 hrs, ovulation at about 1500 hrs, followed by spawning. Egg diameter 0.82 to 0.88 mm, one oil globule. Hatch at 2.14 mm NL, flexion at approximately 6 mm SL. Illustrations of larvae in Fritzsche (1978) and Richards (2005b). Caught by pole and line and as bycatch in a variety of gear including gill nets. Particularly important in the Philippines, Japan, and the Mediterranean Sea. See Yoshida & Nakamura 1965; Rodríguez-Roda 1966; Fritzsche 1978; Uchida 1981; Collette & Nauen 1983; Grudtsev 1992; Yesaki & Arce 1994; Collette & Aadland 1996; Niiya 2001a, b; Schaefer 2001; Richards 2005b.

Auxis thazard (Lacepède 1800), Frigate Tuna.   Probably cosmopolitan in warm waters but relatively few documented occurrences in the Atlantic Ocean. Maximum size 58 cm FL. IGFA all-tackle record 1.72 kg. Longevity four years. Smallest maturing female off the west coast of Thailand 31 to 33 cm FL. Length at 50% maturity in the Gulf of Thailand 34 to 37 cm FL. In the southern Indian Ocean, spawning extends from August to April, north of the equator from January to April at sea surface temperatures of 24°C or higher. Fecundity estimates range from 78,000 to 1.37 million eggs in 31.5- to 44.2-cm females. Egg diameter 0.84 to 0.92 mm, one oil globule. Hatch at 2.32 mm NL, flexion at approximately 6 mm SL. Illustrations of larvae 4.5 mm NL to 25.0 mm SL in Richards (2005b). Caught by pole and line and as by catch in a variety of gear including gill nets. Particularly important in the Philippines and Japan. See Yoshida & Nakamura 1965; Uchida 1981; Collette & Nauen 1983; Grudtsev & Korolevich 1986; Yesaki & Arce 1994; Collette & Aadland 1996; Schaefer 2001; Richards 2005b.

Gasterochisma melampus (Richardson 1845), Butterfly Kingfish.   Circumglobal in southern temperate waters, mostly between 35°S and 50°S. Maximum size 195 cm FL. Originally described in three different genera: as *Gasterochisma melampus* based on a 181-mm juvenile from New Zealand; as *Chenogaster holmbergi* based on a 132-cm adult from Uruguay; and as *Lepidothynnus huttoni* based on a 167-cm adult from New Zealand. One reason for this is that earlier workers were not aware of the dramatic allometric growth changes, particularly in the length of the pelvic fin (Ito et al. 1994). Reproductive biology, eggs, and larvae unknown. Taken as bycatch by tuna longliners fishing for Southern Bluefin Tuna. See Collette & Nauen 1983; Kohno 1984; Ito et al. 1994.

Katsuwonus pelamis (Linnaeus 1758), Skipjack Tuna.   Cosmopolitan in tropical and warm temperate waters within the 15° isotherms. Maximum size 108 cm FL and 34.5 kg. IGFA all-tackle record 20.54 kg. Longevity at least eight years. Sex ratio about 1:1, but fisheries that rely on young, immature fish are dominated by females; older captured fish are mostly male. Size at first maturity 40 to 55 cm FL, depending on area. Spawn several times per season. Spawn in batches at sea surface temperatures of 24°C to 29°C throughout the year in the Caribbean and other equatorial waters, in the Atlantic, Pacific and Indian oceans, and from spring to early fall in subtropical waters, with the spawning season becoming shorter as distance from the equator increases. Models of migration have been proposed, especially from the central Pacific into the eastern Pacific. Apparent courtship involved one fish following another with its snout close to the caudal fin of the lead fish, the following fish displaying dark vertical bars, and the lead fish wobbling from side to side; similar to observations made in large tanks of Pacific Bonito, *Sarda chiliensis,* and Yellowfin Tuna (Magnuson & Prescott 1966; Margulies et al. 2007). Fecundity increases with size but is highly variable, the number of eggs per season in females 41 to 87 cm FL ranges from 80,000 to 2,000,000. Diameter of eggs still within the ovary 0.80 to 1.17 mm, with a single oil globule, 0.22 to 0.45 mm. Hatching at 2.3 to 3.0 mm NL, flexion at 5.0 to 7.0 mm SL. Illustrations of larvae in Sun (1960), Fritzsche (1978); Ambrose (1996), and Richards (2005b). Skipjack make up 59% of the commercial tuna catch and are mostly canned as light meat tuna. They are taken at the surface mostly with purse seines and pole-and-line gear. Majkowski (2007) categorized most stocks as "Moderately Exploited"; Joseph (2009b) concluded that overfishing of Skipjack stocks was not occurring and that the stocks were not overfished. See Sun 1960; Iversen et al. 1970; Erdman 1977; Fritzsche 1978; Collette & Nauen 1983; Matsumoto et al. 1984; Wild & Hampton 1994; Pagavino & Gaertner 1995; Ambrose 1996; Schaefer 2001; Richards 2005b; Majkowski 2007; Joseph 2009a, b.

Thunnus alalunga (Bonnaterre 1788), Albacore.   Cosmopolitan in tropical and temperate waters of all oceans including the Mediterranean Sea but not at the surface between 10°N and 10°S. Maximum size 127 cm FL. IGFA all-tackle record 40.0 kg. Longevity 13 years. Immature Albacore (<80 cm) generally have a sex ratio of 1:1 but males predominate in catches of mature fish. Maturity attained at about 90 to 94 cm FL for females, 94 to 97 cm FL for males. Spawning occurs at sea surface temperatures of 24°C or higher. Use of combined Japanese and U.S. tagging data confirm the frequent westward movement of young Albacore and infrequent eastward movements, in the North Pacific. This corresponds to Albacore life history in which immature fish recruit into fisheries in the western and eastern Pacific and then gradually move nearer to their spawning grounds in the central and western Pacific before maturing. Fecundity increases with size but there is no clear correlation between fork length, ovary weight, and number of eggs. A 20 kg female may produce between 2 and 3 million eggs per season, released in at least two batches. Egg diameter 0.84 to 0.94 mm, one oil globule, 0.24 mm in diameter. Hatching at 2.60 mm NL, flexion at about 6.0 mm SL. Larvae illustrated by Fritzsche (1978) and Richards (2005b). An important fishery exists for this species, which is mainly marketed as canned white meat tuna, the most expensive canned tuna. They are caught by longlining, live-bait fishing, purse seining, and trolling. The North Atlantic stock was considered "Vulnerable" and the South Atlantic stock "Critically Endangered" based on 1996 assessments (IUCN 2009). Majkowski (2007) considered the North Atlantic stock to be "Overexploited"; Joseph (2009b) feels that this stock is in an overfished state and that overfishing is currently taking place. Majkowski (2007) considered the Indian Ocean and North Pacific stocks to be "Fully Exploited." See Fritzsche 1978; Collette & Nauen 1983; Chang et al. 1993; Labelle et al. 1993; Richards 2005b; Majkowski 2007; Ichinokawa et al. 2008; IUCN 2009; Joseph 2009b.

Thunnus albacares (Bonnaterre 1788), Yellowfin Tuna.   Widespread throughout tropical and temperate waters of the world. Maximum size over 200 cm FL. IGFA all-tackle record 176.35 kg. Longevity at least six years. Smallest mature individuals in the Pacific off the Philippines and Central America in the 50 to 60 cm size group at an age of 12 to 15 months. Length at 50% maturity in the eastern Pacific 69 cm for males, 92 cm for females. Spawning occurs throughout the year in the core areas of distribution at sea surface temperatures of 24°C or higher, but peaks are observed in the northern and southern summer months respectively. Spawning occurs almost entirely at night between 2200 and 0600 hrs. Reproductively active Yellowfin spawn almost daily. Courtship behavior in aquaria consisted of one to three males following a female and often flashing vertical bars on the body. Estimated average batch fecundity 2.5 million ova or 67.3 ova/g of body weight. Batch fecundity estimates in the eastern Pacific ranged from 162,918 ova for a 1180-mm female to 8,026,026 ova for a 1460-mm female. Egg diameter 0.90 to 1.04 mm, one oil globule. Incubation 24 to 38 hours at 26°C. Hatching at 2.6 to 2.7 mm TL, flexion at 4.5 to 6.1 mm. Illustrations of larvae and juveniles in Sun (1960), Fritzsche (1978), Ambrose (1996), and Richards (2005b). Yellowfin is the second most important species of tuna for canning, constituting 24.0% of the tuna catch, and is the primary target of the purse-seine fishery in the eastern Pacific. They are also taken primarily by pole-and-line fishing. Considered of "Least Concern" based on a 1996 assessment (IUCN 2009) and "Fully Exploited" by Majkowski (2007), but Joseph (2009b) considered the Indian Ocean stock as being overfished with overfishing currently taking place. See Sun 1960; Fritzsche 1978; Collette & Nauen 1983; Ambrose 1996; Schaefer 1998, 2001; Richards 2005b; Majkowski 2007; Margulies et al. 2007; IUCN 2009; Joseph 2009a, b.

Thunnus atlanticus (Lesson 1830), Blackfin Tuna.   Confined to the western North Atlantic from Cape Cod to southern Brazil in waters of at least 20°C. Maximum size 110 cm FL. IGFA all-tackle record 22.39 kg. Longevity eight years. Males grow larger than females. Length of females at 50% maturity 49.8 cm FL and for males 52.1 cm FL off northeastern Brazil; both sexes mature at age two years. Around Florida, the spawning season extends from April to November with a peak in May, while in the Gulf of Mexico, spawning apparently lasts from June to September. Eggs unknown. Length at flexion about 6 mm SL. Illustrations of larvae and juveniles in Fritzsche (1978) and Richards (2005b). The largest fisheries for Blackfin Tuna are a live-bait and pole fishery off the southeastern coast of Cuba and a handline artisanal fishery off northeastern Brazil but there are also important sports fisheries in Florida and the Bahamas. See Erdman 1977; Fritzsche 1978; Collette & Nauen 1983; Neilson et al. 1994; Freire et al. 2005; Richards 2005b.

Thunnus maccoyii (Castelnau 1872), Southern Bluefin Tuna.   Found throughout the Southern Ocean between 30 and 50°S but migrating to warm waters between northwest Australia and Indonesia for spawning. Maximum size 225 cm FL and 200 kg. IGFA all-tackle record 158 kg. Longevity 20+ years. Sex ratio in catches shows that as juveniles, females outnumber males; the situation is reversed in adults. Maturity can occur at 120 cm FL but more commonly at 130 cm (about eight years old). Age at first maturity five to seven years at 110 to 125 cm FL. Size at 50% maturity 152 cm FL. The spawning season extends throughout the southern summer from about September/October to March at water surface temperatures in excess of 24°C. Once females start spawning, they appear to spawn daily. Migration pattern shown in Caton (1994). An asynchronous indeterminate spawner with annual batch fecundity of 57 ova/g body weight. Fecundity of a 158-cm female with gonads weighing about 1.7 kg each was estimated at about 14 to 15 million eggs. Intraovarian eggs 0.66 to 1.10 mm in diameter with one or two large oil globules. A very important commercial species especially off Australia. The meat is highly prized for the sashimi markets of Japan. Considered "Critically Endangered" by a 1996 assessment (IUCN 2009), "Depleted" by Majkowski (2007), and seriously overfished by Joseph (2009a). See Collette & Nauen 1983; Thorogood 1986; Caton 1994; Farley & Davis 1998; Schaefer 2001; Majkowski 2007; IUCN 2009; Joseph 2009a.

Thunnus obesus (Lowe 1839), Bigeye Tuna.   Worldwide in tropical waters of the Atlantic, Indian, and Pacific oceans but absent from the Mediterranean Sea. Maximum size over 200 cm FL. IGFA all-tackle record 197.31 kg. Maximum age 16 years, although 80% of fish caught in Australia were less than five years. Males tend to predominate in catches over the entire size range but particularly at larger sizes. Minimum length at maturity for females 80 to 102 cm FL and predicted length at 50% maturity 102 to 135 cm in different regions. In the eastern and central Pacific, spawning has been recorded between 15°N and 15°S and between 105° and 175°W during most months when sea surface temperatures exceeded 24°C, with a peak from April through September in the northern hemisphere and between January and March in the southern hemisphere. Spawning is primarily at night between 1900 and 0400 hr. The average mature female spawns every 2.6 days. The estimated mean batch fecundity is 24 ova/g body weight. The number of eggs per spawning has been estimated at 2.9 to 6.3 million. Eggs 1.03 to 1.08 mm in diameter with one oil droplet 0.23 to 0.24 mm in diameter. Length at flexion about 6 mm SL. Illustrations of larvae in Fritzsche (1978) and Richards (2005b). An extremely valuable fishery resource especially for the sashimi market. In the Pacific, Bigeye are exploited by longliners from 40°N to 40°S and by purse seiners from 10°N to 20°S. Bigeye were considered "Vulnerable" but the Pacific stock was considered "Endangered" by a 1996 assessment (IUCN 2009). Majkowski (2007) considered the Atlantic and Indian Ocean stocks to be "Fully Exploited" and both Pacific stocks to be "Overexploited." Joseph (2009b) considered the eastern Pacific stock to be overfished and the Atlantic stock slightly overfished. See Fritzsche 1978; Collette & Nauen 1983; Miyabe 1994; Schaefer 2001; Richards 2005b; Schaefer et al. 2005; Farley et al. 2006; Majkowski 2007; IUCN 2009; Joseph 2009b.

Thunnus orientalis (Temminck & Schlegel 1844), Pacific Bluefin Tuna.   Known from the Gulf of Alaska to southern California and Baja California in the eastern Pacific; from Sakhalin Island in the southern Sea of Okhotsk south to the southeastern Australia and New Zealand in the western Pacific; most abundant near Japan. Maximum length 300 cm FL and maximum weight 555 kg. IGFA all-tackle record 325 kg. Longevity 16 years. Spawning occurs between Japan and the Philippines in April, May, and June, off southern Honshu in July, and in the Sea of Japan in August. A model of migration is presented by Bayliff (1994). The sex ratio is about 1:1. Size at first maturity is 150 cm FL and 60 kg at an age of five years. Batch fecundity increases with length, from about 5 million eggs at 190 cm FL to about 25 million eggs at 240 cm FL. The linear relationship between batch fecundity F and fork length L (cm) is

$$F = 3.2393 \times 10^5 \times L - 5.2057 \times 10^7$$

Broodstock of Pacific Bluefin that were artificially hatched and reared, spawned in captivity. The resulting eggs hatched and the young were reared to the juvenile stage (Sawada et al. 2005). A highly valuable species. See Collette & Nauen 1983; Bayliff 1994; Schaefer 2001; Sawada et al. 2005; Chen et al. 2006.

Thunnus thynnus (Linnaeus 1758), Atlantic Bluefin Tuna.   Genetic differentiation and homing to breeding sites indicates that there are two reproductively isolated stocks although there is considerable transatlantic migration of individuals from both stocks. The western Atlantic stock is found from Labrador and Newfoundland south into the Gulf of Mexico and Caribbean Sea; the eastern Atlantic stock occurs from off Norway south to the Canary Islands and the Mediterranean Sea. Maximum size over 300 cm FL, common to 200 cm. IGFA all-tackle record 678.58 kg. Longevity at least 35 years. Maturity is reached at about four or five years and 115 to 121 cm FL in the Mediterranean Sea. Maturity is delayed in the Gulf of Mexico to age eight to ten years and 200 cm FL. The western Atlantic stock spawns in the Gulf of Mexico from mid-April to early July at temperatures of 22.6 to 27.5°C starting at age eight although most fish first spawn closer to age 12. The eastern Atlantic stock spawns in the Mediterranean Sea from May to August at temperatures of 22.5 to 25.5°C starting at age three and full recruitment is reached by age five. There are distinct behaviors during the spawning period, most noticeably changes in diving times and depths. Estimated relative batch fecundity is greater (more than 90 ova/g of body weight) than estimated for other tunas in the genus *Thunnus*. Egg diameter 1.00 to 1.12 mm, one oil globule, 0.25 to 0.28 mm. Hatching size 2.0 to 3.0 mm TL, flexion at about 6 mm SL. Illustrations of larvae in Fritzsche (1978) and Richards (2005b). A highly valued species for the Japanese sashimi markets that has led to severe overfishing in both the eastern and western Atlantic. Also an important gamefish, particularly in the United States and Canada. The eastern Atlantic stock was considered "Endangered" and the western Atlantic stock "Critically Endangered" by 1996 assessments (IUCN 2009). The two stocks were considered "Overexploited" and "Depleted," respectively by Majkowski (2007), seriously overfished by Joseph (2009a), and "Critically Endangered" by MacKenzie et al. (2009). High priority needs to be given to protecting spawning adults in the Gulf of Mexico and Mediterranean Sea. Large adults in the northern foraging region in the Gulf of Maine and Gulf of St. Lawrence also need protection because this region represents a critical refugium for the largest fishes, so-called western "giants." See Fritzsche 1978; Collette & Nauen 1983; Sissenwine et al. 1998; Corriero et al. 2003; Richards 2005b; Majkowski 2007; Rooker et al. 2007, 2008; Boustany et al. 2008; IUCN 2009; Joseph 2009a; MacKenzie et al. 2009.

Thunnus tonggol (Bleeker 1851), Longtail Tuna.   An epipelagic species that is widely distributed along the tropical shores of the Indo-West Pacific. Maximum size about 130 cm FL. IGFA all-tackle record 35.9 kg. Longevity at least five years. Sex ratio 1:1. Probably spawn more than once a year, perhaps in two spawning seasons in the Gulf of Thailand. Smallest mature female in Thailand was 43 cm FL. Fifty percent of females in the Gulf of Thailand were mature at 39.6 cm. Longtail Tuna from Southeast Asia mature at a smaller size than fish from Australia-Papua New Guinea. Fecundity of fish ranging in size from 43.8 to 49.1 cm varied from 1.2 to 1.9 million eggs. No information available on fertilized eggs but intra-ovarian eggs had a diameter of 1.09 mm and an oil globule of 0.31 to 0.33 mm. There are two major fishing grounds for Longtail Tuna, one off the South China Sea coast of Thailand and Malaysia and the other off countries bordering the North Arabian Sea. Catch of this species is increasing in many areas but landings are frequently confused with Yellowfin Tuna in some regions. Caught mostly with purse seines and gillnets, not with longlines. See Collette & Nauen 1983; Yesaki 1994.

Family Luvaridae (Louvar)

A monotypic cosmopolitan family found in tropical and temperate waters and containing only *Luvarus imperialis* (Rafinesque 1810), Louvar. This is a large, heavily built, blunt-snouted species reaching 2 m TL, commonly 1 m TL. It feeds on jellyfish, ctenophores, and salps and has an extremely long gut, 5 to 11 times standard length. Apparently solitary. Spawns in late spring and summer in the Mediterranean. Estimated total number of eggs from a 173.5-cm FL female from the Gulf of Mexico is 47.5 million. Eggs undescribed. Hatching at 3.5 mm or smaller. Flexion at 6 to 7 mm, length at transformation 9 to 10 mm, larvae pelagic. Pelagic juveniles infrequently collected in coastal waters between 40°N and 90°S. Juveniles go through a dramatic metamorphosis in the shape and size of the head and fins leading to the names hystricinella, astrodermella, and luvarella for three growth stages of juveniles. Occasionally captured with purse seines. See Gottschall & Fitch 1968; Topp & Giradin 1971; Tyler et al. 1989; Farooqui et al. 2005.

ORDER TETRAODONTIFORMES

This order contains the most highly evolved fishes with nine families with about 350 species (Nelson 2006).

Family Molidae (Ocean Sunfishes)

Three cosmopolitan species in three monotypic genera: *Mola mola* (Linnaeus 1758), *Ranzania laevis* (Pénnant 1776), and *Masturus lanceolatus* (Liénard 1840). These highly specialized epipelagic fishes reach enormous sizes of more than 2 m TL and 1000 kg). Ocean sunfishes are considered to be among the most fecund of bony fishes, spawning large, spherical pelagic eggs that measure 1.42 to 1.8 mm in diameter and contain 20 to 40 oil globules. Time to hatching takes at least seven to eight days. Newly hatched larvae are extremely rotund and measure 1.8 mm NL. Flexion does not occur, the tip of the notochord atrophies. Ocean sunfishes have lost the caudal fin; dorsal and anal fin rays grow around the posterior part of the fish and meet in the middle to form what has been termed a clavus (Johnson & Britz 2005).

Masturus lanceolatus (Liénard 1840), Sharptail Mola.   Spawning season and fecundity unknown. Egg diameter 1.8 mm. About 40 oil globules. Hatching size unknown. Transformation takes place at about 5 mm TL. Illustrations of larvae from 2.8 mm TL to 22.5 mm preclaval length in by Lyczkowski-Shultz (2005).

Mola mola (Linnaeus 1758), Ocean Sunfish.   Based on seasonal changes in gonad index and gonad maturation phases, the spawning period off Japan is estimated to be from August to October. Asynchronous oocyte development suggests that Ocean Sunfish are multiple spawners. The ovary of a 150-cm female contained about 300 million eggs. Eggs undescribed but probably similar to other molids and containing multiple oil globules. Larval transformation takes place at about 4 mm TL. Illustrations of larvae from 1.8 mm TL to 11.0 mm TL in Lyczkowski-Shultz (2005). See Schmidt 1921; Lyczkowski-Shultz 2005; Nakatsubo et al. 2007.

Ranzania laevis (Pénant 1776), Slender Mola.   Reproductive biology unknown. Egg diameter 1.42 to 1.65 mm, 20 to 30 oil globules, 0.05 to 0.16 mm in diameter. Hatching size 1.8 mm NL. Length at transformation about 15 mm. Illustrations of eggs and larvae from 1.8 mm TL to 11.0 mm preclaval length were presented by Lyczkowski-Shultz (2005).

CONCLUSIONS

Even the basic knowledge of reproductive biology of most epipelagic fishes is limited (see summary in Table 2.1). Most epipelagic fishes show little evidence of sexual dimorphism except in body size. They are highly fecund and spawn in warm waters, usually 24°C or warmer. Fertilization appears to be external in all oceanic epipelagic fishes. Most eggs have one or more oil globules, enabling them to float at the surface, or filaments that facilitate attachment to floating vegetation. Most species of the Beloniformes have filaments on the eggs, but the most oceanic species have lost these filaments. Important, fast-growing food fishes with widespread spawning sites such as both species of dolphinfishes, Skipjack, and Yellowfin Tuna, seem to still be in relatively stable condition and are probably not in danger of extinction, in spite of high fishing pressures. Long-lived species with restricted spawning sites such as the three species of bluefin tunas are threatened and in need of protection from over-fishing. Efforts to successfully manage pelagic fish stocks are likely to fail until adequate information regarding their reproductive biology becomes available. It may be necessary to implement marine protected areas (MPAs) for pelagic conservation (Game et al. 2009).

These have had some successes with inshore fishes and fishes that live associated with benthic habitats and may be useful in conserving some marine epipelagic species.

ACKNOWLEDGMENTS

Kurt M. Schaefer and William J. Richards read early drafts of the manuscript and provided useful comments, and Maria Jose Juan Jorda read the section on Scombridae and provided many additional references.

REFERENCES

Ambrose DA (1996). Scombroidei. In: Moser HG (ed.), The early stages of fishes in the California Current region. California Cooperative Oceanic Fishery Investigations Atlas 33: 1257–1293.

Angel, MV (1993). Biodiversity of the pelagic ocean. Conservation Biology 7: 760–772.

Arata GF Jr (1954). A contribution to the life history of the swordfish, *Xiphias gladius* Linnaeus, from the south Atlantic coast of the United States and the Gulf of Mexico. Bulletin of Marine Science of the Gulf and Caribbean 4: 183–243.

Bane, GW Jr (1965). The opah (*Lampris regius*), from Puerto Rico. Caribbean Journal of Science 5: 63–66.

Bayliff WH (1994). A review of the biology and fisheries for northern bluefin tuna, *Thunnus thynnus*, in the Pacific Ocean. FAO Fisheries Technical Paper 336(2): 244–295.

Beardsley GL Jr (1967). Age, growth, and reproduction of the dolphin *Coryphaena hippurus*, in the straits of Florida. Copeia 1967: 441–451.

Beardsley GL Jr, Merrett NR, Richards WJ (1975). Synopsis of the biology of the sailfish, *Istiophorus platypterus* (Shaw and Nodder, 1791). NOAA Technical Report NMFS SSRF-675(3): 95–120.

Berkeley SA, Houde ED (1978). Biology of two exploited species of halfbeaks, *Hemiramphus brasiliensis* and *H. balao* from southeast Florida. Bulletin of Marine Science 28: 624–644.

Berkeley SA, Houde ED, Williams F (1975). Fishery and biology of ballyhoo on the southeast Florida coast. University of Miami Sea Grant Special Report 4: 1–15.

Boehlert GW (1984). Scanning electron microscopy. In: Moser HG et al. (eds.), Ontogeny and systematics of fishes. American Society of Ichthyologists and Herpetologists Special Publication 1: 43–48.

Boustany AM, Reeb CA, Block BA (2008). Mitochondrial DNA and electronic tracking reveal population structure of Atlantic bluefin tuna (*Thunnus thynnus*). Marine Biology 156: 13–24.

Boyce DG, Tittensor DP, Worm B (2008). Effects of temperature on global patterns of tuna and billfish richness. Marine Ecology Progress Series 355: 267–276.

Breder CM Jr (1932). On the habits and development of certain Atlantic Synentognathi. Carnegie Institution of Washington Publication 435, Papers from the Tortugas Laboratory 28(1): 1–25.

Breder CM Jr (1938). A contribution to the life histories of Atlantic Ocean flying fishes. Bulletin of the Bingham Oceanographic Collection 6(5): 1–126.

Breder CM Jr, Rasquin P (1952). The sloughing of the melanic area of the dorsal fin, an ontogenetic process in *Tylosurus raphidoma*. Bulletin of the American Museum of Natural History 99: 1–24.

Breder CM Jr, Rasquin P (1954). The nature of post-larval transformation in *Tylosurus acus* (Lacepède). Zoologica, New York 39: 17–30.

Brown-Peterson NJ, Franks JS, Burke AM (2000). Preliminary observations on the reproductive biology of wahoo, *Acanthocybium solandri*, from the northern Gulf of Mexico and Bimini, Bahamas. Proceedings of the Gulf and Caribbean Fisheries Institute 51: 414–427.

Bruun AF (1935). Flying-fishes (Exocoetidae) of the Atlantic. Dana Report 6: 1–106.

Caton AE (1994). Review of aspects of southern bluefin tuna biology, population, and fisheries. FAO Fisheries Technical Paper 336(2): 96–343.

Chang S-K, Liu H-C, Hsu C-C (1993). Estimation of vital parameters for Indian albacore through length-frequency data. Journal of the Fisheries Society of Taiwan 20: 1–13.

Chen C-H (1988). Beloniformes. In: Okiyama M (ed.), *An Atlas of the Early Stage Fishes in Japan*. Tokai University Press, Tokyo, pp. 259–301.

Chen K-S, Crone P, Hsu C-C (2006). Reproductive biology of female Pacific bluefin tuna *Thunnus orientalis* from south-western North Pacific Ocean. Fisheries Science 72: 985–994.

Chiang W-C, Sun C-L, Yeh S-Z, Su W-C, Liu D-C, Chen W-Y (2006). Sex ratios, size at sexual maturity, and spawning season seasonality of sailfish *Istiophorus platypterus* from eastern Taiwan. Bulletin of Marine Science 79: 727–737.

Collette BB (1974). The garfishes (Hemiramphidae). of Australia and New Zealand. Records of the Australian Museum 29: 11–105.

Collette BB (2003). Family Belonidae Bonaparte 1832: needlefishes. California Academy of Sciences Annotated Checklists of Fishes No. 16.

Collette BB (2004). Family Hemiramphidae Gill 1859: halfbeaks. California Academy of Sciences Annotated Checklists, Fishes No. 22.

Collette BB (2005a). Belonidae: needlefishes. In: Richards WJ (ed.), *Early Stages of Atlantic Fishes: An Identification Guide for the Western Central North Atlantic*. CRC Press, Boca Raton, Florida, pp. 909–931.

Collette BB (2005b). Hemiramphidae: halfbeaks. In: Richards WJ (ed.), *Early Stages of Atlantic Fishes: An Identification Guide for the Western Central North Atlantic*. CRC Press, Boca Raton, Florida, pp. 933–953.

Collette BB, Aadland CR (1996). Revision of the frigate tunas (Scombridae, *Auxis*), with descriptions of two new subspecies from the eastern Pacific. Fishery Bulletin 94: 423–441.

Collette BB, Chao LN (1975). Systematics and morphology of the bonitos (*Sarda*) and their relatives (Scombridae, Sardini). Fishery Bulletin 73: 516–625.

Collette BB, Nauen CE (1983). FAO species catalogue. Vol. 2. Scombrids of the world. An annotated and illustrated catalogue of tunas, mackerels, bonitos and related species known to date. FAO Fisheries Synopsis 125(2): 1–37.

Collette BB, Parin NV (1970). Needlefishes (Belonidae) of the eastern Atlantic Ocean. Atlantide Report 11: 7–60.

Collette BB, Russo JL (1985). Morphology, systematics, and biology of the Spanish mackerels (*Scomberomorus*, Scombridae). Fishery Bulletin 82: 545–692.

Collette BB, McGowen GE, Parin NV, Mito S (1984a). Beloniformes: development and relationships. In: Moser HG et al. (eds.), *Ontogeny and Systematics of Fishes*. American Society of Ichthyologists and Herpetologists Special Publication 1, Allen Press, Lawrence, Kansas, pp. 335–354.

Collette BB, Potthoff T, Richards WJ, Ueyanagi S, Russo JL, Nishikawa Y (1984b). Scombroidei: development and relationships. In: Moser HG et al. (eds.), *Ontogeny and Systematics of Fishes*. American Society of Ichthyologists and Herpetologists Special Publication 1, Allen Press, Lawrence, Kansas, pp. 591–620.

Collette BB, Reeb C, Block BA (2001). Systematics of the tunas and mackerels (Scombridae). In: Block BA, Stevens ED (eds.), *Tuna: Physiology, Ecology, and Evolution*. Academic Press, San Diego, pp. 1–33.

Collette BB, McDowell JR, Graves JE (2006). Phylogeny of recent billfishes (Xiphioidei). Bulletin of Marine Science 79: 455–468.

Corriero A, Desantis S, Deflorio M, Acone F, Bridges CR, de la Serna JR, Megalofonou P, De Metrio G (2003). Histological investigation on the ovarian cycle of the bluefin tuna in the western and central Mediterranean. Journal of Fish Biology 63: 108–119.

Cotten N, Comyns BH (2005). Exocoetidae: flyingfishes. In: Richards WJ (ed.), *Early Stages of Atlantic Fishes: An Identification Guide for the Western Central North Atlantic*. CRC Press, Boca Raton, Florida, pp. 955–989.

Cressey RF, Lachner EA (1970). The parasitic copepod diet and life history of diskfishes (Echeneidae). Copeia 1970: 310–318.

de Sylva DP (1975). Synopsis of biological data on the Mediterranean spearfish, *Tetrapturus belone* Rafinesque. NOAA Technical Report NMFS SSRF-675(3): 121–131.

de Sylva DP, Breder PR (1997). Reproduction, gonad histology, and spawning cycles of North Atlantic billfishes (Istiophoridae). Bulletin of Marine Science 60: 668–697.

Die DJ (2006). Are Atlantic marlins overfished or endangered? Some reasons why we may not be able to tell. Bulletin of Marine Science 70: 529–543.

Ditty JG (2005). Coryphaenidae: dolphinfishes. In: Richards WJ (ed.), *Early Stages of Atlantic Fishes: An Identification Guide for the Western Central North Atlantic*. CRC Press, Boca Raton, Florida, pp. 1511–1515.

Ditty JG, Shaw RF, Grimes CB, Cope JS (1994). Larval development, distribution, and abundance of common dolphin *Coryphaena hippurus*, and pompano dolphin, *C. equiselis*. Fishery Bulletin 92: 275–291.

Erdman DS (1977). Spawning patterns of fish from the northeastern Caribbean. In: Stewart HB (ed.), Cooperative investigations of the Caribbean and adjacent regions II. FAO Fisheries Report No. 200: 145–169.

Farley JH, Davis TLO (1998). Reproductive dynamics of southern bluefin tuna, *Thunnus maccoyii*. Fishery Bulletin 96: 223–236.

Farley JH, Clear NP, Leroy B, Davis TLO, McPherson G (2006). Age, growth and preliminary estimates of maturity of bigeye tuna, *Thunnus obesus*, in the Australian region. Australian Journal of Marine and Freshwater Research 57: 713–724.

Farooqui TW, Shaw RF, Lindquist DC. (2005). Luvaridae: louvar. In: Richards WJ (ed.), *Early Stages of Atlantic Fishes: An Identification Guide for the Western Central North Atlantic*. CRC Press, Boca Raton, Florida, pp. 2111–2117.

Freire KMF, Lessa R, Lins-Oliveria JE (2005). Fishery and biology of blackfin tuna *Thunnus atlanticus* off northeastern Brazil. Gulf and Caribbean Research 17: 15–24.

Fritzsche RA (1978). Chaetodontidae through Ophidiidae. Vol. 5, Development of fishes of the mid-Atlantic Bight. U.S. Fish and Wildlife Service FWS/OBS-78/12.

Game ET, Grantham HS, Hobday AJ, Pressey RL, Lombard, AT, Beckley LE, Gjerde K, Bustamente R, Possingham HP, Richardson AJ (2009). Pelagic protected areas: the missing dimension in ocean conservation. Trends in Ecology and Evolution 24: 360–369.

Gehringer JW (1956). Observations on the development of the Atlantic sailfish *Istiophorus americanus* (Cuvier), with notes on an unidentified species of istiophorid. Fishery Bulletin 57: 139–171.

Gehringer JW (1970). Young of the Atlantic sailfish *Istiophorus platypterus*. Fishery Bulletin 68: 177–189.

Gibbs RH Jr, Collette BB (1959). On the identification, distribution, and biology of the dolphins, *Coryphaena hippurus* and *C. equiselis*. Bulletin of Marine Science of the Gulf and Caribbean 9: 117–152.

González-Armas R, Klett-Traulsen A, Hernández-Herrera A (2006). Evidence of billfish reproduction in the southern Gulf of California, Mexico. Bulletin of Marine Science 70: 705–717.

Gottschall DW, Fitch JE (1968). The louvar, *Luvarus imperialis* in the eastern Pacific, with notes on its life history. Copeia 1968: 181–183.

Govoni JJ, West MA, Zivotofsky D, Zivotofsky AZ, Bowser PR, Collette BB (2004). Ontogeny of squamation in Swordfish, *Xiphias gladius*. Copeia 2004: 301–306.

Graham JB, Dickson KA (2000). The evolution of thunniform locomotion and heat conservation in scombrid fishes: new insights based on the morphology of *Allothunnus fallai*. Zoological Journal of the Linnean Society 129: 419–466.

Gray KN, McDowell JR, Collette BB, Graves JE (2009). A molecular phylogeny of the Echeneoidea (Perciformes: Carangoidei). Bulletin of Marine Science 84: 183–198.

Grudtsev ME (1992) Particularites de repartition et caracteristiques biologiques de la melva *Auxis rochei* (Risso) dans les eaux du Sahara. International Commission for the Conservation of Atlantic Tunas Collective Volume of Scientific Papers 39(1): 284–288.

Grudtsev ME, Korolevich LI (1986). Studies of frigate tuna *Auxis thazard* (Lacepede) age and growth in the eastern part of the equatorial Atlantic. International Commission for the Conservation of Atlantic Tunas Collective Volume of Scientific Papers 25: 269–274.

Hardy JD Jr, Collette BB (2005). Scomberesocidae: sauries. In: Richards WJ (ed.), *Early Stages of Atlantic Fishes: An Identification Guide for the Western Central North Atlantic*. CRC Press, Boca Raton, Florida, pp. 905–907.

Hubbs CL, Wisner RL (1980). Revision of the sauries (Pisces, Scomberesocidae), with descriptions of two new genera and one new species. Fishery Bulletin 77: 521–566.

Hyde JR, Humphreys R Jr, Musyl M, Lynn E, Vetter, R (2006). A central North Pacific spawning ground for striped marlin, *Tetrapturus audax*. Bulletin of Marine Science 70: 683–690.

Ichinokawa M, Coan AL Jr, Y. Takeuchi Y (2008). Transoceanic migration rates of young North Pacific albacore, *Thunnus alalunga*, from conventional tagging data. Canadian Journal of Fisheries and Aquatic Science 65: 1681–1691.

IGFA (2009). International Game Fish Association World Record Game Fishes. International Game Fish Association, Dania Beach, FL.

Ito RY, Hawn DR, Collette BB (1994). First record of the butterfly kingfish *Gasterochisma melampus* (Scombridae) from the North Pacific Ocean. Japanese Journal of Ichthyology 40: 482–486.

IUCN (2009). IUCN Red List of threatened species. Version 2009. www.iucnredlist.org (accessed 13 November 2009).

Iversen RTB, Nakamura EL, Gooding RM (1970). Courting behavior in skipjack tuna, *Katsuwonus pelamis*. Transactions of the American Fisheries Society 99: 93.

Jesien RV, Barse AM, Smyth S, Prince ED, Serafy JE (2006). Characterization of the white marlin (*Tetrapturus albidus*) recreational fishery off Maryland and New Jersey. Bulletin of Marine Science 79: 647–657.

Johnson GD, Britz R (2005). Leis' Conundrum: homology of the clavus of ocean sunfishes. 2. Ontogeny of the median fins and axial skeleton of *Ranzania laevis* (Teleostei, Tetraodontiformes, Molidae). Journal of Morphology 266: 11–21.

Joseph J (2009a). Plenty more tuna in the sea? New Scientist 203(2719): 22–23.

Joseph J (2009b). Status of the world fisheries for tuna. International Seafood Sustainability Foundation. www.iss-foundation.org (accessed 15 April 2009).

Khokiattiwong S, Mahon R, Hunte W (2000). Seasonal abundance and reproduction of the fourwing flyingfish, *Hirundichthys affinis*, off Barbados. Environmental Biology of Fishes 59: 43–60.

Khrapkova-Kovalevskaya, NV (1963). Data on reproduction, development and distribution of larvae and young fish of *Oxyporhamphus micropterus* Val. (Pisces, Oxyporhamphidae). Trudy Instituta Okeanologii Akademiya Nauk SSSR 62: 49–61 (in Russian).

Kikawa S (1975). Synopsis of biological data on the shortbill spearfish, *Tetrapturus angustirostris* Tanaka, 1914 in the Indo-Pacific areas. NOAA Technical Report NMFS SSRF-675(3): 39–54.

Kitchell JF, Martell SJD, Walters CJ, Jensen OP, Kaplan IC, Watters J, Essington TE, Boggs CH (2006). Billfishes in an ecosystem context. Bulletin of Marine Science 79: 669–682.

Klawe WL (1966). Observations on the opah, *Lampris regius* (Bonnaterre). Nature 210: 965–966.

Kohno, H (1984). Osteology and systematic position of the butterfly mackerel, *Gasterochisma melampus*. Japanese Journal of Ichthyology 31: 268–286.

Kovalevskaya NV (1982). Reproduction and development of flying fishes of the family Exocoetidae. Voprosy Ikhtiologii 22: 582–587 (in Russian, translated in Journal of Ichthyology 22: 48–54).

Labelle M, Hampton J, Bailey K, Murray T, Fournier DA, Sibert JR (1993). Determination of age and growth of South Pacific albacore (*Thunnus alalunga*) using three methodologies. Fishery Bulletin 91: 649–663.

Lovejoy NR (2000). Reinterpreting recapitulation: systematics of needlefishes and their allies (Teleostei: Beloniformes). Evolution 54: 1349–1362.

Lovejoy NR, Iranpour M, Collette BB (2004). Phylogeny and jaw ontogeny of beloniform fishes. Integrative and Comparative Biology 44: 366–377.

Lyczkowski-Shultz L (2005). Molidae: ocean sunfishes. In: Richards WJ (ed.), *Early Stages of Atlantic Fishes: An Identification Guide for the Western Central North Atlantic*. CRC Press, Boca Raton, Florida, pp. 2457–2465.

MacKenzie BR, Mosegaard H, Rosenberg AA (2009). Impending collapse of bluefin tuna in the northeast Atlantic and Mediterranean. Conservation Letters 2: 25–34.

Magnuson JL, Prescott JH (1966). Courtship, locomotion, feeding, and miscellaneous behaviour of Pacific bonito (*Sarda chiliensis*). Animal Behaviour 14: 54–67.

Majkowski J (2007). Global fishery resources of tuna and tuna-like species. FAO Fisheries Technical Paper 483.

Margulies D, Suter JM, Hunt SL, Olson RJ, Scholey VP, Wexler JB, Nakazawa A (2007). Spawning and early development of captive yellowfin tunas (*Thunnus albacares*). Fishery Bulletin 105: 249–265.

Masurekar VB (1968). Eggs and development stages of *Tylosurus crocodilus* (Lesueur). Journal of the Marine Biological Association of India 9: 70–76.

Mather FJ III, Clark HL, Mason JM Jr (1975). Synopsis of the biology of the white marlin, *Tetrapturus albidus* Poey, (1861). NOAA Technical Report NMFS SSRF-675(3): 55–94.

Matsumoto T, Bayliff WH (2008). A review of the Japanese longline fishery for tunas and billfishes in the eastern Pacific Ocean, 1998–2003. Bulletin of the Inter-American Tropical Tuna Commission 24: 1–187.

Matsumoto WM, Skillman RA, Dizon AE (1984). Synopsis of biological data on skipjack tuna, *Katsuwonus pelamis*. NOAA Technical Report NMFS Circular 451.

McBride RS, Thurman PE (2003). Reproductive biology of *Hemiramphus brasiliensis* and *H. balao* (Hemiramphidae): maturation, spawning frequency, and fecundity. Biological Bulletin, Woods Hole 204: 57–67.

McBride RS, Foushee L, Mahmoudi B (1996). Florida's halfbeak, *Hemiramphus* spp., bait fishery. Marine Fisheries Review 58(1–2): 29–38.

McBride RS, Styer JR, Hudson R (2003). Spawning cycles and habitats for ballyhoo (*Hemiramphus brasiliensis*) and balao (*Hemiramphus balao*) in south Florida. Fishery Bulletin 101: 583–589.

McBride RS, Richardson AK, Maki KL (2008). Age, growth, and mortality of wahoo, *Acanthocybium solandri*, from the Atlantic coast of Florida and the Bahamas. Marine and Freshwater Research 59: 799–807.

McDowell JR, Graves JE (2008). Population structure of striped marlin (*Kajikia audax*) in the Pacific Ocean based on analysis of microsatellite and mitochondrial DNA. Canadian Journal of Fisheries and Aquatic Science 65: 1307–1320.

Meisner AD, Collette BB (1999). Generic relationships of the internally-fertilized southeast Asian halfbeaks (Hemiramphidae: Zenarchopterinae). In: Proceedings of the Fifth Indo-Pacific Fish Conference, Nouméa, 1997. Societe Francaise d'Ichthyologie, pp. 69–76.

Mito S (1958). Eggs and larvae of *Tylosurus melanotus* (Bleeker) (Belonidae). In: Uchida K et al. (eds.), Studies on the eggs, larvae and juveniles of the Japanese fishes. Series 1. Second Laboratory of Fisheries Biology, Fisheries Department, Faculty of Agriculture, Kyushu University, Fukuoka, Japan, p. 22.

Miyabe N (1994). A review of the biology and fisheries for bigeye tuna, *Thunnus obesus*, in the Pacific Ocean. FAO Fisheries Technical Paper 336(2): 207–243.

Mukhacheva VA (1960). Some data on the breeding, development and distribution of saury: *Cololabis saira* (Brevoort). Trudy Instituta Okeanologii 41: 163–174 (in Russian).

Nakajima H, Kawahara H, Takamatsu S (1987). The breeding behavior and the behavior of larvae and juveniles of the sharksucker, *Echeneis naucrates*. Japanese Journal of Ichthyology 34: 66–70.

Nakamura I (1975). Synopsis of the biology of the black marlin, *Makaira indica* (Cuvier), 1831. NOAA Technical Report NMFS SSRF-675(3): 17–27.

Nakamura I (1985). FAO species catalogue. Vol. 5. Billfishes of the world. An annotated and illustrated catalogue of marlins, spearfishes, and swordfishes known to date. FAO Fisheries Synopsis 125(5).

Nakatsubo T, Kawachi M, Mano N, Hirose H (2007). Spawning period of ocean sunfish *Mola mola* in waters of the eastern Kanto Region, Japan. Aquaculture Science 55: 613–618 (in Japanese with English abstract).

Neilson JD, Manickhand-Heileman S, Singh-Renton, S (1994). Assessment of hard parts of blackfin tuna (*Thunnus atlanticus*) for determining age and growth. International Commission for the Conservation of Atlantic Tunas Collective Volume of Scientific Papers 42(2): 369–376.

Nelson JS (2006). *Fishes of the World*, 4th ed. John Wiley & Sons, Hoboken, NJ.

Nesterov AA, Shiganova TA (1976). The eggs and larvae of the Atlantic saury, *Scomberesox saurus* of the North Atlantic. Voprosy Ikhtiologii 16: 315–322 (in Russian, translated in Journal of Ichthyology 16: 277–283).

Nichols JT, Breder CM Jr (1928). An annotated list of the Synentognathi with remarks on their development and relationships. Collected by the Arcturus. Zoologica, New York 8: 423-448.

Niiya Y (2001a). Age, growth, maturation and life of bullet tuna *Auxis rochei* in the Pacific waters off Koch Prefecture. Nippon Suisan Gakkaishi 67(3): 429–437 (in Japanese with English abstract).

Niiya Y (2001b). Maturation cycle and batch fecundity of the bullet tuna *Auxis rochei* off Cape Ashizuri, southwestern Japan. Nippon Suisan Gakkaishi 67: 10–16 (in Japanese with English abstract).

Olney JE (1984). Lampriformes: development and relationships. In: Moser HG et al. (eds.), Ontogeny and systematics of fishes. American Society of Ichthyologists and Herpetologists Special Publication 1, Lawrence, Kansas, pp. 368–379.

Olney JE (2005). Chapter 81. Family Lamprididae. In: Richards WJ (ed.), *Early Stages of Atlantic Fishes: An Identification Guide for the Western Central North Atlantic*. CRC Press, Boca Raton, Florida, pp. 995–997.

Orton GL (1964). The eggs of scomberesocid fishes. Copeia 1964: 144–150.

O'Toole B (1999). Phylogeny of the species of the superfamily Echeneoidea (Perciformes: Carangoidei: Echeneidae, Rachycentridae, and Coryphaenidae) with an interpretation on echeneid hitchhiking behaviour. Canadian Journal of Zoology 80: 596–623.

Oxenford, HA (1999). Biology of the dolphinfish (*Coryphaena hippurus*) in the western central Atlantic: a review. Scientia Marina 63: 277–301.

Oxenford HA, Hunte W (1986). A preliminary investigation of the stock structure of the dolphinfish (*Coryphaena hippurus*) in the eastern Caribbean. Fishery Bulletin 84: 451–460.

Oxenford HA, Mahon R, Hunte W (1995). Distribution and relative abundance of flying fish (Exocoetidae) in the eastern Caribbean. I. Adults. Marine Ecology Progress Series 117: 11–23.

Oxenford HA, Murray PA, Luckhurst BE (2003). The biology of wahoo (*Acanthocybium solandri*) in the western central Atlantic. Gulf and Caribbean Research 15: 33–49.

Pagavino M, Gaertner D (1995). Ajuste de una curva de crecimiento a frecuencias de tallas de atún listado (*Katsuwonus pelamis*) pescado en el mar Caribe suroriental. International Commission for the Conservation of Atlantic Tunas Collective Volume of Scientific Papers 44(2): 303–309.

Palko BJ, Beardsley GL, Richards WJ (1981). Synopsis of the biology of the swordfish, *Xiphias gladius* Linnaeus. NOAA Technical Report NMFS Circular 441.

Palko BJ, Beardsley GL, Richards WJ (1982). Synopsis of the biological data on dolphinfishes, *Coryphaena hippurus* Linnaeus and *Coryphaena equiselis* Linnaeus. NOAA Technical Report NMFS Circular 443.

Parin NV (1970). Ichthyofauna of the epipelagic zone (translated from Russian by M. Ravah). Israel Program for Scientific Translations, Jerusalem.

Parin NV, Shakhovskoy IB (2000). A review of the flying fish genus *Exocoetus* (Exocoetidae) with descriptions of two new species from the Southern Pacific Ocean. Journal of Ichthyology 40, Supplement 1: S31–S63.

Potthoff T, Kelley S (1982). Development of the vertebral column, fins and fin supports, branchiostegal rays, and squamation in the swordfish, *Xiphias gladius*. Fishery Bulletin 80: 161–186.

Randall JE (1960). The living javelin. Sea Frontiers 6: 228–233.

Restrepo V, Prince ED, Scott GB, Uozumi Y (2003). ICCAT stock assessments of Atlantic billfish. Australian Journal of Marine and Freshwater Research 54: 361–367.

Richards WJ (2005a). Echeneidae: remoras. In: Richards WJ (ed.), *Early Stages of Atlantic Fishes: An Identification Guide for the Western Central North Atlantic*. CRC Press, Boca Raton, Florida, pp. 1433–1438.

Richards WJ (2005b). Scombridae: mackerels and tunas. In: Richards WJ (ed.), *Early Stages of Atlantic Fishes: An Identification Guide for the Western Central North Atlantic*. CRC Press, Boca Raton, Florida, pp. 2187–2227.

Richards WJ (2005c). Xiphiidae: swordfish. In: Richards WJ (ed.), *Early Stages of Atlantic Fishes: An Identification Guide for the Western Central North Atlantic*. CRC Press, Boca Raton, Florida, pp. 2241–2243.

Richards WJ, Luthy SA (2005). Istiophoridae: billfishes. In: Richards WJ (ed.), *Early Stages of Atlantic Fishes: An Identification Guide for the Western Central North Atlantic*. CRC Press, Boca Raton, Florida, pp. 2231–2240.

Rivas LR (1975). Synopsis of biological data on blue marlin, *Makaira nigricans* Lacepède, 1802. NOAA Technical Report NMFS SSRF-675(3): 1–16.

Robins CR (1975). Synopsis of biological data on the longbill spearfish, *Tetrapturus pfluegeri* Robins and de Sylva. NOAA Technical Report NMFS SSRF-675(3): 28–38.

Rodríguez-Roda J (1966). Estudio de la bacoreta, *Euthynnus alleteratus* (Raf.), bonito, *Sarda sarda* (Bloch), y melva, *Auxis thazard* (Lac.), capturados por las almadrabas españolas. Investigacion Pesquera, Barcelona 30: 247–292.

Rooker JR, Alvarado Bremer JR, Block BA, Dewar H, de Metrio G, Corriero A, Kraus RT, Prince ED, Rodríguez-Marín E, Secor DH (2007). Life history and stock structure of Atlantic bluefin tuna (*Thunnus thynnus*). Reviews in Fishery Science 15: 265–310.

Rooker JR, Secor DH, de Metrio G, Schloesser R, Block BA, Neilson JD (2008). Natal homing and connectivity in Atlantic bluefin tuna populations. Science 322: 742–744.

Rosen DE, Parenti LR (1981). Relationships of *Oryzias*, and the groups of atherinomorph fishes. American Museum Novitates 2719: 1–25.

Sadovy Y, Liu M (2008). Functional hermaphroditism in teleosts. Fish and Fisheries 9: 1–43.

Sawada Y, Okada T, Miyashita S, Murata O, Kumai H (2005). Completion of the Pacific bluefin tuna *Thunnus orientalis* (Temminck et Schlegel) life cycle. Aquaculture Research 36: 413–421.

Schaefer KM (1998). Reproductive biology of yellowfin tuna (*Thunnus albacares*) in the eastern Pacific Ocean. Bulletin of the Inter-American Tropical Tuna Commission 21: 201–272.

Schaefer KM (2001). Reproductive biology of tunas. In: Block BA, Stevens ED (eds.), *Tuna: Physiology, Ecology, and Evolution*. Fish Physiology 19. Academic Press, San Diego, pp. 25–270.

Schaefer KM, Fuller DW, Miyabe N (2005). Reproductive biology of bigeye tuna (*Thunnus obesus*) in the eastern and central Pacific Ocean. Bulletin of the Inter-American Tropical Tuna Commission 23: 1–31.

Schlesinger G (1909). Zur Phylogenie und Ethologie der Scombresociden. Verhandlungen der Zoologisch-Botanischen Gesellschaft in Wien 59: 302–339.

Schmidt J (1921). New studies of sun-fishes made during the "Dana" expeditions 1920. Nature 107: 76–79.

Schwenke KL, Buckel JA (2008). Age, growth, and reproduction of dolphinfish (*Coryphaena hippurus*) caught off the coast of North Carolina. Fishery Bulletin 106: 82–92.

Shivji MS, Magnussen JE, Beerkircher LR, Hinteregger G, Lee DW, Serafy JE, Prince ED (2006). Validity, identification, and distribution of the roundscale spearfish, *Tetrapturus georgii* (Teleostei: Istiophoridae): morphological and molecular evidence. Bulletin of Marine Science 79: 483–491.

Shomura RS, Williams F, eds. (1975). Proceedings of the International Billfish Symposium Kailua-Kona, Hawaii, 9–12 August 1972. Part 3. Species synopses. NOAA Technical Report NMFS SSRF-675(3): 1–159.

Sissenwine, MP, Mace PM, Powers JE, Scott, GP (1998). A commentary on western Atlantic bluefin tuna assessments. Transactions of the American Fisheries Society 127: 838–855.

Strasburg, DW (1964). Further notes on the identification and biology of echeneid fishes. Pacific Science 18: 51–57.

Sun C-L, Chang Y-J, Tszeng C-C, Su N-J (2009). Reproductive biology of blue marlin (*Makaira nigricans*) in the western Pacific Ocean. Fishery Bulletin 107: 420–432.

Sun Z-G (1960). Larvae and juveniles of tunas, sailfish and swordfish (Thunnidae, Istio-phoridae, Xiphiidae) from the central and western parts of the Pacific Ocean. Trudy In-stituta Okeanologii 41: 175–191 (in Russian).

Theisen TC, Bowen BW, Lanier W, Baldwin JD (2008). High connectivity on a global scale in the pelagic wahoo, *Acanthocybium solandri* (tuna family Scombridae). Molecular Ecology 17: 4233–4247.

Thorogood J (1986). Aspects of the reproductive biology of the southern bluefin tuna (*Thunnus maccoyii*). Fisheries Science 4: 297–315.

Topp RW, Giradin DL (1971). An adult louvar, *Luvarus imperialis* (Pisces, Luvaridae), from the Gulf of Mexico. Copeia 1971: 181–182.

Tyler JC, Johnson GD, Nakamura I, Collette BB (1989). Morphology of *Luvarus imperialis* (Luvaridae) with a phylogenetic analysis of the Acanthuroidei (Pisces). Smithsonian Contributions to Zoology 485.

Uchida RN (1981). Synopsis of biological data on frigate tuna *Auxis thazard*, and bullet tuna, *A. rochei*. NOAA Technical Report NMFS Circular 436.

Ueyanagi S (1960). On the larvae and the spawning areas of the shirokajiki, *Marlina mar-lina* (Jordan & Hill). Report of the Nankai Regional Fisheries Research Laboratory 12: 85–96 (in Japanese with English abstract).

Ueyanagi S (1964). Description and distribution of larvae of five istiophorid species in the Indo-Pacific. Proceedings of the Symposium on Scombroid Fishes, Marine Biological Association of India Symposium Series 1: 499–528.

Ueyanagi S, Wares PG (1975). Synopsis of biological data on striped marlin, *Tetrapturus audax* (Philippi), 1887. NOAA Technical Report NMFS SSRF-675(3): 132–159.

Wang S-P, Sun C-L, Yeh S-Z, Chiang W-C, Su N-J, Chang Y-J, Liu C-H (2006). Length distributions, weight-length relationships, and sex ratios at lengths for the billfishes in Taiwan waters. Bulletin of Marine Science 79: 865–869.

Watanabe H, Yukinawa M, Nakazawa S, Ueyanagi S (1966). On the larva probably refer-able to slender tuna, *Allothunnus fallai* Serventy. Report of the Nankai Regional Fish-eries Research Lab. No. 23: 85–94 (in Japanese with English summary).

Watson W (1996). Beloniformes. In: Moser HG (ed.), The early stages of fishes in the California Current region. California Cooperative Oceanic Fisheries Investigation Atlas No. 33, pp. 625–657.

White Marlin Status Review Team (2002). Atlantic white marlin status review document. Report to the National Marine Fisheries Service, Southeast Regional Office, St. Peters-berg, Florida.

Wild A, Hampton J (1994). A review of the biology and fisheries for skipjack tuna, *Katsu-wonus pelamis*, in the Pacific Ocean. FAO Fisheries Technical Paper 336(2): 1–51.

Wollam MB (1969). Larval wahoo, *Acanthocybium solandri* (Cuvier), (Scombridae) from the Straits of Yucatan and Florida. Florida Department of Natural Resources Leaflet Series 4(12): 1–7.

Wu C-C, Su W-C, Kawasaki T (2001). Reproductive biology of the dolphin fish *Corphaena* [sic] *hippurus* on the east coast of Taiwan. Fisheries Science 67: 784–793.

Yesaki M (1994). A review of the biology and fisheries for longtail tuna (*Thunnus tonggol*) in the Indo-Pacific region. FAO Fisheries Technical Paper 336(2): 370–387.

Yesaki M, Arce F (1994). A review of the *Auxis* fisheries of the Philippines and some aspects of the biology of frigate (*A. thazard*) and bullet (*A. rochei*) tunas in the Indo-Pacific region. FAO Fisheries Technical Paper 336(2): 409–439.

Yoshida HO, Nakamura EL (1965). Notes on schooling behavior, spawning, and morphology of Hawaiian frigate mackerels, *Auxis thazard* and *Auxis rochei*. Copeia 1965: 111–114.

3

Reproduction in Scorpaeniformes

Marta Muñoz

The Scorpaeniformes, or mail-cheeked fishes, are one of the largest and most morphologically diverse teleostean orders with more than 1400 species classified in 24 to 36 families, depending on the taxonomy (e.g., Washington et al. 1984b; Eschmeyer 1998; Nelson 2006). After Cuvier's (1829) initial characterization, phylogenetic relationships of this group have been considered by many authors (e.g., Matsubara 1943, 1955; Yabe 1985; Gill 1988; Shinohara 1994; Imamura 1996). Although it has been traditionally retained as a taxonomic unit, recent studies indicate that the mail-cheeked fish group is polyphyletic (see Shinohara & Imamura 2007 for a review comparing findings of Imamura & Yabe 2002; Miya et al. 2003; Smith & Wheeler 2004). As a result, this chapter is structured according to the two recognized major scorpaeniform lineages: the Cottoidei and the Scorpaenoidei (Washington et al. 1984b; Kendall 1991; Imamura & Shinohara 1998).

As Washington et al. (1984a) has pointed out, the Scorpaeniformes show a wide range of reproductive strategies. Some families spawn individual pelagic eggs (Anoplopomatidae, Congiopodidae, and Triglidae) while others spawn demersal clusters of adhesive eggs (Cyclopteridae, Cottidae, and Agonidae). However, most of them spawn pelagic egg masses embedded in a gelatinous matrix. There are strong trends toward internal fertilization, and the genera *Sebastes* and *Sebasticus* give birth to live young.

Following the excellent discussion about the evolution of piscine viviparity in rockfishes made by Wourms (1991) and summarized in Table 3.1, it seems that the ancestral stock of scorpaeniforms was probably unspecialized. It is reasonable to assume that their reproduction involved a simple form of oviparity, such as broadcast spawning of pelagic, or oviposition of demersal, eggs. This ancestral

TABLE 3.1 Scenario for evolution of viviparity in scorpaeniform fishes, from Wourms (1991)

Cottoids (mostly demersal)

Recent cottoids. Oviparous.

Internal fertilization in some taxa. Demersal eggs spawned in masses. Parental care. Exception
  Comephorus, pelagic and viviparous.

Scorpaenoids (primarily demersal, secondarily pelagic)

Oviparous ovuliparous. Broadcast spawning of individual pelagic eggs. External fertilization.

Oviparous, ovuliparous. Pelagic spawning, groups of eggs enclosed in gelatinous mass. External
  or internal fertilization.

Oviparous, zygoparous or embryoparous. Pelagic spawning, groups of developing eggs or
  embryos enclosed in gelatinous mass. Internal fertilization.

Viviparous. Internal fertilization. Developing embryos enclosed in gelatinous mass that is
  extruded at parturition. Embryonic nutrition, lecithotrophic.

Viviparous. Internal fertilization. Developing embryos not enclosed in gelatinous mass.
  Embryonic nutrition ranges from lecithotrophic to matrotrophic.

stock diverged into two groups: the lineages Cottoidei and Scorpaenoidei. This is in concordance with the reproductive dichotomy found between demersal spawning cottoids and pelagic spawning scorpaenoids, which according to Washington et al. (1984b) led to a reproductive partitioning of shared habitats. The wide distribution and extensive speciation of the two lineages indicate that they are both successful groups, and as Wourms (1991) states, it is reasonable to assume that reproductive modes account, at least in part, for their success.

In this chapter, the diversity and evolutionary trends of reproduction in scorpaeniform fishes will be presented and discussed.

COTTOID LINEAGE

In most species of cottoids, the general aspects of the mode of reproduction are similar. Oviparity is retained, clusters of adhesive demersal eggs are generally laid, and parental care is common. This reproductive style has proved to be highly successful and is retained in modern cottoids (Washington et al. 1984a). However, there are a few exceptions: species of Anoplopomatidae secondarily adopted a pelagic lifestyle and produce pelagic eggs, and some freshwater species are viviparous.

At least one of the two existing species of pelagic anoplopomids, the sablefish, *Anoplopoma fimbria*, is long-lived (up to 113 years; McFarlane & Beamish 1983, 1995), with pelagic spawning at depths of 300 to 500 m, near the edges of the continental slopes (McFarlane & Nagata 1988). Eggs develop at these depths; following hatching, larvae develop near the surface as far offshore as 180 miles (Wing 1997). Females of *A. fimbria* are extremely fecund, producing up to one million eggs (Alderdice et al. 1988; McFarlane & Nagata 1988). This reproductive style of non-guarding, egg-scattering pelagic spawners and nutrient-poor ova produced in high numbers coincides with that of most fishes and seems to be the ancestral condition (Balon 1984).

Nowadays most cottoids are guarders, including Hexagrammidae, Cottidae, Cyclopteridae, Hemitripteridae, and others. However, they display a wide range of different types of fertilization and parental behavior modes. Among cottids, the reproductive mode can be either copulatory or non-copulatory (Breder & Rosen 1966; Munehara et al. 1989).

Non-copulatory refers to the release of both ova and sperm to the external environment. This mode appears to be widespread among the hexagrammids. Where recorded, hexagrammids also exhibit either a promiscuous or polygamic mating system, or both. This has been demonstrated for species of several genera: *Pleurogrammus monopterygius* (Dermott et al. 2007), *Hexagrammos* spp. (Crow et al. 1997), *Oxylebius pictus* (DeMartini 1987), and the polyandric *Ophidion elongatus* (King & Withler 2005). Most of them exhibit clear courtship behavior involving—at least in different species of *Hexagrammos*—rushing, butting, and undulation of the trunk (Munehara et al. 2000). These actions have also been observed for *Pleurogrammus azonus* (Gomelyuk 1988). When a female enters the nest, the male leans his head on the spawning surface in the nest and spasmodically undulates his trunk. After the female has released her eggs within the seaweed bed, the male passes over them, touching his genital pore to the egg mass and releasing sperm. Sneaking by other males was frequently observed by Gomelyuk (1988) following the release of sperm.

With regard to the diversity of cottid fishes, their phylogenic relationships together with their reproductive modes are shown in Figure 3.1 (Abe & Munehara 2009). Among cottids, there are also some non-copulating marine species. In *Hemilepidotus gilberti*, females spawn adhesive demersal eggs on the spawning substratum. Under these conditions sperm release should take place close to the eggs to ensure that many spermatozoa will encounter eggs. However, males typically cannot get close to the egg mass, which is deposited between the spawning substratum and the female's belly (Hayakawa & Munehara 1996). Consequently, they position themselves behind the spawning female and emit semen at a distance from the eggs, raising hypotheses regarding the pattern of sperm motility in this species. Hayakawa & Munehara (1998) noted high motility and longevity of spermatozoa of this species when placed in experimentally provided ovarian fluid, a feature of spermatozoa that is similar to that of copulatory cottids. Subsequently, Hayakawa & Munehara (2001) demonstrated that internal fertilization occurs frequently in natural conditions in *H. gilberti*, although all of the internally fertilized eggs developed abnormally. Together with the fact that this species is regarded, based on morphology, as one of the most primitive species among Cottidae (Yabe 1985), this suggests that copulation of marine cottids may have evolved

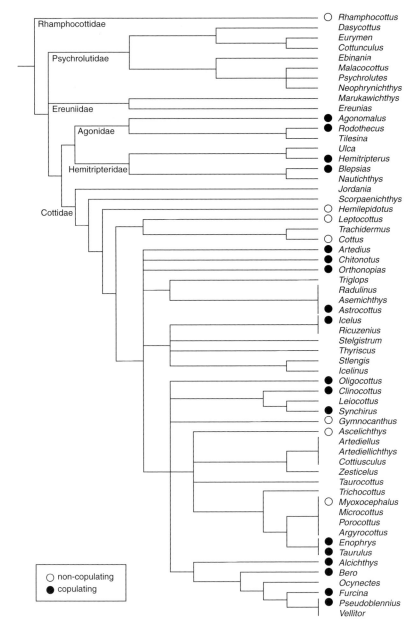

FIGURE 3.1. Phylogeny of cottoid fishes with reproductive mode (non-copulating or copulating). From Abe & Munehara (2009), modified after Yabe (1985). Reprinted with permission from Science Publishers, Enfield, NH.

from the facultative insemination of non-copulatory species, in which the physiological characteristics of spermatozoa are adapted for ovarian fluid conditions.

In copulating species of cottids, spermatozoa are introduced into the female reproductive tract during copulation and associate with ova in the ovarian cavity. However, the entry of spermatozoa into ova and subsequent fertilization occur externally, when the ova come into contact with seawater. This differs from the more typical internal fertilization in fishes because, although sperm enter the ovary and may even enter the micropyle, it does not penetrate the ooplasmic membrane and initiate fertilization and development until eggs are released. This unique reproductive mode is called internal gametic association (IGA) and has been described for several cottid species, including *Alcichthys alcicornis* (Munehara et al. 1989), *Hemitripterus villosus* (Munehara 1992, 1996), *Blepsias cirrhosus* (Munehara et al. 1991), *Radulinopsis taranetzi* (Abe & Munehara 2005), *Bero elegans*, and *Pseudoblennius cottoides* (Koya et al. 1993). Moreover, it is thought to occur in *Artedius harringtoni* (Ragland & Fischer 1987) and the other copulating cottids that have previously been reported to be internal fertilizing species (Munehara et al. 1989), but for which true internal fertilization has not been documented. Hayakawa & Munehara (2001) suggest that the deficiency of calcium ions in the ovarian fluid, which inhibits internal fertilization in copulatory marine sculpins, might have evolved after the establishment of copulatory behavior. It could have developed to preserve eggs until their release to the environment as an alternative to viviparity, since the ovary of these IGA-mode species appears to be unable to supply developing embryos with a suitable environment and enough of essential elements such as oxygen. An excellent review about the reproductive mode in copulating cottoid species has been recently published (Abe & Munehara 2009).

Most of these copulating sculpins transfer sperm by using a large, flexible genital papilla (Ragland & Fischer 1987; Munehara 1988), and copulation can be aided by various morphological features (Munehara 1996). In *Oligocottus snyderi*, the male's first anal ray is enlarged so that it can clasp the female while copulating (Morris 1956). In *Synchirus gilli*, the opposing grip of the male's lower jaw and its pelvic fins are modified to embrace the female body (Krejsa 1964). However, *Blepsias cirrhosus* and *Hemitripterus villosus*, which do not possess such a functional genital organ, have also been inferred to copulate (Munehara et al. 1991). Male *H. villosus* have no intromittent organ and sperm transfer is external, but insemination occurs internally (Munehara 1996). The female first everts the genital papilla and emits gelatinous ovarian fluid from its tip. The male then releases semen in the direction of the gelatinous ovarian fluid, after which the female retracts the gelatinous mass and associated spermatozoa into the ovary, where it is stored. During sperm emission, semen is released at a distance from the gelatinous ovarian fluid, resembling the spawning and ejaculation of *Hemilepidotus gilberti* described above (Hayakawa & Munehara 1996).

In relation with these fertilization tactics, it should be pointed out that spermio-genesis in many cottoids is known to involve atypical paraspermatozoa (i.e., di-morphic spermatozoa) as well as normal euspermatozoa (Hann 1927, 1930; Quinitio et al. 1988; Quinitio 1989; Quinitio & Takahashi 1992; Hayakawa et al. 2002c, 2007; Hayakawa 2007). Parasperm are the nonfertile sperm that are regularly produced along with normal fertile sperm in reproductive males, unlike aberrant sperm, which are irregularly crippled by some errors during spermatogenesis (Healy & Jamieson 1981; Swallow & Wilkinson 2002). It is not always easy to find a biological explanation for the existence of considerable amounts of paraspermato-zoa, but Hayakawa & Munehara (2004) state that in these cottoids, the lump forma-tion of paraspermatozoa can help euspermatozoa move directly toward the eggs by preventing the lateral dispersion of semen. Without paraspermatozoa, semen would have difficulty reaching the egg mass due to lateral dispersion caused by drag before reaching the egg mass (Hayakawa et al. 2002a). Consequently, males with parasper-matozoa can increase the number of euspermatozoa which will reach the egg mass (Hayakawa et al. 2002c, 2004). In addition, parasperm can help eusperm fertilize the egg mass by blocking sperm from other males, when sperm competition arises as a result of several males releasing sperm at the same time (Hayakawa et al. 2002b).

Intraovarian sperm storage has been described in only a few cottids, and no specialized structures within the female reproductive tract have been found asso-ciated with this reproductive feature. However, Koya et al. (2002) have been able to relate some differences in the characteristics of the location of the stored sperm with differences in the storage period. For example, in *Alcichthys alcicornis*, in which sperm only spends a short time in the female ovary, the intraovarian sperm remain freely floating in the ovarian fluid. In contrast, in the marine cottids *Pseudoblennius cottoides*, *P. percoides,* and *Furcina ishikawae,* in which the sperm is maintained in the ovary for a longer period of time, spermatozoa are located in the posterior end of the ovary and are arranged perpendicular to the ovarian wall epithelium (Shinomiya 1985).

Cottoids are one of the most reproductively diverse groups of marine fishes in terms of providing parental care. Most of them produce demersal eggs, which is considered a fundamental step in the evolution of parental care in fish (Potts 1984), since if the embryos are not protected in some way, they are more vulner-able to predation and other environmental hazards. Hexagrammid males guard benthic nests containing several clutches of eggs deposited from multiple females, as has been described for *Ophidion elongatus* (King & Withler 2005), *Hexagram-mos decagrammus* (Crow et al. 1997), or *Pleurogrammus monopterygius* (Dermott et al. 2007). In the cottoid family Psychrolutidae, the blob sculpin *Psychrolutes phrictus* represents the first direct evidence of parental care in an oviparous deep-sea fish. Drazen et al. (2003) observed, with a remote operating vehicle, aggregations of egg-brooding blob sculpins attending nests of large pinkish eggs. Parents often sat directly on, or were otherwise in contact with, the eggs that were free of sediment, suggesting that the adults cleaned or fanned their nest sites. Finally, although freshwater cottids typically exhibit male parental care of demersal eggs, within the marine cottids a greater variety of care patterns has been documented. Strictly paternal care has been found in *Alcichthys alcicornis* (Koya et al. 1994) and *Artedius harringtoni* (Ragland & Fischer 1987). In contrast, Abe & Munehara (2005) showed that *Radulinopsis taranetzi* females practice exclusive maternal egg care. Biparental care with limited maternal care has been observed in *Hemilepidotus gilberti* (Hayakawa & Munehara 1996) and *H. hemilepidotus* (DeMartini & Patten 1979). In these two species, females remain near the eggs for a few days after spawning, while the males continue to care for them until they hatch several weeks later.

As Petersen et al. (2005) have pointed out, the pattern of male parental care and internal fertilization (or at least internal gamete association) is very unusual in fishes (Clutton-Brock 1991). Yet it appears to have evolved multiple times within the cottids. It appears, for example, in the two genera, *Alcichthys* and *Artedius*, which are located in very different lineages within the phylogeny (Yabe 1985). It seems that eggs serve as courtship devices, with males defending eggs and oviposition sites to obtain additional copulations with females (Rohwer 1978; Ragland & Fischer 1987). Munehara et al. (1994) found support for this hypothesis in *Alcichthys alcicornis*, with males commonly providing care to clutches they had not fathered. In this respect, Petersen et al. (op. cit.) point out that the ability of males to continue to obtain matings at a spawning site, and the role of eggs in attracting additional mates, have been suggested as important components of selection pressure for the evolution of paternal care in fishes (Barlow 1964; Jamieson 1995; Petersen 1995; among others).

Among cottoids, egg hiding, as opposed to parental care, is also common. Some cottids, such as *Pseudoblennius cottoides*, *P. percoides*, and *Furcina ishikawae*, deposit eggs in the peribranchial cavity of sea squirts (Uchida 1932), and *Blepsias cirrhosus* injects eggs into the tissue near the gastral cavity of a sponge (Munehara 1991). All of these fishes deposit eggs only on specific invertebrates. Among the family Hemitripteridae, *Hemitripterus americanus* also attaches its eggs near the base of a sponge (Warfel & Merriman 1944), and *H. villosus* deposits them on polychaete tubes (Munehara 1992, 1996). These last two species probably use invertebrates both as spawning substrates and as protection for the eggs from predators. *Pseudoblennius cottoides*, *P. percoides*, *F. ishikawae*, and *B. cirrhosus* eggs may also take additional advantage of available oxygen, as the eggs obtain oxygen for respiration from the seawater passing through the cavities of the invertebrates (Munehara 1992).

Another cottoid family that is a close relative of sculpins is Liparidae. Snailfish, or liparids, are known to lay eggs attached to hydroids (Able & Musick 1976), algae (Detwyler 1963), in empty bivalve shells (DeMartini 1978), and among polychaete tubes or barnacle colonies (Marliave & Peden 1989). In the case of the liparid genus *Careproctus*, the eggs form compact masses within the branchial chambers of lithodid crabs (Somerton & Donaldson 1998). According to these authors and to Yau et al. (2000), such an association may provide snailfish with protection from potential predators, an optimum aerated environment, and even a means of transport toward food falls, at no apparent cost to the crabs.

SCORPAENOID LINEAGE

The reproductive modes of the cottoid sister taxon, Scorpaenoidei, are also extremely varied, consistent with their broad polyphyly as determined by Smith & Wheeler (2004). Most of scorpaenoid species have separate sexes, although some platycephalids (Shinomiya et al. 2003) and caracanthids are hermaphroditic (Cole 2003). Nearly all the families are oviparous and spawn pelagic eggs, but within the subfamily Sebastinae, most species are viviparous. In this respect, there are many examples of intermediate stages between the two basic modes of reproduction, oviparity, and viviparity. These include ovuliparity, zygoparity, and embryoparity as defined by Blackburn et al. (1985) and Wourms et al. (1988). Ovuliparity refers to the release of ova from the female reproductive tract followed by their fertilization within the external environment (i.e., classical oviparity). Zygoparity refers to the oviparous reproductive mode in which fertilized ova are retained within the female reproductive tract for a short period of time before their release. Embryoparity is the pattern of oviparous reproduction in which the embryo is formed and may develop to an advanced state prior to its release from the female reproductive tract. Consequently, the extreme limits of embryoparity can overlap with those of viviparity.

The least specialized mode of reproduction occurs in non-guarding families like Triglidae, Platycephalidae, and Synanceiidae. Most representatives of these three families are ovuliparous broadcast spawners of individual pelagic eggs, and fertilization is external. Members of the family Triglidae are found on sandy and muddy substrates or rubble, using the free pectoral rays for support and to search for food. According to Potts (1984), fish that live on, or in association with, such relatively mobile substrates normally have pelagic eggs, thereby preventing the risk of abrasion or smothering by moving particles. In fact, the eggs of triglids are pelagic and spherical, with a single oil droplet and a diameter over 1 mm (Dulcic et al. 2001; Muñoz et al. 2001, 2002a, 2003). Triglids can emit growling or grunting sounds, related to reproduction, with their swim bladder (Moulton 1963). Aspects of the biology of the triglids suggest that they have a generalist and opportunistic

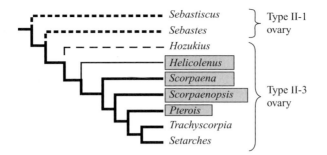

FIGURE 3.2. Phylogenetic tree of scorpaenoids showing their reproductive modes (after Imamura 2005, modified from Koya & Muñoz 2007). Bold dotted line=viviparous. Thin dashed line=, unknown. Thin solid line=zygoparous or embryoparous. Bold line=oviparous. Boxes label those taxa that release the egg mass within a gelatinous matrix.

life style. Triglid growth patterns reveal an extended longevity in most species—16 years for *Chelidonichthys capensis* (McPhail et al. 2001) and 21 years for *Aspitrigla cuculus* Baron (1985)—and many of them exhibit extremely rapid growth before sexual maturity. As Booth (1997) has pointed out, early maturity in long-lived species, with reproduction initiated at a large size, ensures high individual reproductive output.

Another ovuliparous family is Platycephalidae, which seems to contain several species with protandrous sex change, as *Cociella crocodila, Inegocia japonica, Kumococcius rodericensis,* and *Onigocia macrolepis* (Aoyama et al. 1963; Fujii 1970, 1971). In fact, Shinomiya et al. (2003) demonstrated both the existence of this kind of hermaphroditism and of a courtship behavior, at least in *I. japonica.*

According to Wourms (1991), the next phase in the evolutionary scenario of this lineage occurs in Sebastolobine, Scorpaenine, and Pteroine species, and involves a shift from the primitive, unspecialized pattern of spawning individual pelagic eggs to a more specialized pattern of oviparity. Fertilization is still external and the mode of reproduction is ovuliparous, but the small (0.7 to 1.2 mm), spherical to slightly ovoid eggs are now embedded in a large, pelagic, gelatinous matrix (Washington et al. 1984a). Phylogenetic relationships of the scorpaenoids together with their reproductive modes are shown in Figure 3.2 (Koya & Muñoz 2007), where the ovuliparous mode of reproduction of the genera *Setarches, Trachyscorpia, Pterois, Scorpaenopsis* and *Scorpaena* is marked.

Several characteristics of this specialized mode of oviparity distinguish many scorpaenids from the simpler ovuliparous species widely described in the literature. Wourms & Lombardi (1992) have suggested that this specialized mode of

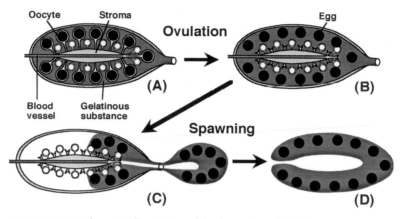

FIGURE 3.3. Schematic illustration of the formation of gelatinous egg masses in the kichiji rockfish. (A) Ovary at the pre-ovulatory period, (B) ovary at post-ovulatory period, (C) spawning, and (D) gelatinous egg mass spawned. From Koya et al. (1995).

oviparity first appeared in the genus *Scorpaena*, so the following discussion focuses on *Scorpaena notata*, and compares the results obtained in our work (Muñoz et al. 2002b, c) with those for other scorpaenids that share this reproductive strategy.

The ovarian stroma of *S. notata* is located in the center of the gonad, and the developing oocytes are connected to it by peduncles extending out into the surrounding lumen. Though the ovarian structure of *S. notata* differs from that of most Teleostei, it is similar to that described for some other scorpaenid species, such as *Dendrochirus brachypterus* (Fishelson 1977, 1978), *Sebastolobus alascanus* (Erickson & Pikitch 1993), *Sebastolobus macrochir* (Koya & Matsubara 1995), *Scorpaenodes littoralis* (Yoneda et al. 2000), *Helicolenus hilgendorfi* (Takano 1989), and *H. dactylopterus dactylopterus* (Muñoz et al. 1999). This type of organization corresponds to the cystovarian type II-3 ovary defined by Takano (1989), and seems to be related to the production of a gelatinous matrix that surrounds the expelled eggs, as shown in Figure 3.3 (Koya et al. 1995).

The peduncles connecting the ovarian follicles to the stroma, the detected paucity and small size of the cortical alveoli, and the thinness of the zona radiata of the oocytes are all characteristics typical of viviparous species (Wourms 1976; Fleger 1977; Takemura et al. 1987). The peduncles are usually considered to be protuberances of placentary or pseudo-placentary connections (Erickson & Pikitch 1993). In oviparous species, their presence may act to prevent oocytes from being too tightly packed (Fishelson 1975) or facilitate the ovulation of mature oocytes directly into the gelatinous mass. The role of the cortical alveoli in the strengthening of the chorion immediately after the fertilization of the egg is well known (Selman & Wallace 1989; West 1990; Iwamatsu et al. 1995; Shibata & Iwamatsu 1996). Their paucity and small size here suggest that less physical protection is needed by the eggs. This is further supported by the presence of a thin zona radiata, which normally provides mechanic protection. In the case of the scorpionfishes, mechanical protection is provided instead by the gelatinous matrix that encloses the eggs.

The testes of *Scorpaena* spp. also show some unusual features not typical in the most basic form of oviparity. In fish, the male germinal epithelium is normally composed of spermatocysts formed when a single clone of primary spermatogonia is enclosed by Sertoli cells. The germ cells develop inside these cysts. At the end of the process, the cysts open and spermatozoa are released into the lobular lumen. However, spermatogenesis among analyzed species of *Scorpaena* does not follow this pattern. Instead, cysts open before spermatogenesis is complete, and therefore sperm maturation finishes in the lobule lumen, without any physical connection with Sertoli cells (Muñoz et al. 2002b; Sàbat et al. 2009). This is semicystic spermatogenesis first described by Mattei et al. (1993) in the *Ophidion* genus, and has rarely been reported in fish.

The spermatozoa of these *Scorpaena* species are classified as primitive, despite exhibiting some modifications related to the specialization of semicystic spermatogenesis that occurs in the genus. As defined by Jamieson & Leung (1991), these are Type I anachrosomal aquasperm, which are characteristic of species using external fertilization. However, while the midpiece is usually shorter than 1 μm in this type of spermatozoon, in *Scorpaena angolensis* (Mattei 1970), *S. notata*, *S. porcus*, and *S. scrofa* it is longer (for example 1.35 μm in *S. notata*, as described by Muñoz et al. 2002b). As the midpiece is made up of mitochondria used to power movement of the flagellum, this extra length provides more energy for the sperm to move (Baccetti & Afzelius 1976) and is generally considered to be an adaptation to internal fertilization (Jamieson 1991; Mattei 1991), since the viscosity of the medium in which fertilization occurs demands more energy supply (Idelman 1967).

As we have seen, there are several reproductive features that place the *Scorpaena* genus, together with other oviparous scorpenids, in a position of having an intermediate reproductive strategy between the simplest ovuliparity and the development of internal fertilization. Among all of the functions attributed to ovarian fluid in fishes, perhaps the most important in the case of scorpenids is keeping the spawn together. If sperm are released onto the grouped mass of eggs within the gelatinous matrix, fertilization is assured and thereby reduces the need for the female to produce large number of eggs, which could explain their relative low fecundity (Muñoz et al. 2005). Observations on scorpionfishes forming pairs during certain times of the year seem to support this hypothesis.

Complex courtship and mating behaviors in oviparous scorpaenids have been described for *Pterois volitans* (Ruiz-Carus et al. 2006), in which males have an elaborate courting display and use their spines in agonistic displays with competing males (Fishelson 1975). Subsequently, females release two mucus-filled egg clusters that dissolve and release the eggs into the water column

At some time in the evolution of this lineage, internal fertilization appeared. According to Wourms (1991), once evolved, its presence was a strong, positive developmental feature that would facilitate the evolution of viviparity.

It is believed that viviparity evolved from oviparity in teleosts, as well as in Scorpaenidae. But despite the repeated suggestion of a scenario in which viviparous genera such as *Sebastes* and *Sebasticus* evolved from oviparous ancestral scorpaenids (such as *Scorpaena*) via embryoparity similar to that of *Helicolenus* (i.e., through the described specialization and a progressively longer retention of embryos), the comparative study of ovarian structures by Koya & Muñoz (2007) does not support this hypothesis. Their analysis suggests that the evolution of the reproductive mode in scorpaenid fishes is based on the primitive dorsal lamella-type ovary (i.e. in which the lamella-like stroma develops from the ovarian hilus located on the dorsal side of the ovary) of *Sebastes* and *Sebasticus*. In these genera, viviparity evolved later, from this ancestral type of ovary. Nevertheless, in the genera that spawn a floating egg mass, such as *Scorpaena* and *Sebastolobus*, the ovarian structure evolved from the primitive one to the previously described central stroma type. Subsequent internal fertilization and embryoparity, such as that of *Helicolenus*, evolved from them.

Ovaries of the dorsal lamella type are the general structure in other families of Scorpaeniformes, although within the family Scorpaenidae, they are only found in viviparous genera. In addition, the ovary of the central stroma type described for many scorpaenid oviparous genera and for the embryoparous *Helicolenus*, has only been observed in scorpaenids and caracanthids (Cole 2003), so it seems to be a structure that originally evolved in this lineage.

The scenario of the evolution of the reproductive mode of scorpaenids, which could be based on the evolution of the above-mentioned ovarian structures, is described as follows (Koya & Muñoz 2007). Current scorpaenids may have evolved from an ancestral clade that had the typical dorsal lamella-type ovary and in which ova were released and fertilized in the external environment. At some time during evolution, copulation behavior developed from this ancestral mode, leading to the development of viviparous species such as *Sebastiscus* and *Sebastes*. However, along a different evolutionary pathway from the ancestral species, other species that spawn floating eggs masses developed, accompanied by a shift in ovarian structure to the central stroma type, which facilitates the formation of the gelatinous matrix (e.g., *Scorpaena, Scorpaenopsis, Scorpaenodes, Pterois, Dendrochirus,* and *Sebastolobus*). The ancestral scorpaenid species possessing an ovary of the dorsal lamella-type and releasing eggs that are fertilized externally may have disappeared following such a bipolarization of reproductive mode. Later, copulation behavior appeared within the group having the central stroma-type ovary. They had not yet evolved to viviparity but stayed in the reproductive mode where they spawn developing embryos embedded in the floating egg mass, as presently seen in *Helicolenus* (Krefft 1961; Sequeira et al. 2003).

Features that suggest a separate evolutionary line for *Sebastes* with viviparity and *Helicolenus* with zygoparity are not limited to their ovarian structures. It is known that the gonadal cycles of males and females in several species of *Sebastes* are not synchronized, and there is a long period of time between spermatogenesis and oogenesis within a copulating pair. This occurs in *Sebastes marinus* and *S. mentella* (Sorokin 1961), *S. pachycephalus pachycephalus* (Shiokawa 1962), *S. serranoides* (Love & Westphal 1981), *S. taczanowskii* (Takemura et al. 1987), and *S. elongatus* (Shaw & Gunderson 2006). The same phenomenon has also been described for *Helicolenus dactylopterus* (Muñoz & Casadevall 2002). In these species, plus others of the same genus like *Sebastes atrovirens* (Sogard et al. 2008) or *Helicolenus lengerichi* (Lisovenko 1979), after copulation—described in depth together with the courtship for *Sebastes inermis* (Shinomiya & Ezaki 1991) and *S. mystinus* (Helvey 1982)—the spermatozoa are retained in the ovaries until the oocytes are mature and fertilization can take place.

However, if the viviparity in scorpaenids evolved from the embryoparity of *Helicolenus*, we would expect that sperm storage structures would also reveal this evolutionary pattern, being simpler in *Helicolenus* and more complex in *Sebastes*. In spite of this, we found that storage of spermatozoa in *Helicolenus* takes place in structures much more complex than those found in viviparous species. In *Sebastiscus* and *Sebastes*, sperm is stored but there are neither specialized ovarian structures nor any unique structural modifications of sperm that can be linked to its storage (Tateishi et al. 1958; Takemura et al. 1987; Eldridge et al. 1991; Takahashi et al. 1991), excepting the invasion by sperm of the follicular epithelia, as described for *Sebastes schlegeli* (Mori et al. 2003). In contrast, in *Helicolenus* there is a crypt (Figure 3.4), probably formed by an epithelial inclusion at the base of the lamellae, which takes up and encloses the sperm (Muñoz et al. 1999, 2000). The stored spermatozoa subsequently seem to be nourished in two ways: through their own energy reserves contained in a large cytoplasm mass that surrounds the sperm head during storage (Vila et al. 2007), and by means of a nutritive contribution from the cryptal epithelium that surrounds the spermatozoa (Muñoz et al. 2002d). Sperm remain protected against the female's immune system by a large number of intercellular junctions among cells of the cryptal epithelium (Vila et al. 2007). This fact has been documented in other scorpaeniform species having intraovarian sperm storage, such as *Alcichthys alcicornis*, in which traces of peroxidase confirm a breakdown of junctional complexes after the spawning period ends (Koya et al. 1997).

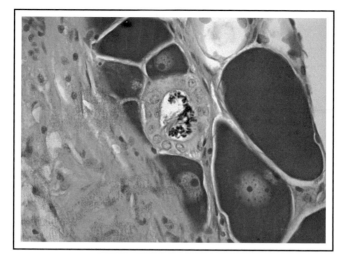

FIGURE 3.4. Intraovarian sperm storage crypt of *Helicolenus dactyopterus*. Epithelial cells surrounding stored sperm are clearly distinguishable.

The differences in ovarian morphology associated with the storage of sperm suggest that viviparous *Sebastiscus* and *Sebastes*, and embryoparous *Helicolenus*, evolved copulation and internal fertilization independantly. This hypothesis is supported by phylogenetic analyses (i.e., Imamura 2004), in which *Sebastiscus* is contained in the most basal clade, *Sebastes* is in the second most basal clade, with the ovary of dorsal lamella type present in both genera, and all genera with ovaries of the central stroma-type are in more recently derived clades. However, it should be pointed out that other recent phylogenetic studies (Smith & Wheeler 2004; Frehlick et al. 2006) consider or suggest that sebastids represent a more derived lineage.

Among viviparous teleosts, members of the genus *Sebastes* are considered relatively primitive with respect to the development of maternal-fetal relationships (Nagahama et al. 1991; Wourms 1991). Gestation is lumenal (i.e., within the ovarian lumen) and the embryos usually develop to term within the egg envelope during a gestation period which is generally longer, compared to the developmental period of oviparous embryos of similar sizes (Yamada & Kusakari 1991). Dygert & Gunderson (1991) suggest that although matrotrophy viviparity (i.e., involving the provision of maternal nutrition for the developing embryo) may be the common mode of reproduction, having been already demonstrated in different species of the genus (Boehlert & Yoklavich 1984; Boehlert et al. 1986; Shimizu et al. 1991; Yoklavich & Boehlert 1991; among others), energy contribution to the development of the embryo varies greatly between species.

Finally, it should be pointed out that most scorpaenids show high fecundities and long reproductive life spans. *Helicolenus dactylopterus* has a fecundity on par with, or even higher than, oviparous scorpaenids, despite the internal fertilization and temporal incubation of the eggs that occurs in this species (Muñoz & Casadevall 2002). In addition, members of the genus *Sebastes* are considered the most fecund of all the teleost viviparous species (Wourms & Lombardi 1992), and that those of the genus *Sebastodes* are also exceptionally fecund (MacGregor 1970). Moreover, their extreme longevity (Pearson & Gunderson 2003; Tsang et al. 2007)—with many species of *Sebastes* reaching maximum ages of 50 to 150 years (Archibald et al. 1981; Love et al. 1990) and a recorded maximum age of 205 years (Cailliet et al. 2001; Munk 2001)—in combination with the evidence that oogenesis continues at advanced ages (de Bruin et al. 2004) reveals a long reproductive life span among members of this group. These traits support the suggestion of Wourms (1991) that scorpaenids have evolved an effective reproductive style that incorporates the best of both possible worlds, namely the fecundity of oviparity combined with the enhanced survival of embryos and larvae conferred by viviparity.

ACKNOWLEDGMENTS

The research was supported in part by contract CTM2006-13964-C03-01 from the "Ministerio de Educación y Cultura" of Spain. I would like to thank reviewers Y. Koya and H. Munehara; their suggestions greatly improved the manuscript.

REFERENCES

Abe T, Munehara H (2005). Spawning and maternal-care behaviors of a copulating sculpin *Radulinopsis taranetzi*. Journal of Fish Biology 67(1): 201–212.

Abe T, Munehara H (2009). Adaptation and evolution of reproductive mode in copulating cottoid species. In: Jamieson B (ed.), *Reproductive Biology and Phylogeny of Fishes (Agnathans and Bony Fishes)*, vol. 8B. Science Publishers, Enfield, New Hampshire, pp. 221–247.

Able KW, Musick JA (1976). Life history ecology and behavior of *Liparis inquilinus* (Pisces Cyclopteridae) associated with the sea scallop *Placopecten magellanicus*. Fishery Bulletin 74(2): 409–421.

Alderdice DF, Jensen JOT, Velsen F (1988). Preliminary trials on incubation of sablefish eggs (*Anoplopoma fimbria*). Aquaculture 69: 271–290.

Aoyama T, Kitajima C, Mizue K (1963). Study of the sex reserval of inegochi, *Cociella crocodila* (Tilesius). Bulletin of Seikai Region Fishery Research Laboratory 29: 11–23.

Archibald CP, Shaw W, Leaman BM (1981). Growth mortality estimates of rockfish (Scorpaenidae) from BC coastal waters 1977–1979. Canadian Technical Reports of Fisheries and Aquatic Sciences 1048.

Baccetti B, Afzelius BA (1976). The biology of the sperm cell. Monographs in Developmental Biology 10: 1–254.

Balon EK (1984). Patterns in the evolution of reproductive styles in fishes. In: Potts GW, Wootton RJ (eds.), *Fish Reproduction: Strategies and Tactics*. Academic Press, London and Orlando, Florida, pp. 36–53.

Barlow GW (1964). Ethology of the Asian teleost *Badis badis*. V. Dynamics of fanning and other parental activities with comments on the behavior of larvae and postlarvae. Zeitschrift für Tierpsychogie 21: 99–123.

Baron J (1985). Les Triglidés (Teleostéens Scorpaeniformes) de la Baie de Douarnenez. 2. La reproduction de: *Eutrigla gurnardus, Trigla lucerna, Trigloporus lastoviza* et *Aspitrigla cuculus*. Cybium 9: 255–281.

Blackburn DG, Evans HE, Vitt LJ (1985). The evolution of fetal nutritional adaptations. Fortschritte der Zoologie 30: 437–439.

Boehlert GW, Yoklavich MM (1984). Reproduction, embryonic energetics and the maternal-fetal relationship in the viviparous genus *Sebastes* (Pisces: Scorpaenidae). The Biological Bulletin 167: 354–370.

Boehlert GW, Kusakari M, Shimizu M, Yamada J (1986). Energetics during embryonic development in kurosoi, *Sebastes schlegeli* Hilgendorf. Journal of Experimental Marine Biology and Ecology 101: 239–256.

Booth AJ (1997). On the life history of the lesser gurnard (Scorpaeniformes: Triglidae) inhabiting the Agulhas Bank, South Africa. Journal of Fish Biology 51: 1155–1173.

Breder CM, Rosen DE (1966). *Modes of Reproduction in Fishes*. TFH Publications, Neptune City, New Jersey.

Cailliet GM, Andrews AH, Burton EJ, Watters DL, Kline DE, Ferry-Graham LA (2001). Age determination and validation studies of marine fishes: do deep-dwellers live longer? Experimental Gerontology 36: 739–764.

Clutton-Brock TH (1991). *The Evolution of Parental Care*. Princeton University Press, Princeton, New Jersey.

Cole KS (2003). Hermaphroditic characteristics of gonad morphology and inferences regarding reproductive biology in *Caracanthus* (Teleostei, Scorpaeniformes). Copeia 2003 (1): 68–80.

Crow KD, Powers DA, Bernardi G (1997). Evidence for multiple maternal contributors in nests of kelp greenling (*Hexagrammus decagrammus*, Hexagrammidae). Copeia 1997: 9–15.

Cuvier G (1829). Le règne animal distribué d'àpres son organisation pour servir de base à l'historie naturelle des animaux et d'introduction à l'anatomie comparée. 2. Nouvelle Édition. Chez Déterville, Paris.

de Bruin JP, Gosden RG, Finch CE, Leaman BM (2004). Ovarian aging in two species of long-lived rockfish, *Sebastes aleutianus* and *S. alutus*. Biology of Reproduction 71: 036–1042.

DeMartini E (1978). Apparent paternal care in *Liparis fucensis* (Pisces: Cyclopteriade). Copeia 1978: 537–539.

DeMartini EE (1987). Paternal defense cannibalism and polygamy: factors influencing the reproductive success of painted greenling (Pisces, Hexagrammidae). Animal Behaviour 35: 1145–1158.

DeMartini EE, Patten BG (1979). Egg guarding and reproductive biology of the red Irish lord, *Hemilepidotus hemilepidotus* (Tilesius). Syesis 12: 41–55.

Dermott SF, Maslenikov KP, Gunderson DR (2007). Annual fecundity, batch fecundity and oocyte atresia of atka mackerel (*Pleurogrammus monopterygius*) in Alaskan waters. Fishery Bulletin 105: 19–29.

Detwyler R (1963). Some aspects of the biology of the seasnail, *Liparis atlanticus* (Jordan and Evermann). Ph.D. thesis, University of New Hampshire, Durham New Hampshire.

Drazen JC, Goffredi SK, Schlining B, Stakes DS (2003). Aggregations of egg-brooding deep-sea fish and cephalopods on the Gorda Escarpment: a reproductive hot spot. The Biological Bulletin 205: 1–7.

Dulcic J, Grubisic L, Katavic I, Skakelja N (2001). Embryonic and larval development of the tub gurnard *Trigla lucerna* (Pisces: Triglidae). Journal of the Marine Biological Association of the United Kingdom 8: 313–316.

Dygert P, Gunderson DR (1991). Energy utilization by embryos during gestation in viviparous copper rockfish, *Sebastes caurinus*. Environmental Biology of Fishes 30: 165–171.

Eldridge MB, Whipple JA, Bowers MJ, Jarvis BM, Gold J (1991). Reproductive performance of yellowtail rockfish *Sebastes flavidus*. Environmental Biology of Fishes 30: 91–102.

Erickson DL, Pikitch EK (1993). A histological description of shortspine thornyhead, *Sebastolobus alascanus*, ovaries: structures associated with the production of gelatinous egg masses. Environmental Biology of Fishes 36: 273–282.

Eschmeyer WN (1998). *Catalog of Fishes*. Vol. 3, *Genera of Fishes, Species and Genera in a Classification*. California Academy of Sciences, San Francisco.

Fishelson L (1975). Ethology and reproduction of pteroid fishes found in the Gulf of Aqaba (Red Sea), especially *Dendrochirus brachypterus* (Cuvier) (Pteroidae: Teleostei). Pubblicazioni della Stazione Zoologica di Napoli 39: 635–656.

Fishelson L (1977). Ultrastructure of the epithelium from the ovary wall of *Dendrochirus brachypterus* (Pteroidae: Teleostei). Cell and Tissue Research 177: 375–381.

Fishelson L (1978). Oogenesis and spawn-formation in the pigmy lion fish *Dendrochirus brachypterus* (Pteroidae). Marine Biology 46: 341–348.

Fleger C (1977). Electron microscopic studies on the development of the chorion of the viviparous teleost *Dermogenys pusillus* (Hemirhamphidae). Cell and Tissue Research 179: 255–270.

Frehlick LJ, Eirín-López J M, Prado A, Su HW (Harvey), Kasinsky HE, Ausió J (2006). Sperm nuclear basic proteins of two closely related species of Scorpaeniform fish (*Sebastes maliger, Sebastolobus* sp.) with different sexual reproduction and the evolution of fish protamines. Journal of Experimental Zoology 305: 277–287.

Fujii T (1970). Hermaphroditism and sex reversal in the fishes of the Platycephalidae. I. Sex reversal of *Onigocia macrolepis* (Bleeker). Japanese Journal of Ichthyology 17: 14–21.

Fujii T (1971). Hermaphroditism and sex reversal in the fishes of the Platycephalidae. II. *Kumococius detrusus* and *Inegocia japonica*. Journal of Ichthyology 18: 109–117.

Gill T (1988). On the classification of the mail-cheeked fishes. Proceedings of the United States National Museum 11: 567–592.

Gomelyuk VE (1988). Spawning behavior of Asian greenling, *Pleurogrammus azonua*, in Peter the Great Gulf. Journal of Ichthyology 28: 82–90.

Hann HM (1927). The history of the germ cell of *Cottus bairdii* Girard. Journal of Morphology and Physiology 43: 427–498.

Hann HM (1930). Variation in spermiogenesis in the teleost family Cottidae. Journal of Morphology and Physiology 50: 393–411.

Hayakawa Y (2007). Parasperm: morphological and functional studies on nonfertile sperm. Ichthyological Research 54: 111–130.

Hayakawa Y, Munehara H (1996). Non-copulatory spawning and female participation during early egg care in a marine sculpin, *Hemilepidotus gilberti*. Ichthyological Research 43: 73–78

Hayakawa Y, Munehara H (1998). The environment for fertilization of the non-copulating marine sculpin, *Hemilepidotus gilberti*. Environmental Biology of Fishes 52: 181–186.

Hayakawa Y, Munehara H (2001). Facultatively internal fertilization and anomalous embryonic development of a non-copulatory sculpin, *Hemilepidotus gilberti* Jordan and Starks (Scorpaeniformes: Cottidae). Journal of Experimental Marine Biology and Ecology 256: 51–58.

Hayakawa Y, Munehara H (2004). Ultrastructural observations of euspermatozoa and paraspermatozoa in a copulatory cottoid fish, *Blepsias cirrhosus*. Journal of Fish Biology 64: 1530–1539.

Hayakawa Y, Akiyama H, Munehara H, Komaru A (2002a). Dimorphic sperm influence semen distribution in a non-copulatory sculpin, *Hemilepidotus gilberti*. Environmental Biology of Fishes 65: 311–317.

Hayakawa Y, Komaru A, Munehara H (2002b). Obstructive role of the dimorphic sperm in a non-copulatory marine sculpin, *Hemilepidotus gilbertii*, to prevent other males' eusperm from fertilization. Environmental Biology of Fishes 64: 419–427.

Hayakawa Y, Komaru A, Munehara H (2002c). Ultrastructural observations of eu- and paraspermiogenesis in the cottid fish, *Hemilepidous gilbertii* (Teleostei: Scorpaeniformes: Cottidae). Journal of Morphology 253: 243–254.

Hayakawa Y, Akiyama R, Munehara H (2004). Antidispersive effect induced by parasperm contained in semen of a cottid fish, *Helmilepidotus giberti*: estimation by models and experiments. Japanese Journal of Ichthyology 51: 31–42.

Hayakawa Y, Kobayashi M, Munehara H (2007). Spermatogenesis involving parasperm production in the marine cottoid fish, *Hemilepidotus gilberti*. Raffles Bulletin of Zoological Sciences 14: 29–35.

Healy JM, Jamieson BGM (1981). An ultrastructural examination of developing and mature paraspermatozoa in *Pyrazus ebeninus* (Mollusca, Gastropoda, Potamididae). Zoomorphology 98: 101–119.

Helvey M (1982). First observations of courtship behaviour in the rockfish genus *Sebastes*. Copeia 1982: 763–770.

Idelman S (1967). Données récentes sur l'infrastructure du spermatozoïde. Annales Biologiques 6: 113–190.

Imamura H (1996). Phylogeny of the family Platycephalidae and related taxa. Species Divers 1: 123–233.

Imamura H (2004). Phylogenetic relationships and new classification of the superfamily Scorpaenoidea (Actinopterygii: Perciformes). Species Divers 9: 1–36.

Imamura H, Shinohara G (1998). Scorpaeniform fish phylogeny: an overview. Bulletin of the Natural Science Museum, Series A 24: 185–212.

Imamura H, Yabe M (2002). Demise of the Scorpaeniformes (Actinopterygii: Percomorpha): an alternative phylogenetic hypothesis. Bulletin of Fisheries Sciences, Hokkaido University 53: 107–128.

Iwamatsu Y, Shibata Y, Kanie T (1995). Changes in chorion proteins induced by the exudate released from the egg cortex at the time of fertilization in the teleost *Oryzias latipes*. Development, Growth, Differentiation 37: 747–759.

Jamieson BGM (1991). *Fish Evolution and Systematics: Evidence from Spermatozoa*. Cambridge University Press, Cambridge.

Jamieson BGM, Leung LKP (1991). Introduction to fish spermatozoa and the micropyle. In: Jamieson BGM (ed.), *Fish Evolution and Systematics: Evidence from Spermatozoa*. Cambridge University Press, Cambridge, pp. 6–72.

Jamieson I (1995). Do female fish prefer to spawn in nests with eggs for reasons of mate choice copying or egg survival? American Naturalist 145: 824–832.

Kendall AW Jr (1991). Systematics and identification of larvae and juveniles of the genus *Sebastes*. Environmental Biology of Fishes 30: 173–190.

King JR, Withler RE (2005). Male nest site fidelity and female serial polyandry in lingcod (*Ophidion elongates*, Hexagrammidae). Molecular Ecology 14: 653–660

Koya Y, Matsubara T (1995). Ultrastructural observations on the inner ovarian epithelia of kichiji rockfish, *Sebastolobus macrochir*, with special reference to the production of gelatinous material surrounding the eggs. Bulletin of the Hokkaido Natural Fisheries Research Institute 59: 1–17.

Koya Y, Muñoz M (2007). Comparative study on ovarian structures in scorpaenids: possible evolutional process of reproductive mode. Ichthyological Research 54: 221–230

Koya Y, Munehara H, Takano K, Takahashi H (1993). Effects of extracellular environments on the motility of spermatozoa in several marine sculpins with internal gametic association. Comparative Biochemistry and Physiology 106: 25–29.

Koya Y, Munehara H, Takano K (1994). Reproductive cycle and spawning ecology in elkhorn sculpin, *Alcicthys alcicornis*. Japanese Journal of Ichthyology 41: 39–45.

Koya Y, Hamatsu T, Matsubara T (1995). Annual reproductive cycle and spawning characteristics of the female kichiji rockfish, *Sebastolobus macrochir*. Fisheries Science 61: 203–208.

Koya Y, Munehara H, Takano K (1997). Sperm storage and degradation in the ovary of a marine copulating sculpin, *Alcichthys alcicornis* (Teleostei: Scorpaeniformes): role of intercellular junctions between inner ovarian epithelial cells. Journal of Morphology 233: 153–163.

Koya Y, Munehara H, Takano K (2002). Sperm storage and motility in the ovary of the marine sculpin, *Alcichthys alcicornis* (Teleostei: Scorpaeniformes) with internal gametic association. Journal of Experimental Zoology 292: 145–155.

Krefft G (1961). A contribution to the reproductive biology of *Helicolenus dactylopterus* (De la Roche 1809) with remarks on the evolution of the Sebastinae. Rapports et Procès

Verbaux des Réunions du Conseil International pour l'Exploration de la Mer 150: 243–244.

Krejsa RJ (1964). Reproductive behavior and sexual dimorphism in the manacled sculpin, *Synchirus gilli* Bean. Copeia 1964: 448–450.

Lisovenko LA (1979). Reproduction of rockfishes (family Scorpaenidac) off the Pacific coast of South America. Journal of Ichthyology 18: 262–268.

Love MS, Westphal WV (1981). Growth, reproduction, and food habits of olive rockfish, *Sebastes serranoides*, off central California. Fishery Bulletin 79: 533–545.

Love MS, Morris P, McCrae M, Collins R (1990). Life history aspects of 19 rockfish species (Scorpaenidae: *Sebastes*) from the southern California Bight. Technical Report of the National Marine Fisheries Service 87. La Jolla, California.

MacGregor JS (1970). Fecundity, multiple spawning, and description of the gonads in *Sebastodes*. Special Scientific Reports US Fish Wildlife Service Fisheries 596, pp. 1–12.

Marliave JB, Peden AE (1989). Larvae of *Liparis fucensis* and *Liparis callyodon*: is the "cottoid bubblemorp" phylogenetically significant? Fishery Bulletin US 87: 735–743.

Matsubara K (1943). Studies on the scorpaenoid fishes from Japan: anatomy, phylogeny and taxonomy. I. Transactions of the Sigenkagaku Kenkyusho 1: 1–170.

Matsubara K (1955). Fish morphology and hierarchy. Ishizaki-shoten, Tokyo.

Mattei X (1970). Spermiogenèse comparée des poissons. In: Baccetti B (ed.), *Comparative Spermatology*. Academic Press, New York, pp. 57–69.

Mattei X (1991). Spermatozoa ultrastructure and taxonomy in fishes. In: Baccetti B (ed.), *Spermatology 20 Years After*. Raven Press, New York, pp. 985–990.

Mattei X, Siau Y, Thiaw OT, Thiam D (1993). Peculiarities in the organization of testis of *Ophidion* sp. (Pisces: Teleostei). Evidence for two types of spermatogenesis in teleost fish. Journal of Fish Biology 43: 931–937.

McFarlane GA, Beamish RJ (1983). Biology of adult sablefish (*Anoplopoma fimbria*) in waters of western Canada. Proceedings of the International Sablefish Symposium, Anchorage, Alaska. Alaska Sea Grant Report 83—08, pp. 59–80.

McFarlane GA, Beamish RJ (1995). Validation of the otolith cross-section method of age determination for sablefish (*Anoplopoma fimbria*) using oxytetracycline. In: Secor DH Dean JM, Campana SE (eds.), *Recent Developments in Fish Otolith Research*. The Belle W. Baruch Library in Marine Science 19, University of South California Press, Columbia, South California, pp. 319–329.

McFarlane GA, Nagata WD (1988). Overview of sablefish mariculture and its potential for industry. Proceedings of the Fourth Alaska Aquaculture Conference. Alaska Sea Grant Report 88-4, pp. 105–120.

McPhail AS, Shipton TA, Sauer WHH, Leslie RW (2001). Aspects of the biology of the Cape Gurnard, *Chelidonichthys capensis* (Scorpaeniformes: Trigilidae) on the Agulhas Bank, South Africa. Vie et Milieu-Life and Environment 51: 217–227.

Miya M, Takeshima H, Endo H, Ishiguro NB, Inoue JG, Mukai T, Satoh TP, Yamaguchi M, Akira K (2003). Major patterns of higher teleostean phylogenies: a new perspective based on 100 complete mitochondrial sequences. Molecular Phylogenetics and Evolution 26: 121–138.

Mori H, Nakagawa M, Soyano K, Koya Y (2003). Annual reproductive cycle of black rock-fish, *Sebastes schlegeli*, in captivity. Fisheries Science 69: 910–923.

Morris RW (1956). Clasping mechanism of the cottid fish *Oligocottus snydery* (Greely). Pacific Science 10: 314–317.

Moulton JM (1963). Acoutic behaviour of fishes. In: Busnel RG (ed.), *Acoustic Behaviour of Animals*. Elsevier, Amsterdam, pp. 655–693.

Munehara H (1988). Spawning and subsequent copulating behavior of elkhorn sculpin, *Alcichthys alcicornis*, in an aquarium. Japanese Journal of Ichthyology 35: 358–364.

Munehara H (1991). Utilization and ecological benefits of a sponge as spawning bed by the little dragon sculpin, *Blepsias cirrhosus*. Japanese Journal of Ichthyology 38: 179–184.

Munehara H (1992). Utilization of polychaete tubes as spawning substrate by the sea raven, *Hemitripterus villosus* (Scorpaeniformes). Environmental Biology of Fishes 33: 395–398.

Munehara H (1996). Sperm transfer during copulation in the marine sculpin *Hemitripterus villosus* (Pisces: Scorpaeniformes) by means of a retractable genital duct and ovarian secretion in females. Copeia 1996: 452–454.

Munehara H, Takano K, Koya Y (1989). Internal gametic association and external fertilization in the elkhorn sculpin, *Alcichthys alcicornis*. Copeia 1989 3: 673–678.

Munehara H, Takano K, Koya Y (1991). The little dragon sculpin, *Blepsias cirrhosus*: another case of internal gametic association and external fertilization. Japanese Journal of Ichthyology 37: 391–394.

Munehara H, Takenaka A, Takenaka O (1994). Alloparental care in the marine sculpin *Alcichtys alcicornis* (Pisces: Cottidae): copulating in conjunction with parental care. Journal of Ethology 12: 115–120.

Munehara H, Kanamoto Z, Miura T (2000). Spawning behaviour and interspecific breeding in three Japanese greenlings (Hexagrammidae). Ichthyological Research 47: 287–292.

Munk KM (2001). Maximum ages of groundfishes in waters off Alaska and British Columbia and considerations of age determination. Alaska Fisheries Research Bulletin 8: 12–21.

Muñoz M, Casadevall M (2002). Reproductive indices and fecundity of *Helicolenus dactylopterus dactylopterus* (Teleostei: Scorpaenidae) in the Catalan Sea (Western Mediterranean). Journal of the Marine Biological Association of the UK 82: 995–1000.

Muñoz M, Cadasevall M, Bonet S (1999). Annual reproductive cycle of *Helicolenus dactylopterus dactylopterus* (Teleostei: Scorpaeniformes) with special reference to the ovaries sperm storage. Journal of the Marine Biological Association of the UK 79: 521–529.

Muñoz M, Casadevall M, Bonet S, Quagio-Grassiotto I (2000). Sperm storage structures in the ovary of *Helicolenus dactylopterus dactylopterus* (Teleostei: Scorpaenidae): an ultrastructural study. Environmental Biology of Fishes 58: 53–59.

Muñoz M, Casadevall M, Bonet S (2001). Gonadal structure and gametogenesis of *Aspitrigla obscura* (Pisces: Triglidae). Italian Journal of Zoology 68: 39–46.

Muñoz M, Casadevall M, Bonet S (2002a). Testicular structure and semicystic spermatogenesis in a specialized ovuliparous species: *Scorpaena notata* (Pisces Scorpaenidae). Acta Zoologica (Stockholm) 83: 213–219.

Muñoz M, Casadevall M, Bonet S (2002b). The ovarian morphology of *Scorpaena notata* shows a specialized mode of oviparity. Journal of Fish Biology 61: 877–887.

Muñoz M, Koya Y, Casadevall M (2002c). Histochemical analysis of sperm storage in *Helicolenus dactylopterus dactylopterus* (Teleostei: Scorpaenidae). Journal of Experimental Zoology 292: 156–164.

Muñoz M, Sàbat M, Malloll S, Casadevall M (2002d). Gonadal structure and gametogenesis of *Trigla lyra* (Pisces: Triglidae). Zoological Studies 41: 412–420.

Muñoz M, Hernández MR, Sàbat M, Casadevall M (2003). Annual reproductive cycle and fecundity of *Aspitrigla obscura* (Teleostei, Triglidae). Vie et Milieu 53(2–3): 123–129.

Muñoz M, Sàbat M, Vila S, Casadevall M (2005). Annual reproductive cycle and fecundity of *Scorpaena notata* (Teleostei: Scorpaenidae). Scientia Marina 69: 555–562.

Nagahama Y, Takemura A, Takano K, Adachi S, Kusakari M (1991). Serum steroid hormone levels in relation to the reproductive cycle of *Sebastes tackzanowskii* and *S. schlegeli*. Environmental Biology of Fishes 30: 31–38.

Nelson JS (2006). *Fishes of the World*. 4th ed. John Wiley & Sons, New York.

Pearson KE, Gunderson DR (2003). Reproductive biology and ecology of shortspine thornyhead rockfish, *Sebastolobus alascanus,* and longspine thornyhead rockfish, *S. altivelis,* from the northeastern Pacific Ocean. Environmental Biology of Fishes 67: 117–136.

Petersen CW (1995). Male mating success and female choice in permanently territorial damselfishes. Bulletin of Marine Science 57: 690–704.

Petersen CW, Mazzoldi C, Zarrella KA, Hale RE (2005). Fertilization mode, sperm characteristics, mate choice and parental care patterns in *Artedius* spp. (Cottidae). Journal of Fish Biology 67: 239–254.

Potts GW (1984). Parental behaviour in temperate marine teleosts with special reference to the development of nest structures. In: Potts GW, Wootton RJ (eds.), *Fish Reproduction: Strategies and Tactics*. Academic Press, London and Orlando, Florida, pp. 223–242.

Quinitio GF (1989). Studies on the functional morphology of the testis in two species of freshwater sculpins. Ph.D. thesis, Hokkaido University, Sapporo, Japan.

Quinitio GF, Takahashi H (1992). An ultrastructural study on the aberrant spermatids in the testis of the river sculpin, *Cottus hangiongensis*. Japanese Journal of Ichthyology 39: 235–241.

Quinitio GF, Takahashi H, Goto A (1988). Annual changes in the testicular activity of the river sculpin, *Cottus hangiongensis* Mori, with emphasis on the occurrence of aberrant spermatids during spermatogenesis. Journal of Fish Biology 33: 871–878.

Ragland HC, Fischer EA (1987). Internal fertilization and male parental care in the scalyhead sculpin, *Artedius harringtoni*. Copeia 1987: 1059–1062.

Rohwer S (1978). Parent cannibalism of offspring and egg raiding as a courtship strategy. American Naturalist 112: 429–440.

Ruiz-Carus R, Matheson Jr RE, Roberts Jr DE, Whitfield PE (2006). The western Pacific red lionfish, *Pterois volitans* (Scorpaenidae), in Florida: evidence for reproduction and parasitism in the first exotic marine fish established in state waters. Biology and Conservation 128: 384–390.

Sàbat M, Lo Nostro F, Casadevall M, Muñoz M (2009). A light and electron microscopic study on the organization of the testis and the semicystic spermatogenesis of the genus *Scorpaena* (Teleostei, Scorpaenidae). Journal of Morphology 270: 662–672.

Selman K, Wallace RA (1989). Cellular aspects of oocyte growth in teleosts. Zoological Science 6: 211–231.

Sequeira V, Figueredo I, Muñoz M, Gordo LS (2003). New approach to the reproductive biology of *Helicolenus dactylopterus*. Journal of Fish Biology 62: 1206–1210.

Shaw FR, Gunderson DR (2006). Life history traits of the greenstriped rockfish, *Sebastes elongates*. California Department of Fish and Game 92: 1–23.

Shibata Y, Iwamatsu T (1996). Evidence for involvement of the exudate released from the egg cortex in the change in chorion proteins at the time of egg activation in *Oryzias latipes*. Zoological Science 13: 271–275.

Shimizu M, Kusakari M, Yoklavich MM, Boehlert GW, Yamada J (1991). Ultrastructure of the epidermis and digestive tract in *Sebastes* embryos with special reference to the uptake of exogenous nutrients. Environmental Biology of Fishes 30: 155–163.

Shinohara G (1994). Comparative morphology and phylogeny of the suborder Hexagrammoidei and related taxa. Memoirs of the Faculty of Fisheries Hokkaido University 41: 1–97.

Shinohara G, Imamura H (2007). Revisiting recent phylogenetic studies of "Scorpaeniformes." Ichthyological Research 54: 92–99.

Shinomiya A (1985). Studies on the reproductive physiology and ecology in three marine cottid fish. Ph.D. thesis, Hokkaido University, Sapporo, Japan.

Shinomiya A, Ezaki O (1991). Mating habits of the rockfish *Sebastes inermis*. Environmental Biology of Fishes 30: 15–22.

Shinomiya A, Yamada M, Sunobe T (2003). Mating system and protandrous sex change in the lizard flathead, *Inegocia japonica* (Platycephalidae). Ichthyological Research 50: 383–386.

Shiokawa T (1962). Studies on habits of coastal fishes in the Amakusa Islands. Part II. Growth and maturity of the purple rockfish, *Sebastes pachycephalus pachycephalus* Temminck et Schlegel. Recent Oceanographic Works, Japan 6: 103–111.

Smith WL, Wheeler WC (2004). Polyphyly of the mail-cheeked fishes (Teleostei: Scorpaeniformes): evidence from mitochondrial and nuclear sequence data. Molecular Phylogenetics and Evolution 32: 627–646.

Sogard SM, Gilbert-Horvath E, Anderson EC, Fishee R, Berkeley SA, Garza JC (2008). Multiple paternity in viviparous kelp rockfish, *Sebastes atrovirens*. Environmental Biology of Fishes 81: 7–13.

Somerton DA, Donaldson W (1998). Parasitism of the golden king crab, *Lithodes aequispinus*, by two species of snailfish genus *Careproctus*. Fishery Bulletin US 96: 871–884.

Sorokin VP (1961). The redfish: gametogenesis and migrations of the *Sebastes marinus* (L) and *Sebastes mentella* Travin. Rapports et Proces-Verbaux des Reunions du Conseil International pour l'Exploration de la Mer 150: 245–250.

Swallow JG, Wilkinson GS (2002). The long and short sperm polymorphism in insects. Biological Reviews 77: 153–182.

Takahashi H, Takano K, Takemura A (1991). Reproductive cycles of *Sebastes*. Environmental Biology of Fishes 30: 23–29.

Takano K (1989). Ovarian structure and gametogenesis. In: Takashima F, Hanyu I (eds.), *Reproductive Biology of Fish and Shellfish*. Midori-shobo, Tokyo, pp. 3–34.

Takemura A, Takano K, Takahashi H (1987). Reproductive cycle of a viviparous fish, the white-edged rockfish, *Sebastes taczanowski*. Bulletin of the Faculty of Fisheries of Hokkaido University 38: 111–125.

Tateishi S, Mizue K, Inao T (1958). Histological study about the ovaries of several kinds of oviviviparous teleost. Bulletion of the Faculty of Fisheries of Nagasaki University 7: 47–50

Tsang WN, Chaillé PM, Collins PM (2007). Growth and reproductive performance in cultured nearshore rockfish (*Sebastes* spp.). Aquaculture 266: 236–245.

Uchida K (1932). On the fish spawning in sea squirt. Kagaku 2: 56–57.

Vila S, Muñoz M, Sabat M, Casadevall M (2007). Annual cycle of stored spermatozoa within the ovaries of *Helicolenus dactylopterus dactylopterus* (Teleostei Scorpaenidae). Journal of Fish Biology 71: 596–609.

Warfel HE, Merriman D (1944). The spawning habits, eggs and larvae of the sea raven *Hemipterus americanus*, in southern New England. Copeia 1944: 197–205.

Washington BB, Eschmeyer WN, Howe KM (1984a). Scorpaeniformes: relationships. In: Moser HG, Richards WJ, Cohen DM, Fahay MP, Kendall Jr AW, Richardson SL (eds.), *Ontogeny and Systematics of Fishes*. American Society of Ichthyologists and Herpetologists, Special Publication 1. Allen Press, Lawrence, Kansas, pp. 438–447.

Washington BB, Moser HG, Laroche WA, Richards WJ (1984b). Scorpaeniformes: development. In: Moser HG, Richards WJ, Cohen DM, Fahay MP, Kendall Jr AW, Richardson SL (eds.), *Ontogeny and Systematics of Fishes*. American Society of Ichthyologists and Herpetologists, Special Publication 1. Allen Press, Lawrence, Kansas, pp. 405–428.

West G (1990). Methods of assessing ovarian development in fishes: a review. Australian Journal of Marine and Freshwater Research 41: 199–222

Wing BL (1997). Distribution of sablefish, *Anoplopoma fimbria*, larvae in the eastern Gulf of Alaska. In: Wilkins M, Saunders M (eds.), Proceedings of the International Symposium on the Biology and Management of Sablefish, *Anoplopoma fimbria*. Seattle, Washington. NOAA National Marine Fisheries Service Technical Report 130, pp. 13–26.

Wourms JP (1976). Annual fish oogenesis. I. Differentiation of the mature oocyte and formation of the primary envelope. Developmental Biology 50: 338–354.

Wourms JP (1991). Reproduction and development of *Sebastes* in the context of the evolution of piscine viviparity. Environmental Biology of Fishes 30: 111–126.

Wourms JP, Lombardi J (1992). Reflections on the evolution of piscine viviparity. American Zoologist 32: 276–293.

Wourms JP, Grove BD, Lombardi J (1988). The maternal embryonic relationship in viviparous fishes. In: Hoar WS, Randall DJ (eds.), *Fish Physiology 11B*. Academic Press, San Diego, pp. 1–134.

Yabe M (1985). Comparative osteology and myology of the superfamily Cottoidea (Pisces: Scorpaeniformes) and its phylogenetic classification. Memoirs of the Faculty of Fisheries of Hokkaido University 32: 1–130.

Yamada J, Kusakari M (1991). Staging and the time course of embryonic development in kurosoi, *Sebastes schlegeli*. Environmental Biology of Fishes 30: 103–110.

Yau C, Collins MA, Everson I (2000). Commensalism between a liparid fish (*Careproctus* sp.) and stone crabs (Lithodidae) photographed *in situ* using a baited camera. Journal of the Marine Biological Association of the UK 80: 379–380.

Yoklavich MM, Boehlert GW (1991). Uptake and utilization of ^{14}C-glycine by embryos of *Sebastes melanops*. Environmental Biology of Fishes 30: 147–153.

Yoneda M, Miura H, Mitsuhashi M, Matsuyama M, Matsuyama S (2000). Sexual maturation, annual reproductive cycle, and spawning periodicity of the shore scorpionfish, *Scorpanenodes littoralis*. Environmental Biology of Fishes 58: 307–319.

4

Parental Care, Oviposition Sites, and Mating Systems of Blennioid Fishes

Philip A. Hastings and Christopher W. Petersen

Mating systems of species result from a complex interaction of phylogenetic constraints and a host of environmental factors (Emlen & Oring 1977; Shuster & Wade 2003). Well-documented, fundamental features of the reproductive biology of fishes affecting their mating systems include mode of fertilization, egg type, and parental care pattern (Gross & Shine 1981). These features evolve relatively slowly in most fishes as they vary little within lineages (Mank et al. 2005). Environmental factors in general are more variable within lineages and consequently are more often responsible for variation in the mating systems within particular lineages. These include the ability of males to sequester females or resources critical to females, as well factors that influence the timing of mating and the distribution of mating sites. In this chapter, we review the fundamental reproductive features of blennioid fishes, as well as the role of various environmental factors on the form of the mating system. We focus on aspects of oviposition sites as the major environmental factor in externally fertilizing species, which includes the majority of blennioids.

Blennioid fishes are common in many coastal regions of the world and prominent members of most reef communities in both tropical and temperate areas. Blennies and other small, benthic fishes have received considerable research attention in recent years as their important roles in coastal communities are becoming more widely evident (Munday & Jones 1998; Depczynski & Bellwood 2003; Smith-Vaniz et al. 2006) and their value as study organisms more widely appreciated (Patzner et al. 2009).

The Blennioidei is a monophyletic lineage of perciform fishes and includes at least 883 species allocated among six families (Springer 1993; Hastings &

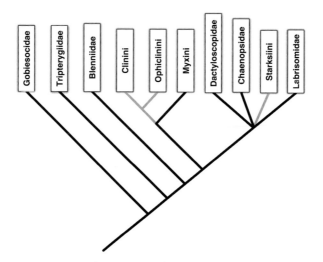

FIGURE 4.1. Character map for fertilization pattern in lineages of blennioid fishes and their sister group the cling-fishes (Gobiesocidae). Solid black = external fertilization of demersal eggs. Solid gray = internal fertilization.

Springer 2009a). The bulk of this diversity is found in the combtooth blennies (Blenniidae) with 387 species (Hastings & Springer 2009a) and the triplefin blennies (Tripterygiidae) with 163 species (Fricke 2009). Both of these lineages are global in distribution. The remaining families are smaller and have more restricted distributions with the Clinidae (85 species) found primarily in temperate oceans and the Chaenopsidae (91), Labrisomidae (109) and Dactyloscopidae (48) found primarily in the Neotropics (Hastings 2009). Five of the six families have evidence of monophyly, while the Labrisomidae lack known morphological synapomorphies (Hastings & Springer 2009b). The phylogenetic relationships of blennioids are largely unresolved. Preliminary phylogenetic analyses based on nuclear and mitochondrial markers (Lin 2009) are largely consistent with the phylogenetic hypothesis proposed by Springer & Orrell (2004) based on selected morphological features (Hastings & Springer 2009b). In that hypothesis, tripterygiids, blenniids, and clinids are sequential sister groups to an unresolved lineage that includes chaenopsids, labrisomids, and, tentatively, dactyloscopids (Figure 4.1).

The biology of blennies, including several aspects of their reproductive biology, has received increased study in recent years (Patzner et al. 2009). In a brief survey of the primary literature, we found 113 journal articles covering some aspect of blenny reproductive biology or reproductive behavior. These papers re-

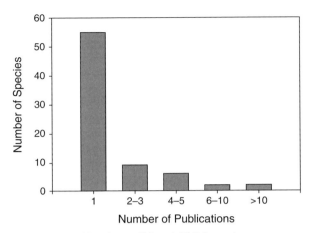

FIGURE 4.2. Numbers of blennioid fish species per quantity of publications covering some aspect of their reproductive biology.

ported on one or more of 74 blennioid species. This represents only 8 percent of the 883 known species; thus, nothing is published concerning the reproductive biology of over 90 percent of blenny species. In addition, the published studies on blennies are heavily biased towards a surprisingly few well-studied species (Figure 4.2). Twenty-seven papers involve some aspect of the reproductive biology of the peacock blenny, *Salaria pavo* (Blenniidae), while 12 concern the reproductive biology of the Azoran rock-pool blenny, *Parablennius parvicornis* (Blenniidae). The vast majority of blennies that have been studied, 55 of 74 species, are treated in a single published study. The distribution of published studies is also biased towards certain families. In particular, it might appear that the reproductive biology of the Clinidae is relatively well known with 27 percent of its species covered in one or more publications. However, most of these are from surveys of specific aspect of their reproductive biology, such as the morphology of male copulatory organs and sperm transmission mechanisms (Fishelson et al. 2006, 2007), details of maternal investment in offspring (Gunn & Thresher, 1993) or general ecology (Prochazka & Griffiths 1992), and do not provide detailed insight into the mating system of most of these species.

In spite of this uneven coverage, several general patterns regarding the reproductive biology and mating systems of blennies emerge. In the present study, we explore the role of selected constraints on the mating systems of blennioids with an emphasis on parental care pattern and oviposition sites. We view these as key features influencing multiple aspects of their mating systems.

FERTILIZATION MODE AND PARENTAL CARE

A fundamental determinant of mating systems in animals is the mode of fertilization (Gross & Shine 1981; Mank et al. 2005). The majority of blennioids are assumed to have external fertilization of demersal eggs deposited on the substrate (Breder & Rosen 1966) or carried on the body of the male (see below). Although unconfirmed for most species, assuming that this feature follows phylogenetic lines, this includes all species of the Tripterygiidae (163 species), Blenniidae (387), Chaenopsidae (91), and Dactyloscopidae (48), most members of the Labrisomidae (78 of 109 species) and a few of the Clinidae (including 9 of the 85 clinid species). Thus, 88 percent (776 of 883) of the species of blennioids probably exhibit external fertilization. In addition to being by far the predominant condition, all available evidence, including the similar condition in the Gobiesocoidei (clingfishes and relatives; Breder & Rosen 1966), the apparent sister group of the Blennioidei, (Hastings & Springer 2009b), indicates that external fertilization of demersal eggs is the plesiomorphic condition in the Blennioidei (Figuvre 4.1).

Internal fertilization (IF) evolved independently at least twice and probably only twice within the Blennioidei (Figure 4.1). IF is seen in two tribes of the Clinidae: the Clinini, with 64 species, and the Ophiclinini, with 12 species. George & Springer (1980) hypothesized that these tribes were sister groups based primarily on the presence of "ovoviviparity" and intromittent organs in males. This relationship was corroborated by earlier (Stepien et al. 1997) and more recent (S. von der Heiden pers. comm.) molecular analyses, implying that internal fertilization evolved once in their common ancestor. Males of both lineages have prominent intromittent organs that represent modifications of the genital papilla, and females retain fertilized eggs within the follicles until hatching (Gunn & Thresher 1993; Fishelson, et al. 2006; Moser 2007). Among blennioids, IF evolved a second time in the unrelated labrisomid tribe Starksiini, a monophyletic lineage that includes two genera, *Starksia* with 30 species, and *Xenomedea* with only one (Rosenblatt & Taylor 1971). All starksiines have putative intromittent organs (Hubbs 1952), and although unconfirmed for most species, *Xenomedea* and at least four species of *Starksia* are confirmed to have internal fertilization (Rosenblatt & Taylor 1971).

Mode of fertilization appears to dictate the pattern of parental care in fishes in general (Ridley 1978; Gross & Shine 1981; Gross & Sargent 1985; Mank et al. 2005), as well as in blennioids. So far as is known, all species of blennioids with external fertilization exhibit male guarding of eggs that are deposited either in the male's territory, in a shelter, or on a modified (cleared) patch of substrate, or are carried on their body (Figure 4.3). Eggs may be deposited on the substrate in a monolayer (e.g., Clarke & Tyler 2003) or in a mass held together with filaments (e.g., Breder 1941). In most cases, parental care takes on the form of egg defense by chasing conspecific and heterospecific egg predators. Predators on the eggs of blennies include a variety of other fish species and invertebrates (e.g., Sunobe 1998; Hirayama et al. 2005), and conspecifics including paternal males themselves (e.g., Ohta & Nakazono 1988; Hamada & Nakazono 1989; Kraak & van der Berghe 1992; Vinyoles et al. 1999).

Parental care in blennies also includes behaviors that appear to increase the survivorship of developing embryos. The most commonly reported behavior of this type is fanning of eggs (e.g., Gibran et al. 2004; Lengkeek & Didderen 2006). In *Aidablennius sphinx,* the frequency of egg fanning is positively correlated with brood size (Kraak & Videler 1991; Oliveira et al. 2000) and with female mate preference (e.g., Neat & Locatello 2002). However, not all paternal blennioids fan eggs (e.g., Petersen 1988). Other parental behaviors of blennies include nest cleaning (e.g., Oliveira et al. 2000), mouthing of eggs to increase circulation and/or remove debris and dead eggs (e.g., Breder 1941; Kraak 1996), and application of antibiotics via the prominent anal-fin glands of male blenniids (Giacomello et al. 2006). Secretions from these glands are attractive to receptive females (Barata et al. 2008). While similar fin glands are present in other lineages of blennioids such as the triplefins (Northcott & James 1996), their function in these latter groups has not been documented.

Importantly, these forms of parental care in blennioids are shareable care in that the per-egg cost of parental care is not additive (Wittenberger 1979, 1981); that is, once a male is caring for eggs, it costs very little more to care for additional eggs as long as they can be accommodated within the spawning site. As a consequence, parental males often continue courting females and may obtain additional clutches of eggs. Thus parental males often guard clutches of eggs that are in different stages of development (see Neat & Lengkeek 2009). However, defense of eggs is costly to males in reduced feeding opportunities because their movements may be restricted to nest sites (e.g., Gonçalves & Almada 1997).

The importance of male egg guarding is reflected in female mate preferences in some blenny species. Females have been demonstrated to prefer mating with males that are already mating with other females (e.g., Petersen 1989) and those males already guarding other clutches of eggs (e.g., Kraak & Videler 1991; Kraak & Groothuis 1994; Kraak & Weissing 1996). In the redlip blenny, females appear to monitor survivorship of eggs in the nests of males and adjust their mating preferences accordingly (Côté & Hunte 1989b). The commonly observed correlations between male body size and female preference (e.g., Hastings 1988a, b, 1992) and male body size and male reproductive success (reviewed in Neat & Lengkeek 2009) have been hypothesized to be a consequence of increased parental abilities (i.e., defense of eggs) by larger males (e.g., Côté & Hunte 1989a, 1993).

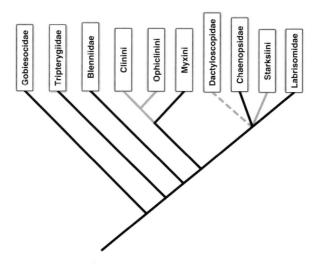

FIGURE 4.3. Character map for parental care pattern in lineages of blennioid fishes and their sister group, the clingfishes (Gobiesocidae). Solid black = male defense of eggs deposited on the substrate. Dashed gray = male egg carrying. Solid gray = female internal gestation.

The sand stargazers (Dactyloscopidae) exhibit a unique form of parental care that may be related to their unusual habitat. These fishes occur on sand and rubble substrates adjacent to or sometimes far from reefs (Hastings & Springer 2009b). They typically burrow in the substrate and exhibit a number of extraordinary features associated with this behavior. These behaviors include upwardly directed mouth and eyes, the lateral line placed high on the body, lip fimbriae preventing sand from entering the buccal cavity, and a modified branchial pump (Todd 1973). One of the most striking features of sand stargazers is their method of parental care. Sand and rubble habitats often lack suitable substrate for egg deposition and egg guarding, and, unique among blennioids (Figure 4.3), dactyloscopid males carry clutches of eggs in one or two balls behind each pectoral fin. The eggs, about 1 mm in diameter, are attached to one another by fine adhesive filaments and a thicker adhesive thread is located in the central portion of each egg mass. Eggs are held in place between the large pectoral fins and the recurved spines in the anterior anal fin of males of species in the genera *Dactyloscopus, Dactylagnus,* and *Myxodagnus* (Böhlke & Chaplin 1968; Petti 1969). Males of some species reportedly carry two distinct egg balls, one behind each pectoral fin (Robins et al. 1999). A single egg mass in *M. opercularis* reportedly included 404 eggs, while one in *Dactyloscopus mundus* contained 3,342 eggs (Petti 1969). Little else is known of this behavior including how eggs are transferred from the female to the male, the number of eggs and number of clutches that are carried by males and consequently how many females mate with single males. It seems likely that this does not represent sharable care, and a male may be constrained to carry the eggs of a single female until they hatch, much like the situation in pipefishes and seahorses (e.g., Berglund et al. 1989) and oral brooders such as cardinalfishes (e.g., Okuda et al. 2003).

Blennies with internal fertilization exhibit a shift from male care of eggs to internal female care and, often, provisioning of embryos (Figure 4.3). This transition from external to internal fertilization in the Starksiini was accompanied by a shift from male parental care, typical of other labrisomid blennies (e.g., Petersen 1988), to female care of internally developing embryos (Rosenblatt & Taylor 1971). In the more well-studied internally fertilizing clinids, care also shifted from male guarding of eggs in the Myxiodini (Stepien 1986) to internal retention of developing eggs by females. Female clinids also contribute significant resources to embryos in addition to those present in the eggs (Veith 1979, 1980). This is clearly not shareable care, although several features of females appear to maximize their fecundity. Chief among these is superfoetation in which clutches of eggs in as many as eight different stages are maintained within the ovary of single females (Gunn & Thresher 1993; Moser 2007).

OVIPOSITION SITES IN EXTERNALLY FERTILIZING BLENNIOIDS

Diversity of Sites

Among blennioids with paternal care of demersal eggs, the variety of spawning sites is immense. These range from temporary territories in the intertidal (Almada et al. 1992; Almada & Santos 1995; Gonçalves & Almada 1998), to holes or crevices (Almada et al. 1994), to species-specific dependence on empty tests of invertebrates (Hastings 1988a, b, 1992). These spawning sites are defended by resident males, and courtship activities are centered around these sites. Courtship may include a variety of visual, chemical, and auditory displays (e.g., Almada et al. 1990; de Jong et al. 2007; Barata et al. 2008). Receptive females approach and enter the nest site and deposit eggs directly on the substrate where they are fertilized by the resident male (and sometimes by nonresident males; see below). Following egg deposition, females leave the spawning site. In some cases entire clutches of eggs are deposited with a single male (e.g., *Ophioblennius macclurei*, Côté & Hunte 1989a), while in others, females may immediately spawn with nearby males (e.g., *Acanthemblemaria macrospilus*, Stephens et al. 1966). After spawning, males continue guarding the oviposition site and consequently the eggs (Petersen 1988) and may also perform one or more egg maintenance behaviors (see above). Parental males often continue courtship, and in many species, successful males may attract multiple mates to oviposition sites (see Neat & Lengkeek 2009). Thus, the typical pattern in blennioids is the combination of paternal care and male courtship, because the main factor limiting male reproductive success is access to females.

In species where males and females differ in habitat utilization, males are often found in either more protected microhabitats such as vacant invertebrate tests as opposed to open surfaces (e.g., Hastings 2002), shallower sites within a rock reef (e.g., Petersen 1988; Almada et al. 1992) or more frequently on living versus dead corals (e.g., Clarke & Tyler 2003).

Because male reproductive success is typically dependent upon defense of an oviposition site, male-male competition for these limited resources (Buchheim & Hixon 1992; Hastings & Galland 2009) may be intense (e.g., Shibata & Kohda 2006), especially where shelters vary in quality as reproductive sites (Hastings 1988a, b). A number of features of blennioids appear to maximize their fighting capabilities, including large body size (Faria & Almada 2001), and large jaws and robust associated muscles (Kotrschal 1988; Hastings 2002).

In a well-studied population of the peacock blenny, *S. pavo*, that occurs in an area of sandy substrate, suitable oviposition sites are limited, being found only in artificial tiles that are defended by large males (Almada et al. 1994, 1995). This shortage of suitable oviposition sites excludes small males from shelter defense and leads to sex role reversal where females actively court males. While this situation is unusual in blennioids, similar sex role shifts have been observed in the blenniid *Petroscirtes breviceps* during the middle of an extended breeding season when free space for egg deposition becomes limiting (Shibata & Kohda 2006).

Sites and Male Reproductive Success

Reproductive success of male blennioids results from a complex interaction of male-male competition for oviposition sites and female preference for various aspects of males and the oviposition sites they defend. Mate choice parameters in blennies were recently reviewed by Neat & Lenkeek (2009). Here we expand their analysis to discuss aspects of oviposition sites in greater detail.

Females typically show preferences for larger males in mating, and this has been demonstrated for several blennies (e.g., *Coralliozetus angelicus*, Hastings 1988b; *Acanthemblemaria crockeri*, Hastings 1988a; *Ophioblennius macclurei*, Côté & Hunte 1989a; *Emblemaria hypacanthus*, Hastings 1991; *Salarias pavo*, Fagundes et al. 2007). This preference may also be reflected in the field, where the reproductive success of male blennies is often correlated with their size for these and many other species (e.g., the above species plus *Fosterygion varium*, Thompson 1986; *Malacoctenus macropus*, Petersen 1988; *Ophioblennius macclurei*, Côté &

Hunte 1989; *Salaria pavo*, Oliveira et al. 1999; *Parablennius parvicornis*, Oliveira et al. 2000; *Aidablennius sphinx*, Neat & Locatello 2002; and Fagundes et al. 2007; but no effect found by Kraak & Videler 1991). In *Aidablennius sphinx*, females lay more eggs, and do so more quickly, with larger males, reflecting either or both a preference for large males and perhaps sperm limitation in small males (Locatello & Neat 2005). In *Ophioblennius macclurei*, females pay a cost of increased aggression from territorial damselfish in order to mate with larger (preferred) males far from their home territory (Reynolds & Côté 1995)

The degree that this pattern of greater reproductive success by larger males is found in field populations may vary over a spawning season (e.g., Oliveira et al. 1999) and depends on the ability of larger males to sequester more successful sites. For example, within populations of *Emblemaria hypacanthus*, Hastings (1992) found that although females preferred larger males in mate choice experiments where other factors were controlled, in the field larger males only obtained higher reproductive success (RS) in a population where nest sites were larger and allowed for high levels of simultaneous egg development. Additionally, preferences for larger males can be masked in populations when larger males are not associated with preferred oviposition sites. Petersen (1988) found that larger males did not have higher RS in *Malacoctenus hubbsi*, but when the effects of oviposition site were statistically removed, larger males were preferred as mates. Females of the angel blenny (*Coralliozetus angelicus*), a barnacle specialist, prefer large males, but at some sites, larger barnacles are lower quality, tending to be more heavily fouled internally where eggs are deposited. Because large, preferred males are forced to occupy especially large shelters (barnacles) and the quality of the largest shelters is below a threshold required by females, the reproductive success of large males is reduced (Hastings 1988b).

In species where the sexes differ in habitat utilization, females often show preferences for males in the direction of the habitat differences (e.g., Petersen 1988). For example, in the carmine triplefin, *Axoclinus storyae*, male territories are at greater depth than female territories, and males defending deeper sites have higher mating success (Petersen 1989). In the chaenopsid *Acanthemblemaria spinosa*, males occupying shallower shelters have increased reproductive success although the sexes do not differ in the depth of shelters occupied; in both sexes larger individuals were found in more shallow shelters (Clarke & Tyler 2003) where their feeding rate is greater because of increased availability of planktonic prey (Clarke 1992). The greater reproductive success of males defending more shallow shelters may be a result of the preference of females for larger males (Clarke & Tyler 2003).

In some blennies, females prefer to mate with males defending protected sites. For example, in the labrisomid *Malacoctenus macropus*, males with the most protected oviposition sites had the highest mating success (Petersen 1988). This suggests that sexual selection through female choice is one of the selective factors favoring males to live in more protected sites.

In species that lay eggs in the open, more vertical (as opposed to more horizontal) surfaces probably have two advantages as oviposition sites, both related to egg development. First, vertical surfaces probably receive more current flow than horizontal surfaces, which should increase oxygenation rates of eggs. Secondly, eggs should be less likely to receive excess amounts of sediments that might reduce egg development rates or increase egg mortality through infection or hypoxia. Species that spawn in open sites often do not fan eggs (e.g., *Tripterygion*, Wirtz 1978; *Malacoctenus*, Petersen 1988), but may provide other forms of parental care, such as guarding eggs and removing dead eggs (Wirtz 1978).

Although some aspects of oviposition sites are preferred by females, these sites can have limited areas for oviposition and thereby constrain male mating success. This appears to limit male reproductive success in several species (e.g., *Emblemaria hypacanthus*, Hastings 1992; *Aidablennius sphynx*, Neat & Locatello 2002). In these species, females lay eggs in a monolayer and, although clutches can be contiguous, they do not overlap. Other species that deposit eggs on exposed surfaces have cryptically colored eggs that are deposited more loosely with algae (Petersen pers. obs.); it is less clear in these species if oviposition substrate ever becomes limiting.

The long-term predictability of the quality of oviposition sites appears to vary substantially among species. For some, such as *M. hubbsi*, sites in the intertidal have decreased success during spring tides when the sites are exposed to air, and this appears to change preferred oviposition sites at those times (Petersen 1988). In response to this temporal variability in oviposition site attractiveness, males in some species appear to assess territory quality indirectly by their mating success. In *M. hubbsi,* males abandon territories within three days when females are kept from mating in those territories (Petersen 1988), and in *Ophioblennius macclurei,* males are more likely to abandon sites after periods of low male reproductive success (Côté & Hunte 1989b). In this second case, males move oviposition sites but stay within the same feeding territories.

In many species of fishes with paternal care, females prefer to mate with males guarding eggs, especially when those eggs or clutches are at early stages of development (Kraak & Videler 1991). In *Ophioblennius atlanticus*, males taking over sites with surviving eggs present realize an increase in mating success over males occupying sites with no eggs (Santos 1995). This has several potential benefits for females, the most obvious one being increased survivorship of eggs due to either dilution of egg predation by the paternal male, increased parental care by the male, or a lower likelihood of abandonment of an oviposition site. In blennioids, this pattern has been suggested by female preferences for mating with males guarding eggs (Kraak & Videler 1991; Kraak & Groothuis 1994; Kraak & Weissing 1996, Fagundes et al. 2007) or mating males (Petersen 1989; Geertjes & Videler 2002). The most intriguing study of this kind is the work of Kraak & van der Berge (1992) on *Aidablennius sphynx*. These authors showed that if a male is given eggs, his future reproductive success increases, and if eggs are removed, his future reproductive success decreases. When nests are empty, females appear to lay a few eggs (1 to 10) in a nest, and if these "test" eggs survive, females will subsequently lay larger clutches with that male.

In some hole nesting combtooth blennies, a positive association between male size and size of the shelter opening has been reported (e.g., Crabtree & Middaugh 1982; Takegaki et al. 2008). Relative fit of males within their shelter entrance in the chaenopsids *Coralliozetus angelicus* and *Acanthemblemaria crockeri* shows considerable variation, ranging from 20 to 98 percent of the shelter opening filled by the resident's head; however, relative fit of males within their shelter entrance was unrelated to their body size or to the number of eggs they defended in either species (Hastings 1988a, b).

Size assortative pairing has been reported in at least two blenniids, *Ophioblennius macclurei* (Côté & Hunte 1989a) and *Istiblennius enosimae* (Sunobe et al. 1995). The preference of large females for large males is not surprising, given that a preference for large males occurs in many blennioids (see above). The important point here is that small females appear to be restricted from mating with large males. In species with relatively open oviposition sites, multiple females spawning simultaneously or consecutively with a male is common (Petersen 1988; Hamada & Nakazono 1989; Petersen 1989; Neat 2001; Geertjes & Videler 2002), and size assortative spawning is not reported in these species. In these species, males maximize mating success by mating with as many females as possible; small females are not excluded from oviposition sites and may mate at the same time that large females are spawning. However, if the number of females a male can mate with is restricted, then preferences for large females can evolve either because these females have qualitatively or quantitatively better offspring, or because their rates of oviposition during individual spawning events are higher. Restrictions on female mates would most likely occur in species where oviposition sites are limited to single females spawning at a time, and spawning is limited either to specific diel or lunar cycles. All of these conditions occur in *Ophioblennius macclurei*; spawning occurs during a limited period at dawn, is concentrated within a lunar cycle (Robertson et al. 1990) and the oviposition sites appear to restrict spawning to one female at a time. With limited opportunities to spawn, we would expect males to differentially select large females. An interesting prediction from this hypothesis is that we would expect male choice and size assortative pairing to be most intense during the peak of spawning in the lunar month. Thus, we would expect sexual selection for male choice to be less intense during other times in the lunar month when fewer females are spawning. Shibata & Kohda (2006) reported a shift in sex roles in *Petroscirtes breviceps* at the middles of an eight-month breeding season when spawning sites were restricted in availability, but they did not report the size of mating pairs.

A second possible reason for size assortative pairing would be if individuals assort in the habitat according to size, and females travel limited distances to find mates, thereby reducing their mating options. Size assortative habitat use has been reported in *Acanthemblemaria spinosa* (Clarke & Tyler 2003), but it is unclear whether this pattern ever leads to size assortative mating.

There is no evidence in any blennioid species that sperm limits fertilization success in the field, and sperm limitation is unlikely to be a constraint on fertilization success in these and similar fishes (Giacomello et al. 2007). Spawning in enclosed areas such as cracks or invertebrates probably leads to very effective fertilization at very low sperm concentrations. Spawning in open areas with greater current flow is more likely to result in sperm dilution and reduced fertilization rates, but no data exist to address this prediction for marine blennioids. However, in the freshwater blenniid *Salaria fluviatilis,* males in riverine populations were currents may disperse sperm have larger testes and release more sperm compared to males from lake populations (Vinyoles et al. 2002; Neat et al. 2003a).

OVIPOSITION SITES AND ALTERNATIVE REPRODUCTIVE TACTICS

Alternative male reproductive tactics (ARTs), including both sneak spawning and female mimicry, are known for several species of blennioids. This topic has been recently reviewed (Taborsky 2001 for fishes, and Oliveira et al. 2001a and Almada & Robalo 2008 for blennioids). Within blennioids, alternative reproductive tactics appear to take three forms: small males sneaking into nests and releasing sperm (e.g., *Axoclinus* spp. and *Tripterygion spp.*, Wirtz 1978; Neat 2001); satellite males that associate with specific nests and their territorial males, do some nest defense, but also sneak spawn (e.g., *Parablennius parvicornis* and *Salaria pavo*; Oliveira et al. 2002; Gonçalves et al. 2003a, b); and males that mimic female behavior and morphology to gain access to oviposition sites in what has been described for fishes as pseudofemale behavior (e.g., *Salaria pavo*, Gonçalves et al. 1996, 2005). Males practicing ARTs often forego investment in accessory gonadal structures associated with mate attraction in territorial males (de Jonge et al. 1989; Oliveira et al. 2001b; Neat et al. 2003b) and often have larger testes (Ruchon et al. 1995; but see Geertjes & Videler 2002 for an exception).

Oliveira et al. (2008) reviewed the nine species of blennioids where ARTs have been documented, which includes only the Tripterygiidae and Blenniidae. However, the absence of reports from the less well-studied Labrisomidae is probably a result of less published work on these species. For example, in the labrisomid *Malacoctenus hubbsi*, one of us has observed individuals adopting pseudofemale behavior, either for multiple days or as a potential season-long tactic. In addition, territorial males that do not attract females early during the breeding period lose their courtship coloration and attempt to enter nearby nests with active spawning (Petersen, unpublished observations). Although these sightings were rare events (only one individual of each type was observed), it suggests that ARTs are more widely distributed across the blennioids than currently reported.

With the recent proliferation of microsatellite primers for a variety of fish species, the genetic confirmation of multiple paternity, and thus the success of ARTs is now possible. Multiple paternity was recently demonstrated within clutches of eggs defended by single males in the blenniid *Scartella cristata* (Mackiewicz et al. 2005). Because male replacement at oviposition sites is well known in blennioids (e.g., Santos 1995; Tyler & Tyler 1999), some cases of males guarding unrelated eggs may represent nest turnover rather than successful sneak fertilizations. Nest turnover should be distinguishable from lost fertilizations to competing males because in the former, eggs fathered by a different male should be the oldest eggs at the oviposition site.

Reports on ARTs in blennioids conform to a pattern of smaller males "making the best of a bad situation," suggesting that they have lower current reproductive success than territorial males, and that the mating tactic is conditional and not an equally successful mating strategy (Ruchon et al. 1995; Gonçalves & Almada 1997). Generally, we expect alternative mating tactics in blennioids to exist when there is a class of males, typically significantly smaller males, that have substantially lower expected mating success than larger territorial males.

Blennioids vary substantially in how physically isolated oviposition sites are within the environment. These range from small holes in virtually all chaenopsids and some blenniids, to the often relatively open oviposition sites in the labrisomids and many of the tripterygiids. Oviposition sites located in holes or crevices, especially when they are relatively small compared to individual adults (i.e., in chaenopsids), often lead to single females spawning with males at any one time, and make it relatively easy for males to defend these sites from conspecific male intruders (Hastings 1986). ARTs have never been reported within the Chaenopsidae, and instead, neighboring males exhibit vigorous courtship from their own shelter when in sight of a spawning pair (Hastings pers. obs.) and females may spawn consecutively with multiple males (Stephens et al. 1966). This is in contrast to open substrate spawners, like tripterygiids, where as many as five females of *Axoclinus storyae* may spawn with a male at once. ARTs have been reported for several of these triplefin species (Wirtz 1978; Neat 2001). Generally, we expect ARTs to be more likely in open substrate spawning species than in hole-nesting species, simply because of increased access to nests by small males, either as a sneaker male making quick intrusions into relatively open oviposition site, or as one of many conspecifics in a nest at once (Neat & Locatello 2002). ARTs are known in some non-blennioid species that are hole nesters, but in these species the hole or crevice is either large enough for multiple spawning partners to be in a nest at once, or males appear to release sperm at a distance from the female and fan the sperm toward the nest with their pectoral fins (Mazzoldi & Rasotto 2002).

The second characteristic that we predict will exist in species with ARTs is that of having a class of males that have a low expectation of mating success for a given time in the reproductive season. In species that live for several years, this will most often consist of younger year classes of males or, in short-lived species with prolonged spawning seasons, those males that have recently recruited to the reproductive population. For example, in the Gulf of California triplefin, *Axoclinus nigricaudus*, during the summer months there are two size classes of males. The smaller size class does not take on male secondary sexual characteristics, does not have a well-developed accessory gland, and can be observed sneak spawning with larger territorial males (Neat 2001; Petersen pers. obs.). In contrast, *Axoclinus storyae* (= *A. carminalis*) appears to be largely an annual species, with breeding restricted to the summer months. All males are mature, have well-developed secondary sexual characteristics including red coloration and developed accessory glands, and attempt to court females to oviposition sites (Petersen 1989, pers. obs.). However, late in the breeding season, small individual males do appear, presumably young of the year from spawnings earlier in the season, and attempt to sneak spawn (Petersen, pers. obs.). Thus, in Gulf of California *Axoclinus* species, variation in body size within a species appears to have a direct relationship with the existence of ARTs, as predicted generally by Taborsky (2001). In species such as hole nesters where there is a large variation in male size and large males can become associated with successful (and limited) oviposition sites, we expect the more likely life-history tactic will be for males to delay maturation, and only the largest males to have well-developed secondary sexual characteristics. However, where size assortative mating provides mating opportunities for young, small males such as in the angel blenny (*Coralliozetus angelicus*), males of all sizes defend shelters from which they court and spawn with similarly sized females (Hastings 1986, 1988b).

Alternative mating tactics have evolved in at least three families of blennies and several times within the blenniids (Oliveira et al. 2001a; Almada & Robalo 2008), so this characteristic appears to have high evolvability in this group, especially where oviposition sites are more exposed and more accessible. ARTs have not been reported in species with readily defensible oviposition sites (i.e., within small shelters), and they may also be absent in internally fertilizing blennioids, but the mating behaviors of these species are largely unknown.

SIZE DIMORPHISM AND PATTERNS
OF MALE REPRODUCTIVE SUCCESS

Blennioid species differ in their pattern of sexual size dimorphism, ranging from males being larger, to sexes similar in size, to females being larger. Because small changes in body size of males and/or females can affect the presence or absence of size dimorphism, it appears that body size dimorphism is a phylogenetically labile feature in blennioids. This is not surprising given the varied nature of growth and mortality in fishes such as blennies and the fact that sexual size dimorphism results from separate but related selection on male and female body size (Blankenhorn 2005). For example, angel blennies are highly dimorphic in body size, with the frequency of males increasing disproportionately in the larger size classes (Hastings 1991); however, this may result from differential mortality of females because they experience greater exposure to predators (Hastings 2002); faster growth of males (not tested); or both (Hastings 1991).

Similar to externally fertilizing clinids (e.g., *Heterostichus rostratus*, Stepien 1986), males of internally fertilizing clinids are the larger sex in several species (Gunn & Thresher 1993). The larger size of males in internally fertilizing species is surprising given persistent selection on gestating females to maximize fecundity. This observation predicts a large male advantage in mating in these fishes, but this has yet to be demonstrated.

Externally fertilizing blennies exhibit a range of dimorphism patterns. Large male size dimorphism is predicted in species where larger males are able to obtain higher reproductive success (RS), which often will be related to their ability to sequester preferred oviposition sites. Stable patterns of female preference with respect to oviposition site should select for larger males that can defend these sites. The intensity of this selection will be proportional to the variance in male reproductive success due to variation in oviposition site. However, several factors may restrict the ability of males to maintain sites that allow them to continue to mate with females and obtain high levels of RS. First, as mentioned earlier, size of oviposition sites in some species may place upper limits on male RS due to limited oviposition site area (e.g., *Emblemaria hypacanthus*, Hastings 1992). Additionally, preferred sites must be predictable and not change in quality over the reproductive season, and must continue to be preferred after males begin defending them. In *Malacoctenus hubbsi*, high spring tides may expose otherwise preferred oviposition sites during the spawning period, reducing the RS of males defending them (Petersen 1988). Similarly, over time, barnacle tests used for oviposition sites may became fouled and thus less preferred as oviposition sites in *Corralliozetus angelicus* (Hastings 1988b). While most studies of male RS have treated oviposition sites as fixed-quality entities, these studies, and those where males change oviposition sites in response to changes in RS (e.g., Côté & Hunte 1989b), suggest that such temporal instability of oviposition-site quality may be common in blennioids.

Under what circumstances will females prefer larger males? We expect that females should select males defending oviposition sites that maximize the fitness of their young. This could occur if their offspring have better survivorship at specific sites or if spawning with particular males has genetic benefits. Although there is some suggestion that females in some species do prefer larger males because of increased survivorship of young guarded by large males (e.g., Giacomello & Rasotto 2005), we cannot discern any a priori predictions of where this association might exist. We cannot, for example, find any reason to predict that variance in egg survivorship should be greater for species that nest in holes versus species that nest in more open oviposition sites.

ANALYSIS OF FIFTEEN SPECIES OF BLENNIOIDS

We examined patterns of sexual size dimorphism (SSD) and the correlation between male size and male mating success in several species of blennioids where we have recorded field data on male mating success and male size. These 15 study species are distributed across four of the six families of blennioids, with most data coming from species from the Gulf of California (13 of 15; Table 4.1).

Data on sexual size dimorphism were either collected from individuals at the same time that mating success data were being collected, or were obtained from museum collections. In all species, immature individuals were excluded from analysis. In chaenopsids, only those individuals whose sex could be accurately identified based on sex-specific genital papillae (Hastings 1991) were included. For several genera (e.g., *Malacoctenus*, *Ophioblennius*), only individuals with gonadal evidence of active reproduction, as assessed by visual inspection of gonads, were included. In *Axoclinus nigricaudus*, SSD was estimated using only year 2+ males and females (i.e., as estimated from size frequency distributions), effectively excluding males employing ARTs.

Data on male mating success in the field was determined in one of two general ways. For species that had readily visible eggs, eggs were censused either once (several chaenopsids) or multiple times up to one month (e.g., *Acanthemblemaria crockeri* and *Ophioblennius* spp.). RS was inferred based on either total area of

TABLE 4.1 Sexual size dimorphism and male reproductive success
in 15 species of blennioid fishes

Family/species	M/F SL	Correlation coefficient (r) for SL vs. RS (N)	p	Source
Labrisomidae				
Malacoctenus tetranemus	1.11*	0.076 (26)	0.714	CP
M. macropus	1.07*	0.554 (19)	0.014	Petersen 1988
M. hubbsi	0.96*	0.205 (36)	0.792	Petersen 1988
M. margaritae	0.94*	0.015 (21)	0.948	CP
Tripterygiidae				
Axoclinus nigricaudus	1.04*	0.26 (29)	0.173	CP
A. storyae	1.00	−0.362 (21)	0.107	Petersen 1989
Blenniidae				
Ophioblennius macclurei	0.98*	0.699 (19)	0.001	CP
O. steindachneri	1.00	−0.04 (8)	0.920	CP
Chaenopsidae				
Acanthemblemaria balanorum	1.06	0.4003 (17)	0.110	PH
A. crockeri	1.10*	0.3469 (56)	0.009	Hastings 1988a
A. macrospilus	1.10*	0.1852 (38)	0.266	PH
Coralliozetus angelicus	1.08*	0.3442 (133)	0.001	Hastings 1988b
C. micropes	1.04*	0.1987 (38)	0.232	PH
C. rosenblatti	1.33*	0.6839 (11)	0.020	PH
Emblemaria hypacanthus	1.10*	0.5232 (34)	0.001	Hastings 1992

M/F SL = male to female ratio of mean standard length (SL).

*indicates male and female sizes are significantly different based on a *t*-test.

r = correlation coefficient of SL vs. RS (N = number of males sampled; RS = reproductive success).

p = significance level for coefficient r.

CP = Petersen unpublished data.

PH = Hastings unpublished data.

eggs oviposited (*Ophioblennius*) or exact counts of eggs (chaenopsids). This number was then run in a correlation against standard length of males guarding the eggs. For more detail on these methods, see Hastings (1988a, b). This technique was used for all chaenopsids and blenniids.

For species with open oviposition sites and more cryptic eggs, we used the techniques reported in Petersen (1988, 1989) for estimating reproductive success. Individuals were censused during the morning spawning period, and the relative mating success of a male was estimating by integrating the amount of time a male spawned with females over a period of weeks. This technique was used for all labrisomids and tripterygiids.

There were substantial differences in SSD among the 15 species, with nine species having larger males, three species having larger females, and two species showing no pattern of SSD (Table 4.1). Although chaenopsids tended to be more likely to show male SSD (six of seven species), generally there was little clear phylogenetic signal with four of the six genera showing interspecific variation in the pattern of SSD.

The general prediction that species where males are larger would have a stronger correlation between male size and mating success was only weakly supported in our data set. Ignoring the phylogenetic signal, five of nine species where there was a significant male SSD showed a positive correlation between male size and male RS. Only one (*Ophioblennius macclueri*) of five species where males were not larger showed a positive correlation between male size and male RS (Fisher exact test $p = 0.168$). Similarly, there was a non-significant trend for SSD to be correlated with the correlation between male size and male RS ($r = 0.453$, $p = 0.104$, 2-tailed).

Although there are slight trends toward a positive correlation between male SSD and a size advantage in male mating success for territorial males in these blennioid species, the pattern is not striking, and clearly other factors are affecting the observed patterns. Within some genera like *Malacoctenus*, the species with the most striking pattern of male SSD showed a notable lack of correlation between RS and male size. While there are many possible explanations for these patterns, here we mention three types of reasons why sexual selection on increased male size is not strictly mirrored in our data set.

First, male and female blennies may have different energy budgets or different survivorship due to differences in their ecology or reproductive behavior. There may be selection for increased male size. But in some species, defense of oviposition sites may change feeding rates, or may change male mortality rates relative to females. Second, our estimates of male RS with size are based on our best censusing of reproductive males. It is possible that males with no mating success are underrepresented in some species, and if they are smaller, these uncounted males would limit our ability to determine positive correlations between reproductive success and male size. Finally, our estimate of short-term reproductive success may not represent the selective differential on male size versus female size. This might be because short-term male RS is not a good predictor of lifetime RS for males, or that species differ in selection on female RS and female size.

With the relatively simple analysis presented here, we cannot distinguish between the possibilities listed above and other potential reasons accounting for lack of a strong correlation between male RS and male SSD. We consider this analysis an initial step in understanding how patterns of SSD and male reproductive success interact with oviposition site and other elements of the reproductive biology of these fishes to produce the rich patterns of mating systems we see in blennioids.

SUMMARY

This brief overview of the factors impinging on the mating systems of blennioid fishes reveals the importance of fertilization mode and parental care pattern in setting the selective arena in which environmental factors act to determine a species' mating system. A key component in the majority of blennioids with external fertilization of demersal eggs is the predictability and defensibility of oviposition sites. These determine to a large extent the options for resource defense by dominant males, the patterns of female mate choice, and the reproductive options of non-territorial males.

REFERENCES

Almada VC, Robalo JI (2008). Phylogenetic analysis of alternative reproductive tactics: problems and possibilities. In: Oliveira RF, Taborsky M, Brockmann HJ (eds.), *Alternative Reproductive Tactics: An Integrative Approach*. Cambridge University Press, Cambridge, pp. 52–62.

Almada VC, Santos RS (1995). Parental care in the rocky intertidal: a case study of adaptation and exaptation in Mediterranean and Atlantic blennies. Reviews of Fish Biology and Fisheries 5:23–37.

Almada VC, Oliveira RF, Barata EN, Gonçalves EJ, Rito AP (1990). Field observations on the behaviour of the breeding males of *Lipophrys pholis* (Pisces: Blenniidae). Portugaliae Zoologica 1: 27–36.

Almada VC, Gonçalves EJ, Oliveira RF, Barata EN (1992). Some features of the territories in the breeding males of the intertidal blenny *Lipophrys pholis* (Pisces: Blenniidae). Journal of the Marine Biological Association, UK 72: 187–197.

Almada VC, Gonçalves EJ, Santos AJ, Baptista C (1994). Breeding ecology and nest aggregations in a population of *Salaria pavo* (Pisces: Blenniidae) in an area where nest sites are very scarce. Journal of Fish Biology 45: 819–830.

Almada VC, Gonçalves E, Oliveira RF, Santos AJ (1995). Courting females: ecological constraints affect sex roles in a natural population of the blenniid fish, *Salaria pavo*. Animal Behaviour 49: 1125–1127.

Barata EN, Serrano RM, Miranda A, Nogueira R, Hubbard PC, Canário AVM (2008). Putative pheromones from the anal glands of male blennies attract females and enhance male reproductive success. Animal Behaviour, 75: 379–389.

Berglund A, Rosenqvist G, Svensson I (1989). Reproductive success of females limited by males in two pipefish species. American Naturalist 133: 506–516.

Blanckenhorn WU (2005). Behavorial causes and consequences of sexual size dimorphism. Ethology 111: 977–1016.

Böhlke JE, Chaplin CG (1968). *Fishes of the Bahamas and Adjacent Tropical Waters*. Livingston Press, Wynnewood, Pennsylvania.

Breder CM (1941). On the reproductive behavior of the sponge blenny, *Paraclinus marmoratus* (Steindachner). Zoologica 26: 233–236.

Breder CM, Rosen DE (1966). *Modes of Reproduction in Fishes*. Natural History Press, Garden City, New York.

Buchheim JR, Hixon MA (1992). Competition for shelter holes in the coral-reef fish, *Acanthemblemaria spinosa* Metzelaar. Journal of Experimental Marine Biology and Ecology 164: 45–54.

Clarke R, Tyler JC (2003). Differential space utilization by male and female spinyhead blennies, *Acanthemblemaria spinosa* (Teleostei: Chaenopsidae). Copeia 2003: 241–247.

Clarke RD (1992). Effects of microhabitat and metabolic rate on food intake, growth and fecundity of two competing coral reef fishes. Coral Reefs 11: 199–205.

Côté IM, Hunte W (1989a). Male and female mate choice in the redlip blenny: why bigger is better. Animal Behaviour 38: 78–88.

Côté IM, Hunte W (1989b). Self-monitoring of reproductive success: nest switching in the redlip blenny (Pisces: Blenniidae). Behavioral Ecology and Sociobiology 24: 403–408.

Côté IM, Hunte W (1993). Female blennies prefer older males. Animal Behaviour 46: 203–205.

Crabtree RE, Middaugh DP (1982). Oyster shell size and the selection of spawning sites by *Chasmodes bosquianus, Hypleurochilus geminatus, Hypsoblennius ionthas* (Pisces, Blenniidae) and *Gobiosoma bosci* (Pisces, Gobiidae) in two South Carolina estuaries. Estuaries 5: 150–155.

de Jonge J, de Ruiter AJH, van den Hurk R (1989). Testis-testicular gland complex of two *Tripterygion* species (Blennioidei, Teleostei): differences between territorial and non-territorial males. Journal of Fish Biology 35: 497–508.

de Jong K, Bouton N, Slabbekoorn H (2007). Azorean rock-pool blennies produce size-dependent calls in a courtship context. Animal Behaviour 74(5): 1285–1292.

Depczynski M, Bellwood DR (2003). The role of crytobenthic reef fishes in coral reef trophodynamics. Marine Ecology Progress Series 256: 183–191.

Emlen ST, Oring LW (1977). Ecology, sexual selection and the evolution of mating systems. Science 197: 215–223.

Fagundes T, Gonçalves DM, Oliveira RF (2007). Female mate choice and mate search tactics in a sex role reversed population of the peacock blenny *Salaria pavo* (Risso, 1810). Journal of Fish Biology 71: 77–89.

Faria C, Almada V (2001). Agonistic behaviour and control of access to hiding places in two intertidal blennies, *Lipophrys pholis* and *Coryphoblennius galerita* (Pisces: Blenniidae). Acta Ethologica 4: 51–58.

Fishelson L, Gon O, Holdengreber V, Delarea Y (2006). Comparative morphology and cytology of the male sperm-transmission organs in viviparous species of clinid fishes (Clinidae: Teleostei, Perciformes). Journal of Morphology 267: 1406–1414.

Fishelson L, Gon O, Holdengreber V, Delarea Y (2007). Comparative spermatogenesis, spermatocytogenesis, and spermatozeugmata formation in males of viviparous species of clinid fishes (Teleostei: Clinidae, Blennioidei). The Anatomical Record 290: 311–323.

Fricke, R (2009). Systematics of the Tripterygiidae. In: Patzner R, Gonçalves E, Hastings P, Kapoor B (eds.), *The Biology of Blennies*. Science Publishing, Enfield, New Hampshire, pp. 37–67.

Geertjes GJ, Videler JJ (2002). A quantitative assessment of the reproductive system of the Mediterranean cave-dwelling triplefin blenny *Tripterygion melanurus*. PSZN: Marine Ecology 23: 327–340.

George A, Springer VG (1980). Revision of the clinid fish tribe Ophiclinini, including five new species, and definition of the family Clinidae. Smithsonian Contributions to Zoology 307: 1–31.

Giacomello E, Rasotto MB (2005). Sexual dimorphism and male mating success in the tentacle blenny *Parablennius tentacularis* (Teleostei: Blenniidae). Marine Biology 147: 1221–1228.

Giacomello E, Marchini D, Rasotto MB (2006). A male sexually dimorphic trait provides antimicrobials to eggs in blenny fish. Biology Letters 2: 330–333.

Giacomello E, Neat F, Rasotto MB (2007). Mechanisms enabling sperm economy in blenniid fishes. Behavioral Ecology and Sociobiology 62: 671–680.

Gibran FG, Santos FB, dos Santos HF, Sabino J (2004). Courtship behavior and spawning in the hairy blenny *Labrisomus nuchipinnis* (Labrisomidae) in southeastern Brazil. Neotropical Ichthyology 2: 163–166.

Gonçalves EJ, Almada VC (1997). Sex differences in resource utilization by the peacock blenny. Journal of Fish Biology 51: 624–633.

Gonçalves EJ, Almada VC (1998). A comparative study of territoriality in intertidal and subtidal blennioids (Teleostei: Blennioidei). Environmental Biology of Fishes 51: 257–264.

Gonçalves EJ, Almada VC, Oliveira RF, Santos AJ (1996). Female mimicry as a mating tactic in males of the blenniid fish *Salaria pavo*. Journal of the Marine Biological Association, UK 76: 529–538.

Gonçalves D, Fagundes T, Oliveira R (2003a). Reproductive behaviour of sneaker males of the peacock blenny. Journal of Fish Biology 63: 528–532.

Gonçalves D, Oliveira RF, Körner K, Schlupp I (2003b). Intersexual copying by sneaker males of the peacock blenny. Animal Behaviour 65: 355–361.

Gonçalves, DM, Matos R, Fagundes T, Oliveira R (2005). Bourgeois males of the peacock blenny, *Salaria pavo*, discriminate female mimics from females? Ethology 111: 559–572.

Gross MR, Sargent RC (1985). The evolution of male and female parental care in fishes. American Zoologist 25: 807–822.

Gross MR, Shine R (1981). Parental care and mode of fertilization in ectothermic vertebrates. Evolution 35: 775–793.

Gunn JS, Thresher RE (1993). Viviparity and the reproductive ecology of clinid fishes (Clinidae) from temperate Australian waters. Environmental Biology of Fishes 31: 323–344.

Hamada H, Nakazono A (1989). Reproductive ecology of the triplefin, *Enneapterygius etheostomus*, with special reference to the occurrence of fish eggs in the digestive tract of the male. Science Bulletin of the Faculty of Agriculture, Kyusu University 43: 127–134.

Hastings PA (1986). Habitat selection, sex ratio and sexual selection in *Coralliozetus angelica* (Blennioidea: Chaenopsidae). In: Uyeno T, Arai R, Taniuchi T, Matsuura K. (eds.),

Indo-Pacific Fish Biology: Proceedings of the Second International Conference on Indo-Pacific Fishes. Ichthyological Society of Japan, pp. 785–793.

Hastings PA (1988a). Correlates of male reproductive success in the brown cheeked blenny, *Acanthemblemaria crockeri* (Blennioidei: Chaenopsidae). Behavioral Ecology and Sociobiology 22: 95–102.

Hastings PA (1988b). Female choice and male reproductive success in the angel blenny, *Coralliozetus angelica* (Teleostei: Chaenopsidae). Animal Behaviour 38: 115–124.

Hastings PA (1991). Ontogeny of sexual dimorphism in the angel blenny, *Coralliozetus angelica* (Blennioidei: Chaenopsidae). Copeia 1991: 969–978.

Hastings PA (1992). Nest-site size as a short-term constraint on the reproductive success of paternal fishes. Environmental Biology of Fishes 34: 213–218.

Hastings PA (2002). Evolution of morphological and behavioral ontogenies in females of a highly dimorphic clade of blennioid fishes. Evolution 58: 1644–1654.

Hastings PA (2009). Biogeography of Neotropical blennies. In: Patzner R, Gonçalves E, Hastings P, Kapoor B (eds.), *The Biology of Blennies*. Science Publishing, Enfield, New Hampshire, pp. 95–118.

Hastings PA, Galland GR (2010). Ontogeny of microhabitat use and two-step recruitment in a specialist reef fish, the Browncheek Blenny (Chaenopsidae). Coral Reefs in 29: 155–164.

Hastings PA, Springer VG (2009a). Systematics of the Blenniidae (Blennioidei). In: Patzner R, Gonçalves E, Hastings P, Kapoor B (eds.), *The Biology of Blennies*. Science Publishing, Enfield, New Hampshire, pp. 69–99.

Hastings PA, Springer VG (2009b). Systematics of the Blennioidei and the included families Chaenopsidae, Clinidae, Labrisomidae and Dactyloscopidae. In: Patzner R, Gonçalves E, Hastings P, Kapoor B (eds.), *The Biology of Blennies*. Science Publishing, Enfield, New Hampshire, pp. 3–30.

Hirayama ST, Shiiba Y, Sakai H, Hashimota, Gushima K (2005). Fish-egg predation by the small clingfish *Pherallodischthys meshimaensis* (Gobiesocide) on the shallow reefs of Kuchierabu-Jima Island, southern Japan. Environmental Biology of Fishes 73: 237–242.

Hubbs CL (1952). A contribution to the classification of blennioid fishes of the family Clinidae, with a partial revision of the eastern Pacific forms. Stanford Ichthyological Bulletin 4: 41–165.

Kotrschal K (1988). A catalogue of skulls and jaws of eastern tropical Pacific blennioid fishes (Blennioidei: Teleostei): A proposed evolutionary sequence of morphological change. Zeitschrift für Zoologisches Systematik und Evolutionsforschung 26: 442–466.

Kraak SBM (1996). A quantitative description of the reproductive biology of the Mediterranean blenny *Aidablennius sphynx* (Teleostei, Blenniidae) in its natural habitat. Environmental Biology of Fishes 46: 329–342.

Kraak SBM, Groothuis TGG (1994). Female preference for nests with eggs is based on the presence of the eggs themselves. Behaviour 131: 189–206.

Kraak SBM, van der Berghe EP (1992). Do female fish assess paternal quality by means of test eggs? Animal Behaviour 43: 865–867.

Kraak SBM, Videler JJ (1991). Mate choice in *Aidablennius sphynx* (Teleostei, Blenniidae): females prefer nests containing more eggs. Behaviour 119: 243–266.

Kraak SBM, Weissing FJ (1996). Female preference for nests with many eggs: a cost-benefit analysis of female choice in fish with paternal care. Behavioral Ecology 7: 353–361.

Lengkeek W, Didderen K (2006). Breeding cycles and reproductive behaviour in the river blenny (*Salaria fluviatilis*). Journal of Fish Biology 69: 1837–1844.

Linn HC (2009). Evolution of the suborder Blennioidei: Phylogeny and phylogeography of a shallow water fish clade. Unpublished Dissertation, University of California, San Diego.

Locatello L, Neat FC (2005). Reproductive allocation in *Aidablennius sphynx* (Teleostei: Blenniidae): females lay more eggs faster when paired with larger males. Journal of Experimental Zoology 303A: 992–926.

Mackiewicz M, Porter BA, Dakin EE, Avise JC (2005). Cuckoldry rates in the molly miller (*Scartella cristata*, Blenniidae) a hole nesting marine fish with alternative reproductive tactics. Marine Biology 148: 213–221.

Mank JE, Promislow DEL, Avise JC (2005). Phylogenetic perspective on the evolution of parental care in ray-finned fishes. Evolution 59: 1570–1578.

Mazzoldi C, Rasotto MB (2002). Alternative male mating tactics in *Gobius niger*. Journal of Fish Biology 61: 157–172.

Moser HG (2007). Reproduction in the viviparous South African clinid fish *Fucomimus mus*. African Journal of Marine Science 29: 423–436.

Munday PL, Jones GP (1998). The ecological implications of small body size among coral-reef fishes. Annual Review of Oceanography and Marine Biology 36: 373–411.

Neat FC (2001). Parasitic spawning in two species of triplefin blenny: contrasts in demography, behaviour and gonadal characteristics. Environmental Biology of Fishes 55: 57–64.

Neat FC, Lengkeek W (2009). Sexual selection in blennies. In: Patzner R, Gonçalves E, Hastings P, Kapoor B (eds.), *The Biology of Blennies*. Science Publishing, Enfield, New Hampshire, pp. 249–278.

Neat FC, Locatello L (2002). No reason to sneak: why males of all sizes can breed in the hole-dwelling blenny, *Aidablennius sphynx* (Teleostei: Blenniidae). Behavioral Ecology and Sociobiology 52: 66–73.

Neat FC, Lengkeek W, Westerbeek P, Laarhoven B, Videler JJ (2003a). Behavioural and morphological differences between lake and river populations of *Salaria fluviatilis*. Journal of Fish Biology 63: 374–387.

Neat FC, Locatello L, Rasotto MB (2003b). Reproductive morphology in relation to alternative reproductive tactics in *Scartella cristata*. Journal of Fish Biology 62: 1381–1391.

Northcott SJ, James MA (1996). Ultrastructure of the glandular epidermis on the fins of male estuarine triplefins *Fosterygion nigripenne*. Journal of Fish Biology 49: 95–107.

Ohta T, Nakazono A (1988). Mating habits, mating system and possible filial cannibalism in the triplefin, *Enneapterygius etheostomus*. In: Choat JH, Barnes D, Borowitzka MA, Coll JC, Davies PJ, Flood P, Hatcher BG, Hopley D, Hutchings PA, Kinsey D, Orme GR, Pichon M, Sale PF, Sammarco P, Wallace CC, Wilkinson C, Wolanski E, Bellwood O (eds.), Proceedings of the Sixth International Coral Reef Symposium. Vol. 2, Contributed Papers. Townsville, Australia, pp. 797–801.

Okuda N, Fukumori K, Yanagisawa Y (2003). Male ornamentation and its condition-dependence in a paternal mouthbrooding cardinalfish with extraordinary sex roles. Journal of Ethology 21: 153–159.

Oliveira RF, Almada VC, Forsgren E, Gonçalves EJ (1999). Temporal variation in male traits, nesting aggregations and mating success in the peacock blenny. Journal of Fish Biology 53: 499–512.

Oliveira RF, Miranda JA, Carvalho N, Gonçalves EJ, Grober MS, Santos RS (2000). Male mating success in the Azorean rock-pool blenny: the effects of body size, male behaviour and nest characteristics. Journal of Fish Biology 57: 1416–1428.

Oliveira RF, Canario AVM, Grober MS (2001a). Male sexual polymorphism, alternative reproductive tactics, and androgens in combtooth blennies (Pisces: Blenniidae). Hormones and Behavior 40: 266–275.

Oliveira RF, Gonçalves EJ, Santos RS (2001b). Gonadal investment of young males in two blenniid fishes with alternative mating tactics. Journal of Fish Biology 59: 459–462.

Oliveira RF, Carvalho N, Miranda J, Gonçalves EJ, Grober M, Santos RS (2002). The relationship between the presence of satellite males and nest-holders' mating success in the Azorean rock-pool blenny *Parablennius sanguinolentus parvicornis*. Ethology 108: 223–235.

Oliveira RF, Gonçalves DM, Ros A (2008). Alternative reproductive tactics in blennies. In: Patzner R. Gonçalves E, Hastings P, Kapoor B (eds.), *The Biology of Blennies*. Science Publishing, Enfield, New Hampshire, pp. 279–308.

Patzner R, Gonçalves E, Hastings P, Kapoor B (2009). *The Biology of Blennies*. Science Publishing, Enfield, New Hampshire.

Petersen CW (1988). Male mating success, sexual size dimorphism, and site fidelity in two species of *Malacoctenus* (Labrisomidae). Environmental Biology of Fishes 21: 173–183.

Petersen CW (1989). Females prefer mating males in the carmine triplefin, *Axoclinus carminalis*, a paternal brood-guarder. Environmental Biology of Fishes 26: 213–221.

Petti JC (1969). Behavioral and morphological adaptations to burrowing of two species of dactyloscopid fishes from the northern Gulf of California. Master's thesis, University of Arizona, Tucson.

Prochazka K, Griffiths CL (1992). Observations on the distribution patterns, behaviour, diets and reproductive cycles of sand-dwelling clinids (Perciformes: Clinidae) from South Africa. Environmental Biology of Fishes 35: 371–379.

Reynolds JD, Côté IM (1995). Direct selection on mate choice: female redlip blennies pay more for better mates. Behavioral Ecology 6: 175–181.

Ridley M (1978). Parental care. Animal Behaviour 26: 904–932.

Robertson DR, Petersen CW, Brawn JD (1990). Lunar reproductive cycles of benthic-brooding reef fishes: reflections of larval-biology or adult-biology? Ecological Monographs 60: 311–329.

Robins CR, Ray C, Douglas J (1999). *A Field Guide to Atlantic Coast Fishes: North America*. Houghton Mifflin and Harcourt, Boston.

Rosenblatt RH, Taylor Jr LR (1971). The Pacific species of the clinid fish tribe Starksiini. Pacific Science 25: 436–463.

Ruchon F, Laugier T, Quignard JP (1995). Alternative male reproductive strategies in the peacock blenny. Journal of Fish Biology 47: 826–840

Santos RS (1995). Allopaternal care in the redlip blenny. Journal of Fish Biology 47: 350–353.

Shibata J, Kohda M (2006). Seasonal sex role changes in the blenniid *Petroscirtes breviceps*, a nest brooder with paternal care. Journal of Fish Biology 69: 203–214.

Shuster SM, Wade MJ (2003). *Mating Systems and Strategies*. Princeton University Press, Princeton, New Jersey.

Smith-Vaniz WF, Jelks HL, Rocha LA (2006). Relevance of cryptic fishes in biodiversity assessments: a case study at Buck Island Reef National Monument, St. Croix. Bulletin of Marine Science 79: 17–48.

Springer VG (1993). Definition of the suborder Blennioidei and its included families (Pisces: Perciformes). Bulletin of Marine Science 52: 472–495.

Springer VG, Orrell TM (2004). A phylogenetic analysis of 147 families of acanthomorph fishes based primarily on dorsal gill-arch muscles and skeleton. Bulletin of the Biological Society of Washington 11: 237–260.

Stephens JS Jr, Hobson ES, Johnson RK (1966). Notes on distribution, behavior and morphological variation in some chaenopsid fishes from the tropical eastern Pacific, with descriptions of two new species, *Acanthemblemaria castroi* and *Coralliozetus springeri*. Copeia 1966: 424–438.

Stepien CA (1986). Life history and larval development of the giant kelpfish, *Heterostichus rostratus* Girard, 1854. Fisheries Bulletin 84: 809–826.

Stepien CA, Dillon AK, Brooks MJ, Chase KL, Hubers AN (1997). The evolution of blennioid fishes based on an analysis of mitochondrial 12S rDNA. In: Kocher TT, Stepien CA (eds.), *Molecular Systematics of Fishes*. Academic Press, San Diego, pp. 245–270.

Sunobe T (1998). Notes on the mating system of *Omobranchus elegans* and *O. fasciolatoceps* (Blenniidae) at Maizuru, Japan. Ichthyological Research 45: 319–321.

Sunobe T, Ohta T, Nakazono A (1995). Mating system and spawning cycle in the blenny *Istiblennius enosimae*, at Kagoshima, Japan. Environmental Biology of Fishes 43: 195–199.

Taborsky M (2001). The evolution of bourgeois, parasitic and cooperative reproductive behaviors in fishes. Journal of Heredity 92: 100–110.

Takegaki T, Matsumoto Y, Tawa A, Miyano T, Natsukari Y (2008). Size-assortative nest preference in a paternal brooding blenny *Rhabdoblennius ellipes* (Jordan & Starks). Journal of Fish Biology 72: 93–102.

Thompson S (1986). Male spawning success and female choice in the mottled triplefin, *Forsterygion varium* (Pisces: Tripterygiidae). Animal Behaviour 34: 580–589.

Todd ES (1973). A preliminary report on the respiratory pump in the Dactyloscopidae. Copeia 1973: 115–119.

Tyler JC, Tyler DM (1999). Natural history of the sea fan blenny, *Emblemariopsis pricei* (Teleostei: Chaenopsidae), in the western Caribbean. Smithsonian Contribution to Zoology 601: 1–24.

Veith WJ (1979). The chemical composition of the follicular fluid of the viviparous teleost *Clinus superciliosus*. Comparative Biochemistry and Physiology 63A: 37–40.

Veith WJ (1980). Viviparity and embryonic adaptations in the teleost *Clinus superciliosus*. Canadian Journal of Zoology 58: 1–12.

Vinyoles D, Côté IM, De Sostoa A (1999). Egg cannibalism in river blennies: the role of natural prey availability. Journal of Fish Biology 55: 1223–1230.

Vinyoles D, Côté IM, de Sostoa A (2002). Nest orientation patterns in *Salaria fluviatilis*. Journal of Fish Biology 61: 405–416.

Wirtz P (1978). The behaviour of the Mediterranean *Tripterygion* species (Pisces, Blennioidei). Zeitschrift für Tierpsychologie 48: 142–174.

Wittenberger JF (1979). The evolution of mating systems in birds and mammals. In: Marler P, Vanderberg J (eds.), *Handbook of Behavioral Neurobiology: Social Behavior and Communication*. Plenum Press, New York, pp. 271–349.

Wittenberger JF (1981). *Animal Social Behavior*. Duxbury Press, Boston.

5

Gonad Morphology in Hermaphroditic Gobies

Kathleen S. Cole

Gobiid fishes (Family Gobiidae, Order Perciformes), consisting of at least 214 genera and 1,400 species (E. Murdy pers. comm.), constitute the second largest vertebrate family (second only to the Cyprinidae) and the largest family of vertebrates occupying marine environments (Nelson 2006). Among species for which there is information on reproductive biology, all are oviparous, having external fertilization in which gametes are released and fertilization occurs in the external environment. Typically, the female oviposits demersal eggs on a spawning surface prepared by the male, and the male guards the embryos until they hatch.

The reproductive anatomy of gobiids, particularly among males, exhibits considerable morphological diversity. In many species, extensive modifications of the sperm duct form specialized, secretion-producing structures referred to as seminal vesicles (Eggert 1931; Young & Fox 1937; Egami 1960; Arai 1964) or more commonly as sperm duct glands (Miller 1984). These structures show considerable taxon-specific variation in morphology (Miller 1984; Fishelson 1989). Testis morphology is also variable across gobiid taxa. Among some gonochoric (i.e., fixed-sex) species, a portion of the testis is made up of an aggregation of Leydig-like cells, which is referred to as the mesorchial gland (Colombo & Burighel 1974). These cells synthesize and secrete steroids, some of which act to attract gravid female conspecifics (Colombo et al. 1980; Belanger et al. 2004). The size of this gland has been shown to relate to mating tactics, with small sneaker males having a much smaller mesorchial gland than larger, conventionally spawning males (Locatello et al. 2002; Rasotto & Mazzoldi 2002). In other species, such as *Pomatoschistus microps*, organized mesorchial glands are absent and steroid production depends on interstitial cells that are embedded in the sperm duct glands.

In two hermaphroditic genera, *Paragobiodon* and *Gobiodon*, Fishelson (1989) has reported that the mesorchial gland is absent during the female phase but develops at the time of sex change.

Among functionally hermaphroditic goby taxa, the reproductive morphology of the gonad proper has become even further modified, substantially in some cases. These modifications vary considerably across hermaphroditic goby taxa. However, both morphological features and their patterns of development and transition throughout life have been shown to be highly conserved within some hermaphroditic clades (i.e., Cole 1990; Cole & Shapiro 1990). This clade-based distribution of reproductive morphological diversity raises the question as to whether hermaphroditism within the Gobiidae has singular or multiple origins.

As one of the two largest vertebrate families, gobiids have undergone considerable adaptive radiation and speciation, particularly in marine environments. Their success seems likely to be based at least in part on a capacity for developing morphological and physiological innovations. This, in turn, would have increased their ability to become morphologically and functionally specialized, leading to increased micropartitioning of complex habitats and the invasion of new niches. A finding of numerous independent origins of hermaphroditism within the taxon would provide additional support for the hypothesis that character innovation has been a major driver of evolutionary diversity within the Gobiidae.

In this chapter, information has been compiled from a variety of sources to examine the biological implications of diverse reproductive morphologies found among hermaphroditic goby taxa. In the first section, the generalized model of the teleost reproductive complex is compared with those of gonochoric (i.e., fixed-sex, or non-hermaphroditic) and functionally hermaphroditic gobiids to characterize differences and identify patterns of reproductive morphology among hermaphroditic goby taxa. In the second section, the distribution patterns of different forms of reproductive morphology among hermaphroditic goby taxa are examined from a phylogenetic perspective to answer two questions: First, are developmental patterns consistent with known phylogenetic relationships within the Gobiidae (i.e., does reproductive morphology accurately predict clade relationships among hermaphroditic gobies)? And second, to what extent may the evolution of different reproductive morphologies across clades represent independent evolutionary events within the Gobiidae?

A GENERALIZED MODEL OF THE REPRODUCTIVE COMPLEX OF EXTERNALLY FERTILIZING TELEOSTS

The following is a brief overview of teleost gonad morphology. For more detailed information, the reader is referred to Hoar (1955), Grier (1981), Nagahama (1983), Maack & Segner (2003), and Parenti & Grier (2004).

The basic morphology of the ovary among externally fertilizing teleosts is one of a bilobed, caudally united structure that is continuous with the oviduct, the latter extending through the body of the genital papilla and terminating at the genital pore. The ovary consists of peripheral tissue layers forming an ovarian wall that encloses and protects ovigerous (i.e., ova-producing) tissue making up the body of the gonad. The ovigerous tissue is typically anchored to the entire inner layer of the ovarian wall and takes the form of lamellar folds projecting into a central lumen.

The ovigerous tissue contains a variety of developmental stages of female sex cells, including oogonia and more mature stages of oocytes. At ovulation, mature oocytes—now termed ova—are released from their surrounding follicle cells into the ovarian lumen. Among externally fertilizing, non-inseminating teleosts, ova move from the ovarian lumen into the common genital sinus, an open region formed by the union of the two ovarian lobes and which has no ovigerous tissue directly associated with the genital sinus wall. From here, ova pass through a relatively short oviduct and are delivered to the outside of the body via the genital pore.

The typical testis of externally fertilizing teleosts consists of a bilobed, caudally united structure that is continuous with a sperm duct, which in turn leads to the genital papilla and terminates at the genital pore. In its simplest form, the testis consists of an external testis wall and, internally, spermatogenic tissue organized into blind-ended, seminiferous lobules (i.e., seminiferous tubules) *sensu* Grier et al. (1980). The terminal apices of these lobules are usually located at the testis lobe periphery. The proximal, open end of each lobule is oriented toward either a medially located sperm sinus; an internal sperm-collecting duct; or a caudally located common collecting region, frequently referred to as the common genital sinus. In mature males, free spermatozoa can be found in the lobule lumina and in the common collecting region(s) into which lobule lumina empty. The common genital sinus of the male reproductive complex is formed by the posterior union of the testis lobes; it contains no directly affixed spermatogenic tissue, and is continuous with the sperm duct.

Individual seminiferous lobules of the testis consist of a lobule wall made up of support cells and of male germ cells in various stages of development. Among teleosts, the seminiferous lobules may be of the restricted or the unrestricted type (Grier et al. 1980). In the unrestricted type testis, the lobule lumen extends the full length of the lobule and is surrounded along its length by developing germ and support cells. In the restricted type testis, the germ and support cells are confined to the distal end of the blind-ended lobule.

THE REPRODUCTIVE COMPLEX
OF GONOCHORIC GOBY TAXA

In general, the female reproductive complex of gonochoric (i.e., non-hermaphroditic, or constant-sex) goby taxa reiterates the basic teleost model of two ovarian lobes united posteriorly to form a common genital sinus. This non-gametogenic region is confluent with the oviduct and provides the pathway by which ova pass from the gonad to the outside environment. Lamellae of ovigerous tissue extend into a central cavity—the ovarian lumen—and within these lamellae various stages of female sex cells are present. Among immature individuals, these include oogonia and primary oocytes. The primary oocytes may consist of a number of developmental stages identified on the basis of cytological features, including chromatin nucleolar, perinucleolar, and cortical alveolar stages. Among reproductively active females, various stages of vitellogenic oocytes will dominate the ovigerous tissue. In most gonochoric gobiid taxa, a bilobed ovary, common genital sinus, and oviduct constitute the reproductive complex. *Schindleria praematura*, a planktonic paedomorphic goby, is an exception to this pattern. Ovigerous tissue forms a ventrally anchored ridge that projects dorsally into a surrounding lumen within the gonad and is surrounded on three sides by the lumen (Thacker & Grier 2005). Within the ovigerous ridge, stromal tissue is limited, and much of the tissue is made up of oogonia and oocytes. This unusual morphology is likely an adaptation to the extremely small size that characterizes this species.

Among males of gonochoric gobiid taxa, the reproductive complex typically exhibits considerable structural modification. Across taxa there is an impressive diversity of morphology, of both the gonad and the accessory gonadal structures (Miller 1984; Fishelson 1991). The bilobed testis may have an associated testicular gland (Stanley et al. 1965; Columbo & Burighel 1974), also referred to as a mesorchial gland (Miller 1984). This glandular structure is embedded within the testis proper, but it is isolated from the spermatogenic tissue and does not contribute to seminal fluid (Seiwald & Patzner 1989); rather, it is exocrine in function, consists primarily of Leydig cells, and provides a source of steroid hormones (Columbo & Burighel 1974; Seiwald & Patzner 1989). It has also been shown in the black goby, *Gobius niger*, to produce sexual pheromones (Columbo et al. 1980; Locatello et al. 2002). In this species, the relative size of the mesorchial gland is large in conventionally spawning males and small in males that parasitize spawning events by adopting a sneaker strategy (Rasotto & Mazzoldi 2002; Immler et al. 2004). The reduction in the latter is hypothesized to prevent conventionally spawning males from recognizing sneakers on the basis of mesorchial-gland-produced, male-specific pheromones. The mesorchial gland is absent in a number of goby species including the common goby, *Pomatoschistus microps*, *Lebetus* sp. (Miller 1984); two-spotted goby, *Gobiusculus flavescens* (Fishelson 1991); pelagic crystal goby,

Crystallogobius linearis; transparent goby, *Aphia minuta* (Caputo et al. 2003); and *Coryphopterus* spp. (Cole unpublished data). When more information is collected, the mseorchial gland may be found to be more commonly absent than present in some gobiid clades.

The most prominent accessory structures associated with the male reproductive complex among gobiid fishes are paired secretory structures that are associated with the sperm duct. These structures have been referred to as seminal vesicles (Eggert 1931; Young & Fox 1937; Weisel 1949) or sperm duct glands (Miller 1984) and exhibit considerable morphological variation in size and architecture among goby taxa (Miller 1984; Fishelson 1991). In most cases they consist of enlarged lobules lined with a simple columnar epithelium that has a complex ultrastructure (Cinquetti 1997) responsible for producing a sialoglycoprotein-rich fluid (Fishelson 1991; Lahnsteiner et al. 1992). In some species the resulting secretions combine with sperm and seminal fluid to produce sperm trails (Marconato et al. 1996; Scaggiante et al. 1999) from which spermatozoa slowly dissolve out of the mucin-based matrix after their release and deposition on a spawning surface. The size of the secretory structures has been shown to vary among males exhibiting different mating strategies. In some species, parental males have large structures associated with conventional spawning and sperm trail production. In contrast, sneaker males have much smaller secretory structures that are used for sperm storage rather than mucin production and a relatively large testis suitable for releasing large bursts of sperm during sneaking events (Immler et al. 2004; Scaggiante et al. 2004; Mazzoldi et al. 2005). Secretory accessory gonadal structures associated with the male reproductive complex are widespread among goby taxa; however, they are reported to be extremely reduced or absent in several species, including *Amblygobius nocturnus*, *Cryptocentrus lutheri*, *Valenciennea strigata*, *V. sexguttata*, *V. muralis* (Mazzoldi et al. 2005), and *Schindleria praematura*. In the latter species, the secretory accessory structure is reduced, the secretory epithelium is restricted to the ventral region, and instead of consisting of secretory lobules, the structure forms as a single chamber (Thacker & Grier 2005).

THE REPRODUCTIVE COMPLEX
OF HERMAPHRODITIC GOBIIDS

In hermaphroditic gobies, gonadal function varies over an individual's lifetime. Consequently, the terms "female," "male," "ovary," and "testis" that are used to describe the reproductive anatomy and sexual state of gonochores have limited application to the hermaphroditic condition. Accordingly, gonadal state in a hermaphroditic individual is better characterized by a combination of morphology and function. In morphological terms, the gonad may be ovariform, ovotestiform, or testiform. In functional terms, the gonad is either ova producing, sperm producing, both ova and sperm producing, or in the absence of any mature gametes, inactive. Note that the term "inactive" as used here equally describes the gonad of immatures and of adults during a non-reproductive phase. Therefore, on its own, describing a gonad as inactive does not distinguish between these two developmental and functional states. The designation of "immature" or "inactive adult" can only be made in the context of other species-specific information such as size or age at first maturity. By applying the same convention, the functional state of an individual can be characterized as male-active (i.e., sperm producing), female-active (i.e., ova producing), bisexually active (i.e., both sperm and ova producing), or inactive (i.e., no mature gamete production).

The manner in which reproductive anatomy has become modified in hermaphroditic goby species appears to vary according to taxon. Among some taxa, morphological modifications are only associated with the ovariform gonad, while in others, elaborate anatomical modifications of various regions of the reproductive complex are associated with either the female-active or male-active phase. Reproductive modifications among hermaphroditic gobiids are exhibited in their least morphologically elaborated form in the goby genus *Coryphopterus*. In *Coryphopterus glaucofraenum*, the first hermaphroditic *Coryphopterus* species to be studied in detail (Cole & Shapiro 1992), most individuals initially develop an ovariform gonad in which no testis tissues or features are visible. (See Figure 5.1A. First maturation is characterized by the onset of a female-active phase and ova production (Figure 5.1B). Subsequent to the female-active phase, the gonad undergoes a transformation to form a secondary testis (Figure 5.1C). To accomplish this, all elements of prior ovarian structure are lost, including the ovigerous tissue and its lamellar configuration around a central lumen, and the gonad is rebuilt internally by the formation of longitudinally oriented and cell-lined channels. These develop into seminiferous lobules which come to entirely replace the preceding ovigerous tissue (Figure 5.1D). Thus, the gonadal tissue of the initial immature and subsequent female-active phases appears strictly ovariform (Figure 5.1A, B), while that of the secondary male is strictly testiform.

Based on the description above, the pattern of reproductive morphology exhibited by *C. glaucofraenum* involves the absence of testis-associated tissues or recognizable male sex cells within the gonad prior to sex change; the *de nouveau* development of seminiferous lobules within previously active ovigerous tissue at the time of transformation; and the complete disappearance of ovarian tissue and architecture in the secondary testis. In this ontogenetic sequence, female-specific and male-specific gonadal tissues only co-occur for a brief period of time during the transition from female to male function. During this period, the gonad takes the form of a nonfunctional ovotestis by virtue of its temporarily coexisting ovariform and testiform features. In comparison with gonochore gobies in which the

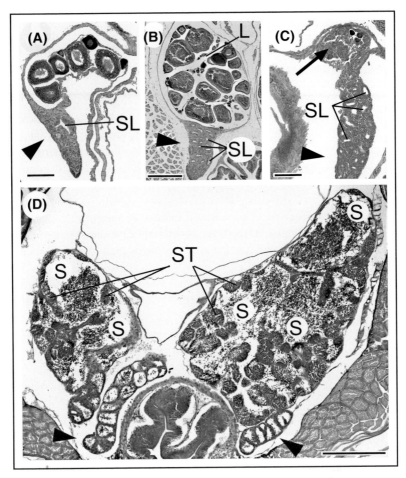

FIGURE 5.1. Histological features shown in transverse section of interim ovotestis gonad morphologies associated with immature and adult *Coryphopterus glaucofraenum*. In all images, dorsal is uppermost.

A. Ovariform gonad of immature, consisting solely of oocyte-bearing tissue, with pAGS (arrowhead) showing cell-bound lumen of future secretory lobule (SL). Scale bar is 100 μ.

B. Female-active gonad with lamellae of ovigerous tissue separated by branches of the central lumen (L) and showing prominent pAGS (arrowhead) with numerous lumina of future secretory lobules (SL). Scale bar is 500 μ.

C. Developing ovotestis (arrow) of transitional fish, with expanding seminiferous lobule lumina (SL) within developing AGS (arrowhead). Scale bar is 100 μ.

D. Reproductive complex of secondary male, including left and right lobes of the secondary testis (ST) made up solely of spermatogenic tissue, including seminiferous lobule lumina filled with sperm (S) and fully developed, actively secreting AGS (arrowhead). Scale bar is 200 μ.

reproductive complex is canalized early into either a male or female pathway, *C. glaucofraenum* expresses an ontogenetic sequence of gonadal morphologies comprising inactive ovariform, functional ovariform, transitional nonfunctional ovotestiform, and functional testiform, respectively.

During the immature and female-active phases, *C. glaucofraenum* does not exhibit any morphological indications of future spermatogenic capacity within the ovigerous tissue. However, there is a novel anatomical feature associated with the gonad proper that is necessary for the development of the reproductive anatomy of the male-active phase. In many gobiid taxa, and in all known hermaphroditic representatives, the male reproductive complex includes accessory structures that are lobular in architecture and secretory in function (i.e., sperm duct glands *sensu* Miller 1984). In hermaphroditic species, these structures usually develop at the time of sex change from female to male function. Among immature and ova-producing *C. glaucofraenum*, a small, ventro-lateral tissue mass located just anterior to the common genital sinus is associated with each gonadal lobe. These accessory masses, which are in close proximity to, but unassociated with, the ovigerous tissue (see arrowhead in Figure 5.1A), are made up of clusters of cells, some of which form a contiguous, single-celled layer around developing channels (Figure 5.1A, B).

In five additional *Coryphopterus* species for which morphological changes associated with a shift from female to male function have been tracked, these tissue masses are universally present among immature and female-active individuals and remain in a stable, undifferentiated state during the ovariform phase of the gonad (Cole & Robertson 1988; Cole & Shapiro 1990, 1992). During seminiferous lobule formation within the transforming gametogenic tissue, the accessory tissue masses undergo cell proliferation, rapid growth, and the development of numerous, epithelial-lined lobules (Figure 5.1C). With continued growth, the developing lobulated structures expand into the body cavity, the lining epithelium becomes columnar, and secretions produced by this epithelium start to be discharged into the lobule lumina (Figure 5.1D)(Cole & Robertson 1988; Cole & Shapiro 1992).

In terms of overall appearance, internal architecture, and secretory activities, the fully developed secretory structures of male-active fish among hermaphroditic *Coryphopterus* spp. appear identical to sperm duct glands associated with the male reproductive complex of gonochore goby taxa (Miller 1984, 1992). Among gonochore taxa, however, lobular secretory structures arise from the wall of the sperm duct, form early in development well before the initiation of sperm production, and are present only in males. The lobular secretory structures of hermaphroditic *Coryphopterus* spp., in contrast, arise from the lateral wall of the gonadal lobes and their precursors are present in all immatures and female-active adults. Because the lobular secretory structures of secondary male *Coryphopterus* spp. have no direct association with the gonoduct, they have been termed accessory gonadal structures, or AGS (Cole & Robertson 1988). The use of this term allows for a distinction to be made between the sperm duct glands *sensu* Miller (1984) that are associated only with the gonoduct of the male reproductive complex among gonochore goby taxa, and the gonadal wall derived structures of hermaphroditic *Coryphopterus* spp. which have the same differentiated appearance and apparent function as sperm duct glands. As AGS arise from the initially undifferentiated tissue masses associated with the ovariform gonad, the latter have been termed *precursive* accessory gonadal structures, or pAGS. The combination of pAGS and AGS that is characteristic of the *Coryphopterus* spp. reproductive complex of different functional states has not been described for any gonochore goby taxa and therefore represents a novel feature of reproductive morphology associated with some hermaphroditic gobies.

The ontogenetic pattern of gonad morphology expressed by *C. glaucofraenum*, in which pAGS associated with the ovariform gonadal phase develop into AGS and male and female gametogenic tissues only co-occur during a brief transitional phase associated with sex change, is also characteristic for all other *Coryphopterus* species that have been examined, including *C. personatus*, *C. dicrus*, *C. hyalinus*, and *C. lipernes* (Cole & Robertson 1988; Cole & Shapiro 1990). In four additional species, *C. alloides*, *C. eidolon*, *C. thrix*, and *C. urospilus*, morphogenic transformations of the gonad have not been tracked, but pAGS have been present in all examined immatures and female-active adults (Cole & Shapiro 1990).

The ovariform–nonfunctional transitional ovotestis–testiform pattern of gonad development exhibited by *Coryphopterus* spp. is also found among species of three other hermaphroditic gobiid genera, including *Rhinogobiops* (previously *Coryphopterus*) *nicholsii* (Cole 1983), *Lophogobius cyprinoides*, and *L. cristulatus* (Cole 1990; Cole unpublished data), and in two *Fusigobius* species, *F. neophytus*, and *F. signipinnis* (Cole 1990; Cole herein). A similar ovariform and testiform morphology has been found in both *L. cyprinoides* and *L. cristulatus* (Figure 5.2A–C and D–F, respectively), including the presence of pAGS associated with the immature and adult ovariform gonad (Figure 5.2A and D, respectively), and in *F. neophytus* (Figure 5.3A, B) and *F. signipinnis* (Figure 5.3C, D). In all four species, the gonad among immatures and ova-producing adults is entirely ovariform, and among sperm-producing adults is entirely testiform. Collectively, these four genera sharing a similar pattern of gonad morphology are referred to here as the Coryphopterus group. The form of gonad development characterized by the expression of a transient, nonfunctional ovotestis between ovariform and testiform phases as described above is termed here an *interim ovotestis* developmental pattern. This pattern involves a relatively straightforward process of tissue replacement and restructuring with no functional overlap in gamete production,

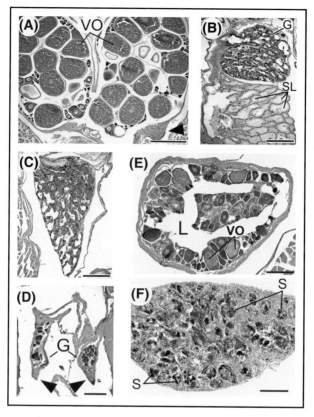

FIGURE 5.2. Histological features shown in transverse section of interim ovotestis gonad morphologies associated with differing reproductive states in *Lophogobius cyprinoides* and *L. crystulatus*. In all images, dorsal is uppermost.

A. *L. cyprinoides* female-active gonad made up entirely of ovigerous tissue containing
   vitellogenic oocytes (VO), with pAGS (arrowhead) of right lobe visible in this plane
   of cut. Scale bar is 500 µ.

B. *L. cyprinoides* reproductive complex of secondary male consisting of a gonad (G)
   made up solely of spermatogenic tissue and an enlarged, actively-secreting AGS consisting of enlarged secretory lobules (SL) filled with secretory material. Scale bar is
   500 µ.

C. Detail of gonad shown in (B) illustrating absence of oocyte-bearing tissue. Scale bar
   is 500 µ.

D. *L. cristulatus* ovariform gonad (G) of immature, consisting solely of oocyte-bearing
   tissue, with pAGS (arrowheads). Scale bar is 200 µ.

E. *L. cristulatus* female-active gonad made up entirely of ovigerous tissue containing
   vitellogenic oocytes (VO) surrounding a central lumen (L). Scale bar is 200 µ.

F. *L. cristulatus* secondary testis lobe made up solely of spermatogenic tissue, including
   seminiferous lobule lumina filled with sperm (S). Scale bar is 100 µ.

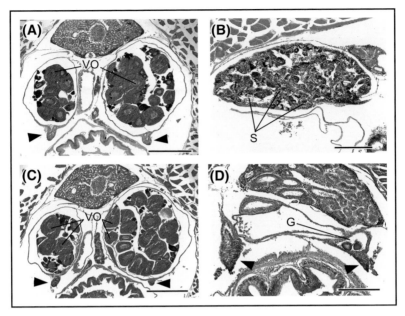

FIGURE 5.3. Histological features shown in transverse section of interim ovotestis gonad morphologies associated with differing reproductive states in *Fusigobius neophytus*, *F. signipinnis*, and *F.* sp.5 (DFH).

A. *F. neophytus* female-active gonad made up entirely of ovigerous tissue containing vitellogenic oocytes (VO), with pAGS (arrowheads). Dorsal is uppermost; scale bar is 500 μ.

B. *F. neophytus* secondary testis lobe made up solely of spermatogenic tissue including seminiferous lobule lumina filled with sperm (S). Dorsal is to the right; scale bar is 200 μ.

C. *F. signipinnis* female-active gonad made up entirely of ovigerous tissue containing vitellogenic oocytes (VO), with pAGS (arrowhead). Dorsal is uppermost. Scale bar is 200 μ.

D. *F.* sp.5 (DFH) (probably *F. melacron*) ovariform gonad (G) of immature, consisting solely of oocyte-bearing tissue, and with associated pAGS (arrowhead). Dorsal is uppermost. Scale bar is 200 μ.

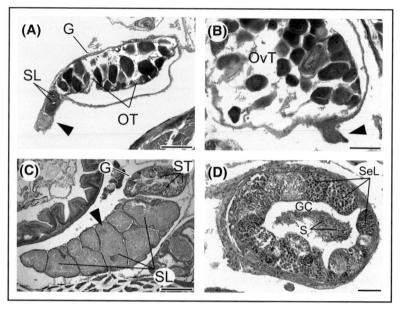

FIGURE 5.4. Histological features shown in transverse section of modified interim ovotestis gonad morphologies associated with differing reproductive states in *Elacatinus multifasciatus*.

A. Ovariform gonad (G) of immature, consisting solely of oocyte-bearing tissue (OT), and with associated pAGS (arrowhead) with cell-bound lumen of future secretory lobule (SL). Dorsal is to the right; scale bar is 500 μ.

B. Female-active gonad made up entirely of ovigerous tissue (OvT) containing vitellogenic oocytes, with associated pAGS (arrowhead). Dorsal is uppermost; scale bar is 500 μ.

C. Reproductive complex of secondary male consisting of a gonad (G) made up entirely of spermatogenic tissue (ST) and an enlarged, actively secreting AGS (arrowhead) consisting of greatly enlarged secretory lobules (SL) filled with secretory material. Dorsal is uppermost; scale bar is 200 μ.

D. Detail of spermatogenic gonad of fish shown in (C), illustrating a peripheral organization of seminiferous lobules (SeL), the absence of oocyte-bearing tissue, and the retention of a central gonadal lumen (GC) for the collection and egress of sperm (S). Dorsal is uppermost; scale bar is 50 μ.

thereby representing a minimally modified example of anatomical transformation among functional hermaphroditic goby taxa.

In the western Atlantic genus *Elacatinus* (recently removed from *Gobiosoma*)(Hoese & Reader 2001), a single species, *Gobiosoma multifasciatus* (Steindachner, 1876), now classified as *Elacatinus multifasciatus* (Steindachner 1876), is confirmed as a protogynous hermaphrodite (Robertson & Justines 1982). As with the Coryphopterus group, the only indication of future sperm-producing potential among immature and ova-producing individuals is the presence of pAGS (see Figure 5.4A, B)(Cole 2008). In the process of sex change, pAGS develop into AGS, and the ovigerous tissue is completely replaced by seminiferous lobules (Figure 5.4C); however, the former ovarian lumen persists (Figure 5.4C, D). Free spermatozoa pass from seminiferous lobules into the gonadal lumen, which now functions as a common collection region that is continuous with the common genital sinus. The Elacatinus pattern of gonad development is essentially the same as that of the Coryphopterus group, except for the retention of the gonadal lumen. Because of this difference, and to distinguish the *E. multifasciatus* pattern from that of the Coryphopterus group, *E. multifasciatus* is described as having a *modified interim ovotestis* developmental pattern, which constitutes a second pattern of hermaphroditic gonad morphology.

Eviota is a speciose gobiid genus found throughout the tropical Indo-West Pacific (Lachner & Karnella 1980; Karnella & Lachner 1981; Jewett & Lachner 1983; Gill & Jewett 2004). *Eviota epiphanes* is hermaphroditic (Cole 1990) and along with three other species, *E. afelei*, *E. disrupta*, and *E. fasciola*, exhibits a somewhat different pattern of gonad morphology than that described for the Coryphopterus group or for *Elacatinus multifasciatus*. The Eviota pattern is characterized by the presence of an ovotestis among all individuals, regardless of sexual function (see Figure 5.5). Immatures and female-active adults all have a predominantly ovariform gonad with oocytes and support cells organized into lamellae surrounding a central lumen (Figure 5.5A, B, C), pAGS or partially or fully developed AGS (Figure 5.5B, C), and clusters of small, haematoxylin-staining cells constituting spermatocysts scattered throughout the gonadal tissue (Figure 5.5A, B, C). A shift to male function involves a number of changes, including the proliferation of spermatocysts and surrounding somatic cells to become organized into seminiferous lobules, an overall reduction of oogenic tissue consisting of oocytes and support cells, the differentiation of pAGS into AGS, and the subsequent increase in AGS size accompanied by the initiation of secretory activity (Figure 5.5D). However, following a transition to male function, the Eviota gonad retains both a central lumen and numerous healthy, early-stage oocytes scattered among the interstices of the seminiferous lobules (Figure 5.5D).

Therefore, the Eviota pattern is one of a universally present ovotestis such that hermaphroditic capabilities are clearly evident within the gonad morphology of

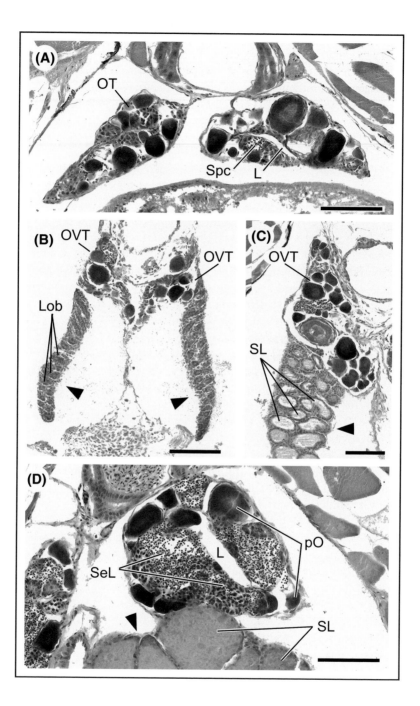

FIGURE 5.5. Histological features shown in transverse section of persistent integrated ovotestis morphologies associated with differing reproductive states in *Eviota epiphanes*. In all images, dorsal is uppermost and scale bar is 100 μ.

A. Integrated ovotestis of immature consisting of oocyte-bearing tissue (OT) and clusters of spermatocytes (SpC), surrounding a central lumen (L).

B. More posterior section of the reproductive complex shown in (A) showing the integrated ovotestis (OVT) and large, fully developed AGS (arrowhead) made up of numerous compressed lobules (Lob).

C. Reproductive complex of female-active adult (no vitellogenic oocytes present in this section) showing integrated ovotestis (OVT) and AGS (arrowhead) with secretory lobules (SL), expanded lobule lumina relative to (B), and containing small amounts of secretions.

D. Integrated ovotestis of male-active adult with sperm-filled seminiferous lobules (SeL) interspersed with pre-vitellogenic oocytes (pO) and retained central gonadal lumen (L). The dorsal-most portion of the AGS (arrowhead), with secretory lobules (SL) filled with secretions, is visible in the lower part of the image.

all individuals. This differs from the Coryphopterus group and *E. multifasciatus* patterns, in which the ovotestis is a transient feature and only the presence of pAGS among immatures and female-active adults indicates hermaphroditic capabilities. An ovotestis in which male and female components are intermixed, as is the case with *Eviota*, is frequently referred to as an undelimited ovotestis (Sadovy & Shapiro 1987). The intermingled distribution of male and female gametocytes observed in all phases of sexual function in *Eviota* produces a third reproductive morphology pattern, the *persistent integrated ovotestis* pattern, which is distinct from the Coryphopterus group and *E. multifasciatus* patterns characterized by a transient interim ovotestis pattern.

The genus *Lythrypnus* has a distribution that includes the eastern central Pacific, eastern Pacific, southeast Pacific, and the western Atlantic (Böhlke & Robins 1960; Greenfield 1988; Bussing 1990). The generalized pattern of gonad morphology in *Lythrypnus* is similar among examined species (Cole 1988; St. Mary 1998). Among immatures and female-active adults, the gonad is usually an ovotestis that is primarily ovarian in both architecture and sex cell type. In at least one species, *L. nesiotes*, rather than having pAGS, nonsecreting but differentiated AGS are associated with the immature ovotestis (see Figure 5.6A). The extent of AGS development among differing gonadal phases has not been reported for the remaining species (St. Mary 1993, 1998, 2000). Among several species, a small number of either pure females (i.e., completely ovarian gonad) or pure males (i.e., completely testicular gonad) may be present (St. Mary 1993, 2000). During the inactive ovariform and female-active phases, male components of the ovotestis consist of small clusters of spermatocytes confined to two regions, one dorsal-most and one ventral-most, in the gonad (Figure 5.6B). With reallocation of function from ova to sperm

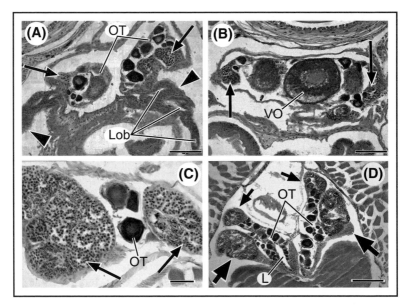

FIGURE 5.6. Histological features shown in transverse section of persistent regionalized ovotestis morphologies associated with differing reproductive states in *Lythrypnus nesiotes*.

A. Ovotestis of immature, consisting of oocyte-bearing tissue (OT) and regionalized spermatocyte-bearing tissue (arrows) with partially developed AGS (arrowhead) showing early lobule formation (Lob). Dorsal is uppermost; scale bar is 100 μ.

B. Female-active ovotestis with late-stage vitellogenic oocyte (VO) in the central ovigerous region, flanked dorsally and ventrally by spermatocyte-bearing tissue regions (arrows). Dorsal is to the right; scale bar is 100 μ.

C. Male-active ovotestis showing enlarged dorsal and ventral sperm-producing regions (arrows) and centrally located, regressed oocyte-bearing tissue (OT). Dorsal is to the right; scale bar is 50 μ.

D. Male-active ovotestis showing resulting lobe formation from expansion of the dorsal (small arrows) and ventral (large arrows) sperm-producing regions linked proximally by the regressed oocyte-bearing tissue region (OT) surrounding a central gonadal lumen (L). Dorsal is uppermost; scale bar is 200 μ.

production, the spermatocyte-occupied regions expand with newly formed semi-niferous lobules coming to constitute the majority of the gonad, and the oviger-ous portion becomes reduced to a relatively small and central region of the ovotestis (Figure 5.6C). In *L. nesiotes*, the independent growth pattern of the two antipodal regions of testis-function tissue results in a bifurcation of each of the gonadal lobes, with the central oogenic region anchoring dorsal and ventral pro-jections of spermatogenic tissue (Figure 5.6D).

The pattern of *Lythrypnus* gonad morphology is novel, having a localized con-centration of male gametocytes and spermatogenic tissues rather than an intermin-gled pattern like that of the persistent integrated ovotestis of *Eviota*. Among imma-tures and ova-producing individuals, the precursive sperm-producing tissues are situated in two separate regions. Moreover, in the male-functioning ovotestis, each of the two regions of precursive spermatogenic tissue develops independently to form two sublobes of spermatogenic tissues connected by a reduced region of tissue consisting of early-stage oocytes. The initial location and subsequent proliferation pattern of spermatocyte-bearing tissues consequently result in a *regionalized* (rather than integrated) *ovotestis* pattern. As a consequence, the gonad morphology of *Lythrypnus* exhibits a fourth pattern: that of a *regionalized persistent ovotestis* rather than an integrated persistent ovotestis pattern like that of *Eviota*.

A fifth pattern of hermaphroditic gonad morphology is shared by three other hermaphroditic genera: *Trimma*, *Priolepis*, and *Bryaninops*. *Trimma* is a speciose Indo-Pacific gobiine genus, while *Priolepis* and *Bryaninops* are less speciose. *Bryaninops* is widespread throughout the Indo-Pacific (Larson 1985), while *Pri-olepis* is found in the Indo-Pacific, the eastern Pacific, and the western Atlantic (Winterbottom & Burridge 1992; Nogawa & Endo 2007). Hermaphroditism is widespread among examined species of these three genera, and all share a num-ber of features of gonad morphology. As illustrated by *Bryaninops loki*, the go-nad develops at the outset as a partitioned structure (see Figure 5.7). Among im-matures and female-active adults, a larger region of oocyte-bearing tissue is organized into a lamellar configuration surrounding a central lumen, and the gonads of both of these functional states have undifferentiated pAGS (Figure 5.7A). Slightly more anteriorly, a smaller region extending along the ventral go-nad periphery from a medial to a ventro-lateral position appears as a thickened portion of the gonad wall. This region exhibits increased cell complexity and is separated from the ovigerous region by a thin boundary layer of connective tis-sue (Figures 5.7B, 5.8B), making this a delimited ovotestis (Sadovy & Shapiro 1987). In transitional individuals, the ova-producing region becomes reduced in size, the ventral region expands and becomes organized into seminiferous lob-ules, and secretory lobules develop within the pAGS in association with their transformation into AGS. In male-functioning adults, the spermatogenic tissue is expanded to occupy much of the gonadal body, and the lobule epithelium of

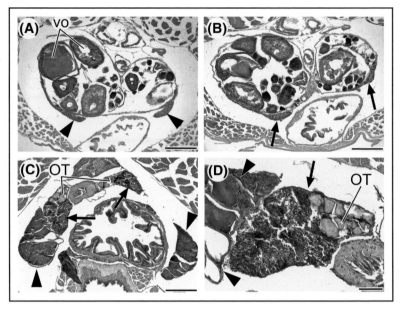

FIGURE 5.7. Histological features shown in transverse section of persistent partitioned ovotestis morphologies associated with differing reproductive states in *Bryaninops loki*.

A. Female-active ovotestis with late-stage vitellogenic oocytes (VO) and prominent, undifferentiated pAGS (arrowhead) posterior to spermatogenic tissue region. Dorsal is uppermost; scale bar is 200 μ.

B. Slightly more anterior section of female-active ovotestis shown in (A), showing future spermatogenic tissue region (arrows). Dorsal is uppermost; scale bar is 200 μ.

C. Reproductive complex with male-active ovotestis showing enlarged sperm-producing tissue regions (arrows for left and right lobes), regressed oocyte-bearing tissue region (OT), and AGS (arrowheads) filled with colloidal material. Dorsal is uppermost; scale bar is 200 μ.

D. Detail of dorsal portion of reproductive complex shown in (C), with oocyte-bearing tissue (OT) dorsal-most (to the right in the image), sperm-producing tissue (arrow) consisting of seminiferous lobules filled with sperm, and the dorsal-most portion of the enlarged AGS (arrowheads) to the left in the image. Dorsal is to the right; scale bar is 50 μ.

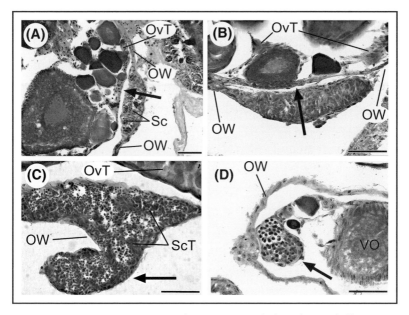

FIGURE 5.8. Persistent partitioned ovotestis morphology (A, B, C) illustrating the separation of oocyte-bearing and spermatocyte-bearing tissues within a partitioned ovotestis of three hermaphroditic goby genera as compared to regionalized persistent ovotestis morphology (D). Scale bar is 50 μ.

A. *Trimma taylori* female-active ovotestis with one region made up of ovigerous tissue (OvT), here containing vitellogenic oocytes, and a small, clearly separated region (see arrow at point of separation) located medially in the ovotestis made up of clusters of spermatocysts (Sc) containing spermatocytes, both enclosed by the ovotestis wall (OW). Dorsal is uppermost.

B. *Bryaninops loki* female-active ovotestis consisting ovigerous tissue (OvT) separated (see arrow at point of separation) from a spermatocyte-bearing region located medially in the ovotestis (arrow). OW is ovotestis wall; dorsal is to the right.

C. *Priolepis hipoliti* early developmental stage spermatocyte-bearing tissue (ScT) organized into a cohesive mass (arrow) within the female-active ovotestis, showing a broad separation from the ovigerous tissue (OvT). OW is ovotestis wall; dorsal is to the right.

D. *Lythrypnus nesiotes* female-active ovotestis with late-stage vitellogenic oocytes (VO) in the ovigerous region and showing a clear demarcation from the ventral spermatocyte-bearing tissue region (arrow). OW is ovotestis wall; dorsal is to the right.

the AGS is actively secreting (Figure 5.7C, D)(Cole 1990). In *Trimma* (Figure 5.8A), *Bryaninops* (Figure 5.8B), and *Priolepis* (Figure 5.8C), all examined individuals exhibit a partitioned ovotestis (Cole 1990, herein), a pattern of gonad morphology that is referred to here as a *persistent partitioned ovotestis* pattern. This pattern is most similar to that of the regionalized ovotestis pattern of *Lythrypnus* spp. (Figure 5.8D).

In one *Priolepis* species, *P. hipoliti*, a histological survey of gonad morphology by Cole (1990) revealed an unusual size-based sequence of ovotestis morphology. Small-sized adults exhibited a predominance of ovigerous tissue and a partially developed AGS showing early secretory lobule development. Large-sized adults had a reduced ovigerous region, a greatly expanded and active spermatogenic region, and fully developed and secreting AGS. Among intermediate-sized adults, both portions of the gonad contained healthy, mature gametes and the AGS were fully developed. The appearance of simultaneous mature gamete production among mid-sized adults indicates that this species appears likely to have an intervening functionally bisexual phase between female- and male-active phases. Pending experimental verification, *P. hipoloti* is tentatively identified as an *interim* (functional) *bisexual hermaphrodite*. A similar bisexual gonad containing mature gametes of both sexes has also been found in its congener, *P. eugenius*, which shows a pattern similar to that of *P. hipoliti* (see Figure 5.9). Among the smallest immatures, the gonad is already partitioned into a dorsal oocyte-bearing region and a ventral spermatocyte-bearing region (Figure 5.9A) and has large and fully differentiated AGS associated with the gonoduct (Figure 5.9B). Among some female-active adults, the ova-producing region makes up the majority of the ovotestis, while the inactive spermatocyte-bearing region consisting of fully differentiated, small-sized seminiferous lobules is relatively contracted (Figure 5.9C). Among other adults, the oocyte and spermatocyte-bearing regions are fully developed, and both regions include mature gametes (Figure 5.9D, E). Among female-active and putatively bisexually active adults, the AGS remain relatively small and shows no signs of secretory activity (Figure 5.9C, D). Among a third group of adults, the male-active ovotestis consists of an enlarged sperm-producing region, a relatively regressed oocyte-bearing tissue region, and extremely enlarged and actively secreting AGS (Figure 5.9F). The persistent partitioned ovotestis pattern exhibited by *P. eugenius* coupled with morphological evidence of an interim functional simultaneous hermaphroditic phase is unique among hermaphroditic gobies, excepting its congener, *P. hipoliti*. The possibility of such a combination of sexual functions has not been investigated among other *Priolepis* species and may be common or even synapomorphic for the genus.

The greatest complexity of reproductive anatomy among hermaphroditic gobiid taxa reported to date is in the Indo-Pacific genus *Gobiodon*. Among immatures and female-active adult *G. oculolineatus*, the gonad is ovariform and has no visible spermatogenic tissue (Cole 2009), a feature that has also been found in *G. okinawae* (Cole & Hoese 2001). Histological features of the tripartite secondary ovotestis are illustrated in Figure 5.10. Among male-active adults, the gonad proper is an ovotestis (Figure 5.10A, B, C) that exhibits three contiguous, intergraded regions: a dorsal gametogenic region, a ventro-lateral lobular region, and a medial stromal region (Figure 5.10B). The gametogenic region is made up of seminiferous lobules and healthy perinucleolar oocytes situated in lobule interstices, which resembles the persistent integrated ovotestis condition of *Eviota* (Figure 5.10C). However, the ventro-lateral region of the gonad consists of enlarged lobules filled with eosinophilic colloidal secretions and is separated from the gametogenic region by medially located stromal tissue that gradually merges with each (Figure 5.10B, D). The tripartite ovotestis configuration formed by regionalization of gametogenic tissue, stromal tissue, and secretion-storing lobules observed within *G. oculolineatus* and *G. okinawae* represents a novel gonad organization that has also been found in several other *Gobiodon* species (Cole unpublished data). It also represents a new pattern in which the adult ovotestis develops secondarily from an ovariform gonad, making it a secondary ovotestis. Because of these features, *Gobiodon* exhibits a sixth gonad morphology pattern, the *tripartite secondary ovotestis* pattern.

In addition to the lobular region of the ovotestis in which colloid secretions accumulate, the *Gobiodon* reproductive complex also includes a pair of lobular structures that are directly associated with the gonoduct. These accessory gonoduct-associated structures, or AGdS, are the production sites for colloidal secretions and are presumably the source for similar appearing material that occupies the secretion-accumulating lobule region of the ovotestis. In all previously described hermaphroditic goby taxa, the accessory structures responsible for secretion production arise from precursive tissues associated with the gonadal wall and develop with the onset of the male-active phase. In contrast, the AGdS of *Gobiodon* are clearly associated with the gonoduct and are fully differentiated in all individuals regardless of their functional phase. The presence of AGdS in *Gobiodon* represents a substantial departure from the secretory AGS found in all other hermaphroditic gobies.

Based on ontogenetic origin, morphology, and connectivity, the AGdS found in all examined *Gobiodon* species appear homologous with the sperm duct glands *sensu* Miller (1984) that are associated with the male reproductive complex of most gonochore goby species. AGdS have been found in *G. okinawae* (Cole & Hoese 2001), *G. oculolineatus* (Cole 2009), and four other *Gobiodon* species (Cole unpublished data), and they represent a novel feature among hermaphroditic gobies. The universal presence of AGdS among all *G. okinawae* and *G. oculolineatus* individuals, regardless of sexual state, constitutes an additional novel feature of the reproductive complex not reported for either gonochoric or other hermaphroditic

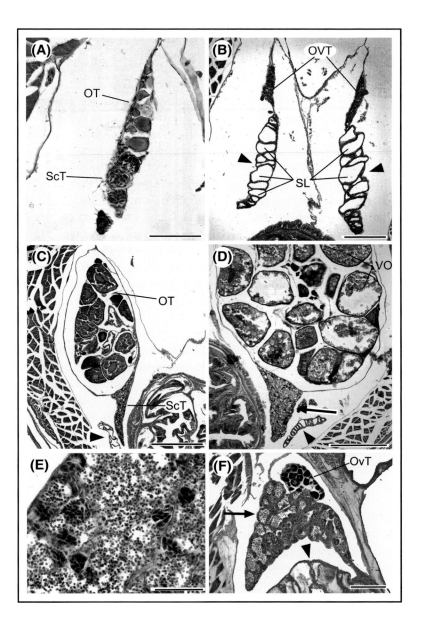

FIGURE 5.9. Histological features shown in transverse section of persistent partitioned ovotestis morphologies associated with differing reproductive states in *Priolepis eugenia*, including functional bisexual phase. In all images, dorsal is uppermost.

A. Ovotestis of immature consisting of oocyte-bearing tissue (OT) dorsally and spermatocyte-bearing tissue (ScT) ventrally. Scale bar is 200 μ.

B. More posterior section of reproductive complex shown in (A), showing ovotestis (OvT) and large, fully developed AGS (arrowhead) made up of broad-lumina, secretory lobules (SL). Scale bar is 100 μ.

C. Reproductive complex of female-active adult showing enlargement of the active oocyte-bearing tissue region (OT) dorsally, inactive spermatocyte-bearing tissue (ScT) ventrally, and fully differentiated but small-sized AGS (arrowhead). Scale bar is 500 μ.

D. Reproductive complex with simultaneous ova and sperm-producing ovotestis, with late-stage vitellogenic oocytes (VO), seminiferous lobules filled with sperm (arrow), and fully differentiated but small-sized AGS (arrowhead). Scale bar is 500 μ.

E. Detail of the sperm-producing region of the ovotestis shown in (D), with sperm-filled seminiferous lobules. Scale bar is 50 μ.

F. Male-active ovotestis showing enlarged sperm-producing region (arrow) and regressed ova-producing region (OvT). The dorsal-most portion of the enlarged, fully developed and secreting AGS (arrowhead) is visible at the bottom of the image. Scale bar is 200 μ.

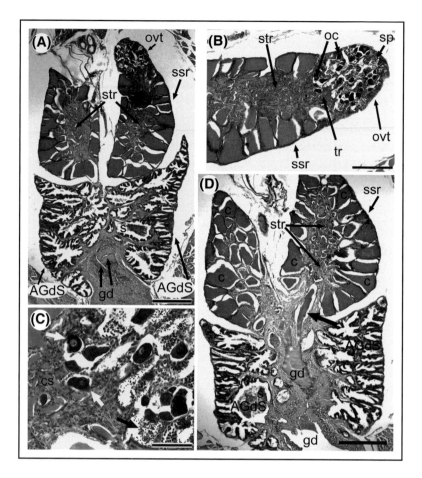

FIGURE 5.10. Histological features shown in transverse section of tripartite secondary ovotestis of *Gobiodon oculolineatus*.

A. Section through posterior region of the reproductive complex showing compressed lumen of gonoduct (gd), AGdS with secretions (s) within secretory lobules, secretion storing lobule region (ssr) of the ovotestis with lobules filled with colloidal material, stromal region (str) of the ovotestis, and posterior-most gametogenic region of the right ovotestis lobe (ovt). Dorsal is uppermost; scale bar is 500 μ.

B. Enlarged view of ovotestis lobe shown in image (A), showing peripherally arranged storage lobules, central stromal core (str) with smaller colloid-filled lumina and loosely packed stromal cells, transitional region (tr) consisting of stromal tissue and early-stage oocytes, and spermiated ovotestis (ovt)-containing oocytes (oc), developing male germ cells and spermatozoa (sp). Dorsal is to the right; scale bar is 250 μ.

C. Enlarged view of the transitional and gametogenic regions of ovotestis lobe detailed in image (B), showing colloid-filled lumina (cs), perinucleolar stage oocytes each surrounded by a ring of follicular cells (white arrows), and cell-delimited lumina containing spermatids (black arrow). Dorsal is to the right; scale bar is 100 μ.

D. More caudal transverse section of the reproductive complex shown in image (A), showing storage lobules (ssr) filled with colloidal material (c) and central stromal region (str) making up entirety of ovotestis (i.e., no gametogenic tissue present); AGdS with secretions (s) in some secretory lobule lumina; portions of the central channel of the gonoduct (gd) moving in and out of the plane of cut; and region of colloid-filled central channel of gonoduct that is continuous with the secretion-storing region (black arrow). Dorsal is uppermost; scale bar is 500. (From Cole 2009; reproduced with permission from Springer Science and Business Media.)

goby taxa. Lastly, the development of a specialized storage region within the gonad proper among all male-active *Gobiodon* (Cole & Hoese 2001; Cole 2008, unpublished data) constitutes a third novel reproductive feature that sets *Gobiodon* apart from other hermaphroditic goby genera so far discussed.

The presence of AGdS and a specialized lobular secretion-storing region within the gonad proper has been reported for only one other genus, *Paragobiodon*, which is also the proposed sister group of *Gobiodon* (Harold et al. 2008). There are few details available regarding the morphology of the reproductive complex in this group. However, in two species, *P. rivulatus* and *P. echinocephalus*, male-active fish also exhibit an enlarged pair of AGdS associated with the gonoduct where secretions are produced, and an enlarged lobular region of the ventral portion of the gonad where secretions are stored (Cole 1990, unpublished data).

The developmental pattern of gonad morphology characterized by *Gobiodon* does not resemble any other pattern among hermaphroditic gobies (except *Paragobiodon*) described here, and both the storage lobule and stromal regions of the gonad, and the derivation of the AGdS from the gonoduct, are so far unique among known hermaphroditic goby taxa.

SUMMARY OF REPRODUCTIVE MORPHOLOGY PATTERNS

Hermaphroditic gobies exhibit a range of reproductive morphologies that are organized into distinctive patterns which show a high degree of taxon-specificity (see Table 5.1 and Figures 5.11 and 5.12). These include (i) the interim ovotestis pattern of the Coryphopterus group (Figure 5. 11A, B, C); (ii) the modified interim ovotestis pattern of *Elacatinus multifasciatus* (Figure 5.11A, B, D); (iii) the persistent integrated ovotestis pattern of *Eviota* (Figure 5.11E, F, G); (iv) the regionalized persistent ovotestis of *Lythrypnus*; (v) the persistent partitioned ovotestis of *Trimma* and *Bryaninops* (Figure 5.11H, I, J) and the specialized pattern of *Priolepis hipoliti* and *P. eugenia*, with an interim functionally bisexual ovotestis (Figure 5.11K, L, M); and (vi) the tripartite secondary ovotestis of *Gobiodon* (Figure 5.12). In four of the first five reproductive morphology patterns listed above (i.e., excepting *Lythrypnus*), the AGS arise from tissues associated with the gonadal wall and only become fully differentiated and active at the onset of male function. In the sixth pattern, shown by *Gobiodon*, secretory structures (AGdS) are associated with the gonoduct rather than the gonadal wall and may be homologous with sperm duct glands *sensu* Miller (1984). However, the AGdS of *Gobiodon* differ from the male-specific sperm duct glands of gonochore goby taxa in that they occur as fully differentiated structures of the reproductive complex irrespective of reproductive state. How these different patterns of reproductive morphology and development may reflect evolutionary patterns associated

TABLE 5.1 Taxon-specific features of ovotestis and male-active gonad morphology
in hermaphroditic gobies

Taxon	Ovotestis duration	Ovotestis gametogenic tissue distribution	Male-active gonad features
Coryphopterus *Lophogobius* *Rhinogobiops* *Fusigobius*	Interim	Integrated	Tissue completely spermatogenic; no retained ovariform features
Elacatinus multifasciatus	Interim	Integrated	Tissue completely spermatogenic; retained gonadal lumen
Eviota	Persistent	Integrated	Male-dominant ovotestis
Lythrypnus	Persistent	Regionalized	Spermatogenic tissue dorsally and ventrally, forming dorsal and ventral lobes
Trimma *Priolepis* *Bryaninops*	Persistent	Partitioned	Male-dominant ovotestis
Gobiodon	Secondary	Integrated	Tripartite gonad with gametogenic, stromal, and storage lobule regions

with the development of functional hermaphroditism within the Gobiidae is the topic of the next section.

DISTRIBUTION PATTERNS OF HERMAPHRODITISM AND PHYLOGENETIC RELATIONSHIPS WITHIN THE GOBIIDAE

The frequent appearance of hermaphroditic sexual patterns, coupled with taxon-linked diversity in reproductive morphology among gobiid fishes, suggests that hermaphroditism may have multiple independent origins within the family. To address this possibility requires an examination of the distribution of hermaphroditic sexual patterns among gobiid taxa within a phylogenetic context. Morphology and molecular-based analyses, either separately or in combination, have consistently identified a number of monophyletic gobiid clades, including the gobiines, sycidiines, oxudercines, amblyopines, and gobionellines (Hoese 1984; Hoese & Gill 1993; Pezold 1993). More recently, the Gobiidae has been expanded to include its former sister taxa, the microdesmids (Wang et al. 2001), the ptereleotrids and schindleriids (Thacker 2003), and the kraemeriids (Thacker 2009), as well as being contracted to exclude the sycidiines, oxydurcines, amblyopines, and gobionellines (Thacker 2009)(refer to Figure 5.13). Clearly, the determination of

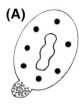 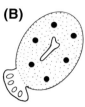 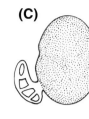 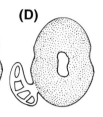

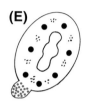 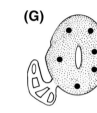

  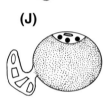

 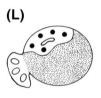

FIGURE 5.11. Progression of gonadal morphologies from female-active (left), to transitional (center), to male-active (right) for four taxon-specific patterns. White background represents oocyte-bearing tissue; large black dots are oocytes; medium-size gray dots represent pAGS tissue; light stipple is early-stage, non-active spermatogenic tissue; dark stipple is male-active spermatogenic tissue.

A. *Coryphopterus* ovariform female-active gonad with oocytes surrounding a central lumen and small channels visible in pAGS.

B. *Coryphopterus* interim ovotestis with regressing lumen, developing spermatocyte-bearing tissue, and developing AGS.

C. *Coryphopterus* secondary testis made up entirely of spermatogenic tissue, no retained gonadal lumen, and fully developed AGS.

D. *Elacatinus multifasciatus* secondary testis made up entirely of spermatogenic tissue, with retained gonadal lumen and fully developed AGS.

E. *Eviota* persistent integrated ovotestis of female-active individual, with oocytes surrounding a central lumen, small clusters of spermatocytes distributed throughout the ovotestis, and pAGS.

F. *Eviota* ovotestis of transforming fish, with developing but not yet mature spermatocyte-bearing tissue, a partially regressed lumen, oocytes (pre-vitellogenic), and developing AGS.

G. *Eviota* ovotestis of a male-active adult, consisting primarily of spermatogenic tissue along, with a small number of non-vitellogenic oocytes, reduced persistent gonadal lumen, and fully developed AGS.

H. *Trimma* persistent partitioned ovotestis of female-active individual, with a dorsal region consisting of oocytes surrounding a central lumen, a ventral region of partially developed but not yet mature spermatocyte-bearing tissue, and pAGS.

I. *Trimma* ovotestis of transforming fish, with a reduced oocyte-bearing region, an expanding region of spermatocyte-bearing tissue, and developing AGS.

J. *Trimma* persistent partitioned ovotestis of a male-active adult, consisting primarily of the expanded region of spermatogenic tissue, a much-reduced region of oocyte-bearing tissue, and fully developed AGS.

K. *Priolepis* persistent partitioned ovotestis of female-active individual identical to that of *Trimma*, with a dorsal oogenic region, a ventral spermatocyte-bearing tissue region, and pAGS.

L. *Priolepis* bisexual ovotestis, with a dorsal oogenic region, a ventral spermatogenic region, and small, fully developed AGS.

M. *Priolepis* persistent partitioned ovotestis of a male-active adult, consisting primarily of the expanded region of spermatogenic tissue, a much-reduced region of oocyte-bearing tissue, and fully developed AGS.

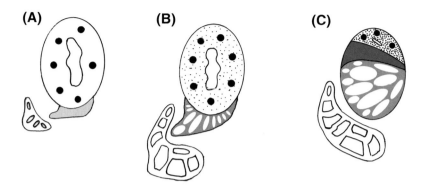

FIGURE 5.12. Progression of gonadal morphologies for *Gobiodon*, from female-active (left), to transitional (center), to male-active (right). White background represents oocyte-bearing tissue; large black dots are oocytes; light stipple is early stage, non-active spermatocyte-bearing tissue; dark stipple is male-active spermatogenic tissue; light gray background represents region of future storage lobules; medium gray is tissue surrounding developing storage lobules (white ovals); dark gray is stromal tissue region.

A. *Gobiodon* ovariform female-active gonad with oocytes surrounding a central lumen; a ventrally located, densely cellular region where storage lobules will later develop; and small-sized, fully developed AGdS associated with the gonoduct.

B. *Gobiodon* integrated ovotestis of transforming fish, with developing but not yet mature spermatocyte-bearing tissue interspersed with pre-vitellogenic oocytes; a developing, ventrally located storage lobular region; and enlarged AGdS.

C. *Gobiodon* tripartite secondary ovotestis of male-active adult, consisting dorsally of an integrated ovotestis region, centrally of a stromal region, and ventrally of a fully developed storage lobular region; and AGdS fully developed.

relationships within the gobiid and gobioid fishes remains challenging, with proposed phylogenies no doubt continuing to undergo revision for the foreseeable future. To date, all reported hermaphroditic gobiid taxa are found within the Gobiidae subfamily, Gobiinae, *sensu* Pezold (1993).

In Thacker's (2003) phylogenetic analysis, all of the hermaphroditic goby genera that were sampled, including *Coryphopterus, Lophogobius, Fusigobius, Priolepis, Elacatinus* (formerly *Gobiosoma*), *Eviota, and Gobiodon*, were placed within a monophyletic clade. In a redefined Gobiidae, which includes the Gobiinae *sensu* Pezold (1993), a number of clades were recovered (Thacker 2009). In this latest phylogeny (Figure 5.13), the deepest split separates the combined Trimma-Priolepis and Coryphopterus-Lophogobius (plus *Fusigobius signipinnis*) groups from all of the remaining examined gobiid taxa. The latter are further subdivided by a split that places one known hermaphroditic genus, *Elacatinus* (formerly *Gobiosoma*), in a monophyletic taxon well separated from the remaining three hermaphroditic goby genera included in that study. These three hermaphroditic genera further resolved into two constituent clades including *Eviota* plus *Gobiodon,* and *Fusigobius neophytus*, respectively.

If hermaphroditism has a singular point of origin within the Gobiidae, the dispersed distribution of hermaphroditic taxa across the family as illustrated in Figure 5.13 indicates that sexual lability has been lost in many more descendent lineages that it has been retained. Alternatively, hermaphroditism has arisen independently at least twice and possibly up to five times within the family (see vertical black arrows in Figure 5.13). In the latter instance, hermaphroditism may have arisen independently in each of five groups: the Coryphopterus-Lophobogius-*F. signipinnis* group; the Trimma-Priolepis-Bryaninops group; the Eviota-Gobiodon group; the Gobiosoma group *sensu* Rüber et al. (2003), which includes *Elacatinus*; and the group containing *F. neophytus sensu* Thacker (2009).

The widespread occurrence of hermaphroditism and shared interim ovotestis pattern of reproductive morphology within the Coryphopterus group, including *Coryphopterus, Lophogobius, Rhinogobiops nicholsii*, and *Fusigobius signipinnis*, supports the hypothesis of a recent common hermaphroditic ancestor proposed for these genera (Thacker & Cole 2002).

Trimma, Priolepis, and *Bryaninops* exhibit a different reproductive morphology pattern, characterized by a persistent partitioned ovotestis. On the basis of a number of morphological autapomorphies, Winterbottom & Emery (1981) have proposed that *Trimma* and *Priolepis* along with *Paratrimma* and *Trimmatom* form a monophyletic clade. In Thacker (2009), *Priolepis, Trimmatom*, and *Trimma* form a monophyletic clade that is sister to a group including *Coryphopterus, Lophogobius, Kraemeria*, and *Fusigobius signipinnis*. Reproductive morphology as described here supports both a close relationship between *Trimma, Priolepis*, and *Bryaninops* (reproductive morphology for *Paratrimma, Trimmatom*, and *Kraemeria* have not

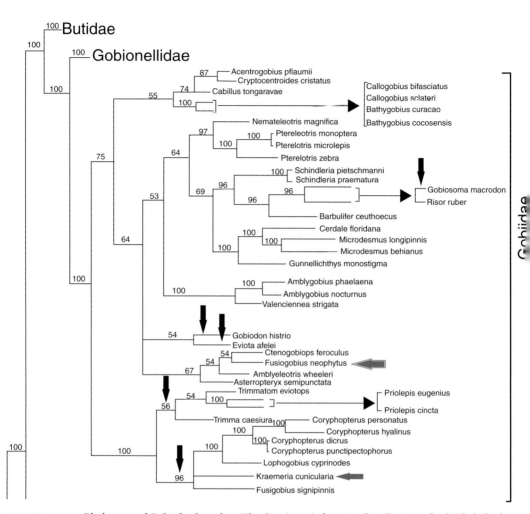

FIGURE 5.13. Phylogeny of Gobiidae based on Thacker (2009) showing distribution of gobiid clades having one or more hermaphroditic members. Black vertical arrows indicate basal region of clades proposed to have an independently evolved hermaphroditic sexual pattern, based on reproductive morphology. Horizontal, white-outlined gray arrow indicates questioned placement of *Fusigobius neophytus*. Horizontal gray arrow indicates questioned placement of the genus *Kraemeria*, for which available information on gonad morphology indicates an absence of hermaphroditic features (Langston unpublished data, Cole unpublished data) and for which hermaphroditism has not been reported.

yet been reported on), and some phylogenetic separation from *Coryphopterus*, *Lophogobius*, *Rhinogobiops*, and *Fusigobius signipinnis*. There are currently no published phylogenies for the placement of *Bryaninops* within the Gobiidae. Larson (1985, 1990) proposed a close relationship between *Bryaninops* and another soft-coral associated goby genus, *Pleurosicya*. Herler (2007) commented on unpublished data that indicates that *Bryaninops* and *Pleurosicya* are somewhat distant from *Gobiodon* and *Paragobiodon*. Based on shared reproductive morphological traits described here, there is strong support for a close relationship between *Bryaninops* and the *Trimma-Priolepis* clade.

Compared with the persistent partitioned and transient unpartitioned ovotestis patterns described above, *Lythrypnus* exhibits a somewhat specialized condition. During the female-active phase, spermatocytes are present and concentrated in two small, localized regions at the dorsal and ventral apices of the gonad (St. Mary 1998, Drilling & Grober 2005). During the male-active phase, the testicular tissue expands from these two locations to take over most of the gonad, relegating the oogenic tissue to a small peripheral region in which early-stage oocytes are retained. The localized ovotestis pattern of *Lythrypnus* has not yet been described for any other hermaphroditic gobies. *Lythrypnus* was not included in either of the two phylogenies of Thacker (2003, 2009), nor has any sister group been proposed, although Böhlke & Robins (1960) noted that superficially, small specimens of *Lythrypnus elasson* sometimes resemble *Priolepis* (*Quisquilius*) *hipoliti*. Based on its persistent regionalized ovotestis morphology, *Lythrypnus* may be most closely related to the Trimma-Priolepis-Bryaninops group, which exhibits a persistent partitioned ovotestis.

The persistent integrated pattern of reproductive morphology of *Eviota* is distinct from that of the previously discussed Coryphopterus and Trimma-Priolepis-Bryaninops groups and of *Lythrypnus*. A comparative analysis by Winterbottom & Emery (1981) showed that *Eviota* shares none of the morphological autapomorphies that unite *Priolepis* and *Trimma*, thereby disassociating *Eviota* from the latter. Several recent molecular analyses indicate that *Eviota* and *Gobiodon* (and by implication, *Paragobiodon*, based on Harold et al. 2008) are closely related but do not share a recent common ancestor with either of the Coryphopterus-Lophogobius or Trimma-Priolepis-Bryaninops groups (Thacker 2003, 2009). A proposed close relationship between *Eviota* and *Gobiodon* is not well supported by reproductive morphology. *Eviota* exhibits a persistent integrated ovotestis pattern in which the ovotestis is present regardless of sexual state. In *Gobiodon*, the gonad is initially ovariform, with an ovotestis only evident among male-active individuals. Moreover, the gonad of male-active adults is highly modified into three functional regions resulting in a tripartite secondary ovotestis pattern. Lastly, secretory functions of the *Gobiodon* reproductive complex are provided by AGdS rather than secretory AGS, which are so far unique to *Gobiodon* and *Paragobiodon* among hermaphroditic gobies. If *Eviota* and *Gobiodon* are in fact sister taxa and hermaphroditism is a shared ancestral condition, the reproductive morphology has subsequently become highly modified in the *Gobiodon/Paragobiodon* lineage. Alternatively, and more likely, the novel features of reproductive morphology found in the latter reflect developmental processes different from that of all other hermaphroditic goby taxa, including *Eviota*, indicating that hermaphroditism in *Gobiodon* spp. (and putatively *Paragobiodon*) evolved independently from the other clades discussed here.

The Gobiosoma group underwent a period of rapid speciation and adaptive radiation early in its evolutionary history (Rüber et al. 2003). In a departure from other hermaphroditic goby clades in which hermaphroditism is widespread, functional hermaphroditism within the speciose Gobiosoma group has been reported for only a single species, *Elacatinus multifasciatus* (Robertson & Justines 1982). However, a number of species within the closely related genera, *Elacatinus*, *Gobiosoma* and *Elacatinus sensu* Rüber et al. (2003) show evidence of a possible ancestral hermaphroditic condition. In several species, immature females exhibit the transient development of pAGS that disappear by first maturity (Cole 2008). If hermaphroditism is an ancestral state within the Gobiosoma group, sexual lability may have played an important role in its adaptive radiation. Rüber et al. (2003) have suggested that behavioral diversification and microhabitat specialization have been instrumental in the rapid evolution and high rate of speciation within some of the more speciose Gobiosoma group clades. As suggested in Cole (2008), the successful exploitation of patchily distributed microniches in this group may have been successful, in part, due to labile sexual patterns that could counter potential constraints on reproductive opportunities. If, as the transient expression of pAGS suggests, hermaphroditism was a basal trait of the Gobiosoma group, it been secondarily lost among most of the descendent lineages.

From a phylogenetic perspective, *Fusigobius* presents a challenge. Molecular analyses by Thacker (2003, 2009) indicate that *Fusigobius*, for which the species *F. neophytus* and *F. signipinnis* were sequenced, is polyphyletic. In both studies, *F. neophytus* falls into a clade with *Asterropteryx*, *Amblyeleotris*, and *Ctenogobiops* (see white-outlined, horizontal gray arrow in Figure 5.13). The second species, *F. signipinnis*, appears to be distant from *F. neophytus*, although its position changes between the two studies. In one study (Thacker 2003), *F. signipinnis* is placed in a clade along with *Ptereleotris*, *Gunnellichthys*, and *Schindleria*, while in the second (Thacker 2009) it comes out as part of the Coryphopterus-Lophogobius group. Gonad morphology in *F. neophytus* supports this move, as it shares all of the distinctive features of the female-active and male-active gonad of *Coryphopterus*, *Rhinogobiops nicholsii*, and *Lophogobius*. These features include pAGS associated with the gonad of immatures and ova-producing adults; the absence of testicular tissues or male sex cells associated with the ovariform gonad; and the reciprocal absence of ovarian features or female sex cells associated with the gonad of sperm-producing adults (Cole 1990). It is of interest to note that *F. signipinnis*, currently placed distant to *F. neophytus*, shows the same gonad morphology pattern as *F. neophytus*. This argues for a much closer association, and probable monophyly, for these two species and a shared ancestry with the Coryphopterus-Lophogobius-*Rhinogobiops nicholsii* group.

IMPLICATIONS OF FUNCTIONAL HERMAPHRODITISM IN GOBIID EVOLUTION

Within the gobiids, hermaphroditism appears to have arisen independently on several occasions. In one clade consisting of *Coryphopterus, Lophogobius, Rhinogobiops*, and *Fusigobius signipinnis*, all exhibit a unidirectional pattern of hermaphroditism involving a single shift from female to male. None of the above taxa are particularly speciose. *Fusigobius* is an Indo-Pacific genus with 11 described species (Randall 2001). *Coryphopterus* is a western Atlantic and eastern Pacific genus having 16 species including two newly described (Victor 2008). *Lophogobius* has three described species (Hoese 1995), although the validity of the current placement of *L. bleekeri* within this genus is in doubt (Hoese pers. comm.). All are relative habitat generalists and are primarily associated with coral reef habitats, except for *Rhinogobiops nicholsii*, which occupies rock and sand habitats of the northeastern Pacific (Eschmeyer et al. 1983), and *Lophogobius cyprinoides*, which is generally found in grass beds, mangroves, tidal creeks, brackish coastal waters, and inland bays of the western Atlantic (Böhlke & Chaplin 1993).

Compared with serial hermaphroditism, in which sexual function can shift back and forth to suit the existing social conditions, a unidirectional form of hermaphroditism appears at first glance to provide more limited adaptive advantages. However, unidirectional hermaphroditism is well suited for these species having broad habitat tolerances and an associated ubiquitous distribution of conspecifics. Under these conditions, the probability of finding a mate is likely high. Consequently, the evolution of unidirectional hermaphroditism in the Coryphopterus group most likely reflects an adaptation associated with the reproductive biology and social organization characteristics of this clade. For one species within the Coryphopterus group, *Rhinogobiops nicholsii*, high-quality nest sites for demersal eggs have been shown to be limited (Breitburg 1987; Kroon et al. 2000). Consequently, reproductive success corresponds to successful nest-site defense, for which large males have a competitive advantage. In having limited access to nest sites, small males are likely to have lower reproductive success than both large males and small adult females, leading to strong selection for size-based protogyny (Ghiselin 1969; Berglund 1990).

A second clade *sensu* Thacker (2009) includes the hermaphroditic genera *Trimma* and *Priolepis*, both of which are relatively speciose genera. *Priolepis* consists of 35 described species (Nogawa & Endo 2007) and an unknown number of undescribed species (Winterbottom pers. comm.), while *Trimma* consists of 58 described species and an estimated 30 to 40 as-of-yet undescribed species (Winterbottom & Southcott 2008). *Trimma* species either occur in hovering aggregations in caves and around overhangs or live more solitarily in reef crevices, while *Priolepis* species form small social groups or live solitarily in small reef holes and crevices (Winterbottom pers. comm.). Species of both genera exhibit serial hermaphroditism, and among some *Priolepis* species, a combination of serial hermaphroditism and simultaneous bisexuality may exist (Cole 1990, herein). The degree to which species of these two genera exhibit habitat segregation among congeners is unclear due to their small size, cryptic lifestyle, and the limited number of studies focusing on this topic. However, serial hermaphroditism is clearly adaptive in instances where mating opportunities may be diminished due to limited mobility relative to encounter rates with conspecifics. This circumstance seems more likely to apply to crevice-dwelling and more solitary species of *Trimma* and *Priolepis* than to the aggregating species of *Trimma*, some of which are also serially hermaphroditic (Sunobe & Nakazono 1993). However, serial hermaphroditism may be adaptive for some *Trimma* species for other reasons. *Trimma* is generally at the small end of the goby size spectrum and therefore faces considerable fecundity constraints on the part of females. In addition, many members of this genus may be very short-lived, as has been reported for *Trimma naso* (Winterbottom & Southcott 2008). These factors introduce considerable unpredictability into potential lifetime reproductive success for both sexes, regardless of the degree of mate access. Consequently, serial hermaphroditism may have developed in response to unpredictable conditions or, alternatively, the development of serial hermaphroditism was an ancestral condition and supported the subsequent evolution of small size, abbreviated life history patterns, or both.

Among the more cryptic and solitary-living *Priolepis* species, life span information is not available, but their larger size suggests a longer life span than that of *Trimma*. In terms of collection numbers, *Priolepis* is not encountered as frequently in rotenone collections as *Trimma* (Winterbottom pers. comm.; Cole pers. obs.) or other common benthic gobies of the western Atlantic (Cole unpublished data). Serial hermaphroditism in *Priolepis* may have evolved as a successful counterbalance to locally low population densities and unpredictability in lifetime reproductive success due to a strong potential for low encounter rates among adults. One recently discovered *Priolepis* species was collected from relatively deep (400m) waters of the Red Sea (Goren & Baranes 1995). If there are further discoveries of deepwater *Priolepis* species and these species prove also to be serially hermaphroditic, this form of flexible sexual pattern may have been instrumental in a bathydemersal radiation within this genus.

Bryaninops is an Indo-Pacific genus of nine described species, all of which are very small sized and form obligate associations with gorgonian and antipatharian corals (Okiyama & Tsukamoto 1989). Most species are microhabitat specialists, often with highly species-specific coral associations (Herler 2007). Due to their small size, associated limited mobility, and highly specific habitat needs, species of *Bryaninops* likely experience the greatest within-population spatial isolation found among hermaphroditic gobies, in which mating opportunities are limited to conspecifics occupying the same sea whip. In addition, their extremely small size imposes significant limitations on female fecundity and they may experience an abbreviated lifespan, which seems to be a common condition among small-sized fishes (Depczynski & Bellwood 2005). *Bryaninops* also expresses serial hermaphroditism and has a reproductive morphology that places it close to *Trimma* and *Priolepis*. To date, this genus has not been included in any phylogenetic analyses of gobiid fishes, so the nature of its relationship to *Trimma* and *Priolepis* remains to be tested. The expression of serial hermaphroditism within *Bryaninops* seems almost obligatory to sustain such a constrained lifestyle. The question as to whether serial hermaphroditism arose prior to, concomitant with, or subsequent to the development of this extreme lifestyle remains unanswered in the absence of any hypotheses of phylogenetic relationships for this genus relative to other gobiid taxa.

Eviota is a tropical Indo-Pacific genus consisting of 50 described species (Gill pers. comm.) and probably at least a similar number of undescribed species (Jewett & Lachner 1983). *Eviota* constitutes a genus of small-sized fishes, many of which have extensive distributions and are abundant within their range (Lachner & Karnella 1980; Gill & Jewett 2004). A number of species occur intertidally (Cole 1990; Gill & Jewett 2004) or subtidally in coral or rock reef crevices, and often in pairs (Sunobe 1988). In a recent Red Sea study, Herler (2007) found that *Eviota* species exhibited a much higher degree of niche segregation among congeners relative to heterogeneric species, indicating strong niche partitioning among coexisting *Eviota* species. The specific nature of the hermaphroditic pattern in *Eviota* remains unclear, but the persistent integrated ovotestis pattern of reproductive morphology suggests a capacity for serial hermaphroditism like that of *Trimma, Priolepis*, and *Bryaninops*. As pointed out earlier, the evolution of serial hermaphroditism in microhabitat specialists may offer the advantage of increased mating opportunities among adult conspecifics that are otherwise constrained by small size, limited mobility, highly specific microhabitat requirements, and/or a possibly short life span.

Gobiodon and *Paragobiodon* are proposed sister taxa (Harold et al. 2008) which share a similar pattern of reproductive morphology. *Gobiodon* consists of more than 30 described species, of which 19 are currently recognized as valid (Harold et al. 2008), while *Paragobiodon* has 6 described species (Akihito et al. 2002). Both *Gobiodon* and *Paragobiodon* are obligate coral-dwelling taxa, often with highly species-specific coral associations (Munday et al. 1997, 1999; Herler 2007). In a recent Red Sea study, *Gobiodon* and *Paragobiodon* were found to constitute some of the most habitat-specialized reef fishes examined (Herler 2007). As proposed for *Eviota*, the evolution of serial hermaphroditism in *Gobiodon* and *Paragobiodon* appears highly adaptive under the constraints of small size, limited mobility, and highly specific microhabitat requirements. However, in the absence of a known sister group for the Gobiodon/Paragobiodon clade, it cannot be determined whether a labile sexual pattern preceded, occurred concomitantly with, or followed the development of such an extreme form of obligate coral-dwelling lifestyle.

The Gobiosoma group of American seven-spined gobies (Gobiidae, Gobiosomatini *sensu* Rüber et al. [2003]) is a western Atlantic clade. Within this relatively speciose taxon, *E. multifasciatus* is the sole known hermaphroditic representative and exhibits a unidirectional protogynous pattern. *Elacatinus multifasciatus* is fairly abundant in intertidal and shallow subtidal habitats, occupying coral rubble and sponges and frequently associating with rock-boring sea urchins (Böhlke & Chaplin 1993; Toller 2005). The sexual pattern of the Gobiosoma group is of interest due to the occurrence of functional hermaphroditism in only a single species within the genus, which is unique among hermaphroditic goby taxa. However, it has been proposed (Cole 2008) that hermaphroditism may be an ancestral condition within the Gobiosoma group that has subsequently been lost in most of the descendent species, excepting *E. multifasciatus*. This hypothesis developed from the observation that there is a transient expression of pAGS prior to first maturity during the immature phase among females of several species closely related to *E. multifasciatus*. Many species within the monophyletic Gobiosoma group are microniche specialists, which has been proposed as an important factor in the adaptive radiation and rapid speciation of this group early in its evolutionary history (Rüber et al. 2003). Functional hermaphroditism may have been instrumental in supporting the success of microniche specialization and subsequent adaptive radiation within this group. By correlation, functional hermaphroditism may have also played a similar role in other speciose genera such as *Trimma*, *Eviota*, and *Priolepis*.

The goby genus *Lythrypnus* contains approximately 20 small-sized species and has a distribution that includes the eastern central Pacific, eastern Pacific, southeast Pacific, and the western Atlantic (Böhlke & Robins 1960; Greenfield 1988; Bussing 1990). All examined species appear to have a serial hermaphroditic sexual pattern (St. Mary 2000). The regionalized ovotestis of *Lythrypnus* exhibits both similarities and differences with that of the partitioned ovotestis of *Trimma*, *Priolepis*, and *Bryaninops*. If *Lythrypnus* is closely related to the Trimma-Priolepis clade (the placement of *Bryaninops* being currently unresolved), differences in the configuration of tissues associated with reproductive function between these two groups may represent a divergence from the ancestral reproductive morphology pattern on the part of either or both. Alternatively, as hypothesized by St. Mary (1998), differences between these two groups could reflect independent origins for hermaphroditism.

In summary, within the Gobiidae, functional hermaphroditism has likely arisen several times within different clades, under differing sets of selective pressures, and with different forms of expression. The development of labile sexuality among numerous goby taxa has also likely played a significant interactive role in the coevolution, and possible convergence, of widely varying social and mating systems and life histories among goby species. In addition, hermaphroditic sexual patterns may have played a central role in the evolution of species-rich genera among gobiids, and therefore contributed to the remarkable diversity of this group. Currently, the phylogenetic resolution of constituent taxa of the Gobiidae suffers from many sampling gaps that remain to be filled. Given the size of this taxon, the likelihood is that many more species—and even genera—remain to be discovered. With the information provided here, gobiid fishes offer an exciting and promising model taxon for the study of evolution processes and adaptive radiation among speciose vertebrate taxa, and of the inter-related evolutionary patterns of speciation, life history, reproductive biology, and sexual expression among marine fishes.

SUMMARY

Reproductive morphology among hermaphroditic goby taxa can be characterized by a number of differing anatomical and ontogenetic patterns that show a high degree of taxon specificity. In some instances, the gonad is either completely ovariform or completely testiform, according to sexual function. An ovotestis, defined here by the co-occurrence of sex-specific reproductive tissues with no ascribed functional connotation, may be present only briefly as an interim form. This transient ovotestis presence results from the brief temporal overlap of sex-specific reproductive tissues during the transition phase between sexual functions (e.g., Coryphopterus group). Among other hermaphroditic taxa, the ovotestis is a more durable feature that persists throughout life (e.g., Eviota) or may form secondarily from an ovariform state (e.g., Gobiodon). In the persistent ovotestis pattern, gametogenesis is typically limited to only ova or sperm production at any one time although an interim, possibly functional, bisexual ovotestis has been documented for Priolepis hipoliti and P. eugenia. Within the ovotestis, the organization of sex-specific reproductive tissues also demonstrates several different patterns. The distribution of oocyte-bearing and spermatocyte-bearing tissues may be integrated (e.g., Eviota, Gobiodon), regionalized (e.g., Lythrypnus), or partitioned (e.g., Trimma, Priolepis, Bryaninops).

Another morphological feature associated with the reproductive complex of hermaphroditic gobies is the elaboration of accessory structures. In most instances, the function of differentiated accessory structures is secretory and appears to be analogous to sperm duct glands of gonochore goby taxa. These secretory accessory structures may arise from the sperm duct (as with AGdS of *Gobiodon*) but more typically develop from precursive or partially developed tissues associated with the gonadal wall (pAGS and AGS, respectively). In *Gobiodon*, an additional elaborated region, which is located directly within the gonad, develops into storage lobules for secretions produced by the AGdS.

The diversity of reproductive morphologies exhibited by hermaphroditic goby taxa is due to a number of novel morphological traits not found in gonochoric goby taxa. Some of these traits may be shared among many, or all, hermaphroditic goby taxa. For example, the functional equivalent of sperm duct glands, the secretory AGS, arise from precursive gonad-associated tissues. These precursive tissues and their subsequent differentiation into gonad-associated AGS represent a novel anatomical feature found among most hermaphroditic goby species. In this review, *Gobiodon* is the only taxon that does not have pAGS that develop into gonad-associated AGS. Instead, their secretory accessory structures (AGdS) arise directly from the gonoduct and therefore may be homologous with gonochore sperm duct glands. However, the AGdS of *Gobiodon* are not entirely consistent with sperm duct glands of gonochore species. In the former, the AGdS are present and fully developed in immatures and in both female-active and male-active adults, while sperm duct glands of gonochoric goby taxa are found only in the male reproductive complex. Consequently, the AGdS of *Gobiodon* may represent a novel reproductive character that is unique to a limited number of gobiine species.

A number of other Gobiodon pattern novelties include the development of a secondary ovotestis during the male-active phase and the partitioning of the secondary ovotestis into gametogenic, stromal, and storage lobule regions. This tripartite form of ovotestis, as well as the presence of storage lobules which dominate the posterior region of the gonad, represent novel features that are currently unique to *Gobiodon*.

A comparison of distribution patterns for different hermaphroditic reproductive morphologies and phylogenetic relationships among hermaphroditic goby taxa indicates a strong concordance. Hermaphroditic reproductive morphology accurately predicts the majority of currently hypothesized clade relationships within the Gobiidae. And where phylogenetic relationships are still unresolved, shared patterns of reproductive morphology (i.e., *Eviota* and *Gobiodon*; *Bryaninops* and *Trimma-Priolepis*) may be phylogenetically informative. The combination of shared reproductive morphologies found among closely related goby taxa and differing reproductive morphologies among distantly related taxa supports a hypothesis of multiple independent origins for hermaphroditism within the

Gobiidae. This finding also argues for a significant capacity for morphological and physiological innovation within the Gobiidae, traits that have likely been instrumental in the frequent and extensive adaptive radiation that is characteristic of this speciose family.

ACKNOWLEDGMENTS

Portions of this chapter benefitted greatly from enlightening conversations with A.C. Gill, D.W. Greenfield, A.S. Harold, D.F. Hoese, H.K. Larson, P.L. Munday, L.R. Parenti, F. Pezold, C.E. Thacker, J.L. Van Tassel, and R. Winterbottom. Specimens were generously provided by D.W. Greenfield, D.F. Hoese, R.C. Langston, H.K. Larson, D.R. Robertson, and R. Winterbottom. Additional assistance and support in the field were kindly provided by L. Orsak and J. Masey (Christensen Research Institute, Papua, New Guinea); A. Hoggett, J. Leis, S. Reader, and L.Vail (Lizard Island Research Station, Australia); and L. Bell, P. Colin (Coral Reef Research Foundation, Palau), and Y. Sadovy (University of Hong Kong). Microphotography imaging assistance was provided by S. Raredon and B. Vine, and artwork by S. Mondon. Cladogram imaging assistance was kindly and patiently provided by S. Raredon. Portions of the research reported on in this chapter were supported by grants from the Smithsonian Institution through the National Museum of Natural History's Caribbean Coral Reef Ecosystems Program (CCRE Contribution no. 830), a Curatorial Fellowship from the Australian Museum, and from funds provided by the University of Hawaii at Mānoa.

REFERENCES

Akihito SK, Ikeda Y, Sugiyama K (2002). Suborder Gobioidei. In: Nakabo, T (ed.), *Fishes of Japan with Pictorial Keys to the Species*, 2nd edition. Tokai University Press, Tokyo, pp. 12139–12168.

Arai R (1964). Sex characters of Japanese gobioid fishes (I). Bulletin of National Science Museum, Tokyo 7: 295–306.

Belanger AJ, Arbuckle WJ, Corkum LD, Gammom DB, Li W, Scott AP, Zielinski S (2004). Behavioural and electrophysiological responses by reproductive female *Neogobius melanostomus* to odours released by conspecific males. Journal of Fish Biology 65: 933–946.

Berglund A (1990). Sequential hermaphroditism and the size-advantage hypothesis: an experimental test. Animal Behaviour 39: 426–433.

Böhlke JE, Chaplin CCG (1993). *Fishes of the Bahamas and Adjacent Tropical Waters*. University of Texas Press, Austin.

Böhlke JE, Robins CR (1960). Western Atlantic gobioid fishes of the genus *Lythrypnus*, with notes on *Quisquilius hipoliti* and *Germannia pallens*. Proceedings of the National Academy of Sciences, Philadephia 112: 73–101.

Breitburg DL (1987). Interspecific competition and the abundance of nest sites: factors affecting sexual selection. Ecology 68: 1844–1855.

Bussing WA (1990). New species of gobiid fishes of the genera *Lythrypnus*, *Elacatinus* and *Chriolepis* from the eastern tropical Pacific. Revista de Biologia Tropical 38: 99–118.

Caputo V, Mesa ML, Candi G, Cerioni PN (2003). The reproductive biology of the crystal goby with a comparison to that of the transparent goby. Journal of Fish Biology 62: 375–385.

Cinquette R (1997). Histochemical, enzyme histochemical and ultrastructural investigation on the sperm duct glands of *Padogobius martensi* (Pisces, Gobiidae). Journal of Fish Biology 50: 978–991.

Cole KS (1983). Protogynous hermaphroditism in a temperate zone territorial marine goby, *Coryphopterus nicholsii*. Copeia 1983: 809–812.

Cole KS (1988). Predicting the potential for sex change on the basis of ovarian structure in gobiid fishes. Copeia 1988: 1082–1086.

Cole KS (1990). Patterns of gonad structure in hermaphroditic gobies (Teleostei: Gobiidae). Environmental Biology of Fishes 28: 125–142.

Cole KS (2008). Transient ontogenetic expression of hermaphroditic gonad morphology within the Gobiosoma group of the Neotropical seven-spined gobies (Teleostei: Gobiidae). Marine Biology 154: 943–951.

Cole KS (2009). Modifications of the reproductive complex and implications for the reproductive biology of *Gobiodon oculolineatus* (Teleostei: Gobiidae). Environmental Biology of Fishes 84: 261–273.

Cole KS, Hoese DF (2001). Gonad morphology, colony demography and evidence for hermaphroditism in *Gobiodon okinawae* (Teleostei: Gobiidae). Environmental Biology of Fishes 61: 161–173.

Cole KS, Robertson DR (1988). Protogyny in a Caribbean reef goby, *Coryphopterus personatus*: gonad ontogeny and social influences on sex change. Bulletin of Marine Sciences 42: 317–333.

Cole KS, Shapiro DY (1990). Gonad structure and hermaphroditism in the gobiid genus *Coryphopterus* (Teleostei: Gobiidae). Copeia 1990: 996–1003.

Cole KS, Shapiro DY (1992). Gonadal structure and population characteristics of the protogynous goby *Coryphopterus glaucofraenum*. Marine Biology 113: 1–9.

Colombo L, Burighel P (1974). Fine structure of the testicular gland of the black goby, *Gobius jozo* L. Cell and Tissue Research 154: 39–49.

Colombo L, Marconato A, Belvedere PC, Frisco C (1980). Endocrinology of teleost reproduction: a testicular steroid pheromone in the black goby, *Gobius jozo* L. Bollettino di Zoologia 47: 355–364.

Depczynski M, Bellwood DR (2005). Shortest recorded vertebrate lifespan found in a coral reef fish. Current Biology 15: R288.

Drilling CC, Grober MS (2005). An initial description of alternative male reproductive phenotypes in the bluebanded goby, *Lythrypnus dalli* (Teleostei: Gobiidae). Environmental Biology of Fishes 72: 361–372.

Egami N (1960). Comparative morphology of the sex characters in several species of Japanese gobies, with reference to the effects of sex steroids on the characters. Journal of the Faculty of Science, University of Tokyo, Section IV 9: 67–100.

Eggert B (1931). Die Geschlechtsorgane der Gobiiformes und Blenniiformes. Zeitschrift Wissenschaftliche Zoologie 139: 249–517.

Eschmeyer WN, Herald ES, Hammann H (1983). *A Field Guide to the Pacific Coast Fishes of North America*. Houghton Mifflin Company, Boston.

Fishelson L (1989). Bisexuality and pedogenesis in gobies (Gobiidae: Teleostei) and other fish, or: why so many little fish in tropical seas? Senckenbergiana Maritima 20: 147–160.

Fishelson L (1991). Comparative cytology and morphology of seminal vesicles in male gobiid fishes. Japanese Journal of Ichthyology 38: 17–30.

Ghiselin MT (1969). The evolution of hermaphroditism among animals. The Quarterly Review of Biology 44: 189–208.

Gill AC, Jewett SL (2004). *Eviota hoesei* and *E. readerae*, new species of fish from the Southwest Pacific, with comments on the identity of *E. corneliae* Fricke (Perciformes: Gobiidae). Records of the Australian Museum 56: 2235–2240.

Goren M, Baranes A (1995). *Priolepis goldshmidtae* (Gobiidae), a new species from the deep water of the northern Gulf of Aqaba, Red Sea. Cybium 19: 343–347.

Greenfield DW (1988). A review of the *Lythrypnus mowbrayi* complex (Pisces: Gobiidae), with the description of a new species. Copeia 1988: 460–470.

Grier H, Linton JR, Leatherland JF, de Vlaming VL (1980). Structural evidence for two different testicular types in teleost fishes. American Journal of Anatomy 159: 331–345.

Grier HJ (1981). Cellular organization of the testis and spermatogenesis in fishes. American Zoologist 21: 345–357.

Harold AS, Winterbottom R, Munday PL, Chapman RW (2008). Phylogenetic relationships of Indo-Pacific coral gobies of the genus *Gobiodon* (Teleostei: Gobiidae), based on morphological and molecular data. Bulletin of Marine Science 82: 119–136.

Herler J (2007). Microhabitats and ecomorphology of coral- and coral rock-associated gobiid fish (Teleostei: Gobiidae) in the northern Red Sea. Marine Ecology 28, Supplement 1: 82–94.

Hoar WS (1955). Reproduction in teleost fish. Memoirs of the Society for Endocrinology 4: 5–24.

Hoese DF (1984). Gobioidei: relationships. In: Moser HG, Richards WJ, Cohen DM, Fahay MP, Kendall AW, Richardson SL (eds.), *Ontogeny and Systematics of Fishes*. American Society of Ichthyologists and Herpetologists Special Publication 1, Gainesville, Florida, pp. 588–591.

Hoese DF (1995). Gobiidae. Gobios, chanquetes y guasetas. In: Fischer W, Krupp F, Schneider W, Sommer C, Carpenter KE, Niem V (eds.), *Guia FAO para Identificación de Especies para lo Fines de la Pesca*. Pacifico Centro-Oriental, 3 Vols. FAO, Rome, pp. 1129–1135.

Hoese DF, Gill AC (1993). Phylogenetic relationships of eleotridid fishes (Perciformes: Gobioidei). Bulletin of Marine Science 52: 415–440.

Hoese DF, Reader S (2001). A preliminary review of the Eastern Pacific species of *Elacatinus* (Perciformes: Gobiidae). Revista De Biologia Tropical 49, Supplement 1: 157–167.

Immler S, Mazzoldi C, Rassoto MB (2004). From sneaker to parental male: change of reproductive traits in the black goby, *Gobius niger* (Teleostei: Gobiidae). Journal of Experimental Zoology 301A: 177–185.

Jewett SL, Lachner EA (1983). Seven new species of the Indo-Pacific genus *Eviota* (Pisces: Gobiidae). Proceedings of the Biological Society of Washington 96: 780–806.

Karnella SJ, Lachner EA (1981). Three new species of the *Eviota epiphanies* group having vertical trunk bars (Pisces: Gobiidae). Proceedings of the Biological Society of Washington 94: 264–275.

Kroon FJ, de Graaf M, Liley NR (2000). Social organisation and competition for refuges and nest sites in *Coryphopterus nicholsii* (Gobiidae), a temperate protogynous reef fish. Environmental Biology of Fishes 57: 401–411.

Lachner EA, Karnella SJ (1980). Fishes of the Indo-Pacific genus *Eviota* with descriptions of eight new species (Teleostei: Gobiidae). Smithsonian Contributions to Zoology 315: 1–136.

Lahnsteiner F, Seiwald M, Patzner RA, Ferrero EA (1992). The seminal vesicles of the male grass goby, *Zosterisessor ophiocephalus* (Teleostei: Gobiidae): Fine structure and histochemistry. Zoomorphology 111: 239–248.

Larson HK (1985). A revision of the gobiid genus *Bryaninops* (Pisces), with a description of six new species. The Beagle: Occasional Papers of the Northern Territory Museum of Arts and Sciences 2: 57–93.

Larson HK (1990). A revision of the gobiid genera *Pleurosicya* and *Luposicya*, with descriptions of eight new species of *Pleurosicya*. The Beagle: Records of the Northern Territory Museum of Arts and Sciences 7: 1–53.

Locatello L Mazzoldi C Rasotto MB (2002) Ejaculate of sneaker males is pheromonally inconspicuous in the black goby, *Gobius niger* (Teleostei: Gobiidae). Journal of Experimental Zoology 293: 601–605.

Maack G, Segner H (2003). Morphological development of the gonads in zebrafish. Journal of Fish Biology 62: 895–906.

Marconato A, Rasotto MB, Mazzoldi C (1996). On the mechanism of sperm release in three gobiid fishes (Teleostei: Gobiidae). Environmental Biology of Fishes 46: 321–327.

Mazzoldi C Petersen CW Rasotto MB (2005). The influence of mating system on seminal vesicle variability among gobies (Teleostei: Gobiidae). Journal of Zoological Systematics and Evolutionary Research 43: 307–314.

Miller PJ (1984). The tokology of gobioid fishes. In: Potts GW, Wootton RJ (eds.), *Fish Reproduction: Strategies and Tactics*. Academic Press, London, pp 119–153.

Miller PJ (1992). The sperm duct gland: a visceral synapomorphy for gobioid fishes. Copeia 1992: 253–256.

Munday PL, Harold AS, Winterbottom R (1999). Guide to the coral-dwelling gobies, genus *Gobiodon* (Gobiidae) from Papua New Guinea and the Great Barrier Reef. Revue Française d'Aquariologie 26: 49–54.

Munday PL, Jones GP, Caley MJ (1997). Habitat specialisation and the distribution and abundance of coral-dwelling gobies. Marine Ecology Progress Series 152: 227–239.

Nagahama Y (1983). The functional morphology of teleost gonads. In: Hoar WS, Randall DJ, Donaldson EM (eds.), *Fish Physiology*, vol. IXA. Academic Press, New York, pp. 223–275.

Nelson JS (2006). *Fishes of the World*. John Wiley & Sons, Inc., Hoboken, New Jersey.

Nogawa Y, Endo H (2007). A new species of the genus *Priolepis* (Perciformes: Gobiidae) from Tosa Bay, Japan. Bulletin of the National Museum of Nature and Science, Series A, Supplement 1: 153–161.

Okiyama M, Tsukamoto Y (1989). Sea whip goby, *Bryaninops yongei*, collected from outer shelf off Miyakojima, East China Sea. Ichthyological Research 36: 1341–8998.

Parenti LR, Grier HJ (2004). Evolution and phylogeny of gonad morphology in bony fishes. Integrative and Comparative Biology 44: 333–348.

Pezold F (1993). Evidence for monophyletic Gobiinae. Copeia 1993: 634–643.

Randall JE (2001). Five new Indo-Pacific gobiid fishes of the genus *Coryphopterus*. Zoological Studies 40: 206–225.

Rasotto MB, Mazzoldi C (2002). Male traits associated with alternative reproductive tactics in *Gobius niger*. Journal of Fish Biology 61: 173–184.

Robertson DR, Justines G (1982). Protogynous hermaphroditism and gonochorism in four Caribbean reef gobies. Environmental Biology of Fishes 7: 137–142.

Rüber L, Van Tassell JL, Zardoya R (2003). Rapid speciation and ecological divergence in the American seven-spined gobies (Gobiidae: Gobiosomatini) inferred from a molecular phylogeny. Evolution 57: 1584–1598.

Sadovy Y, Shapiro DY (1987). Criteria for the diagnosis of hermaphroditism in fishes. Copeia 1987: 136–156.

Scaggiante M, Grober MS, Lorenzi V, Rasotto MB (2004). Changes along the male reproductive axis in response to social context in a gonochoristic gobiid, *Zosterisessor ophiocephalus* (Teleostei: Gobiidae), with alternative mating tactics. Hormones and Behavior 46: 607–617.

Scaggiante M, Mazzoldi C, Petersen C W, Rasotto MB (1999). Sperm competition and mode of fertilization in the grass goby *Zosterisessor ohiocephalus* (Teleostei: Gobiidae). Journal of Experimental Zoology 283: 81–90.

Seiwald M, Patzner RA (1989). Histological, fine-structural and histochemical differences in the testicular glands of gobiid and blenniid fishes. Journal of Fish Biology 35: 631–640.

Stanley H, Chieffi G, Boote V (1965). Histological and histochemical observations on the testis of *Gobius paganellus*. Zeitschrift für Zellforschung 65: 350–362.

St. Mary CM (1993). Novel sexual patterns in two simultaneous hermaphroditic gobies, *Lythrypnus dalli* and *Lythrypnus zebra*. Copeia 1993: 1062–1072.

St. Mary CM (1998). Characteristic gonad structure in the gobiid genus *Lythrypnus dalli* with comparison to other hermaphroditic gobies. Copeia 1998: 720–724.

St. Mary CM (2000). Sex allocation in *Lythrypnus* (Gobiidae): variations on a hermaphroditic theme. Environmental Biology of Fishes 58: 321–333.

Sunobe T (1988). A new gobiid fish of the genus *Eviota* from Cape Santa, Japan. Japanese Journal of Ichthyology 35: 278–281.

Sunobe T, Nakazono A (1993). Sex change in both directions by alternation of social dominance in *Trimma okinawae* (Pisces: Gobiidae). Ethology 94: 339–345.

Thacker CE (2003). Molecular phylogeny of the gobioid fishes (Teleostei: Perciformes: Gobioidei). Molecular Phylogenetics and Evolution 26: 354–368.

Thacker CE (2009). Phylogeny of Gobioidei and placement within Acanthomorpha, with a new classification and investigation of diversification and character evolution. Copeia 2009: 93–104.

Thacker CE, Cole KS (2002). Phylogeny and evolution of the gobiid genus *Corypohopterus*. Bulletin of Marine Science 70: 837–850.

Thacker CE, Grier H (2005) Unusual gonad structure in the paedomorphic teleost *Schindleria praematura* (Teleostei: Gobioidei): A comparison with other gobioid fishes. Journal of Fish Biology 66: 378–391.

Toller W (2005). F7 Final Report—October 1, 2003 to September 30 2005, Recreational Fisheries Habitat Assessment Project Study 3. Patterns of habitat utilization by reef fish on St. Croix. Division of Fish and Wildlife, Department of Planning and Natural Resources, Government of the U.S. Virgin Islands.

Victor BC (2008). Redescription of *Coryphopterus tortugae* (Jordan) and a new allied species *Coryphopterus bal* (Perciformes: Gobiidae: Gobiinae) from the tropical eastern Atlantic Ocean. Journal of the Ocean Science Foundation 1: 1–19.

Wang H-Y, Tsai M-P, Dean J, Lee S-C (2001). Molecular phylogeny of gobioid fishes (Perciformes: Gobioidei) based on mitochondrial 12S rRNA sequences. Molecular Phylogenetics and Evolution 20: 390–408.

Weisel GF (1949). The seminal vesicles and testes of *Gillichthys*, a marine teleost. Copeia 1949: 101–110.

Winterbottom R, Burridge M (1992). Revision of *Egglestonichthys* and of *Priolepis* species possessing a transverse pattern of cheek papillae (Teleostei: Gobiidae), with a discussion of relationships. Canadian Journal of Zoology 70: 1934–1946.

Winterbottom R, Emery AR (1981). A new genus and two new species of gobiid fishes (Perciformes) from the Chagos Archipelago, Central Indian Ocean. Environmental Biology of Fishes 6: 139–149.

Winterbottom R, Southcott L (2008). Short lifespan and high mortality in the western Pacific coral reef goby *Trimma nasa*. Marine Ecology Progress Series 366: 203–208.

Young RT, Fox DL (1937). The seminal vesicles of the goby, with preliminary chemical and physiological studies of the vesicular fluid. Proceedings of the National Academy of Sciences 23: 461–467.

PART TWO

Processes

6

Gonad Development
in Hermaphroditic Gobies

Kathleen S. Cole

Among hermaphroditic goby taxa (Perciformes, Gobiidae), considerable variability in the composition and configuration of gametogenic tissue within the gonad proper is coupled with a diversity of accessory structures of the reproductive complex (Cole 1990, 2009, this volume). Such diversity prompts the question as to how gonad ontogeny and morphogenesis may have become modified to produce such an impressive array of anatomical complexity. In the past decade, major advances in gene expression research have substantially improved our understanding of how cells and tissues first differentiate and then become organized to form the teleost reproductive complex. Genes and gene products such as *vasa*, *olvas*, *nanos*, *DMY/Dmrt1*, *SOX/STY*, *FOXl2*, and the enzyme aromatase figure prominently in these studies. While the number of fish species examined has been relatively limited, and mostly freshwater (i.e., zebrafish, *Danio rerio* and medaka, *Oryzias latipes*), collectively they have provided a working model and testable framework upon which to investigate early reproductive ontogeny in other teleosts. Studies of early ontogeny of the reproductive system in gobiids, although few in number, show numerous similarities with existing teleost model species. Therefore, what is known about the development of the reproductive complex in other teleosts may provide insights into the origins of reproductive morphological diversity found among hermaphroditic goby taxa. This chapter, consisting of three sections, addresses that possibility. The first section provides a brief overview of hermaphroditic gonad morphology among gobiid fishes. For a more detailed coverage, the reader is referred to Chapter 5 (Cole) of this volume. The second section reviews various aspects of early cell differentiation and tissue formation associated with the ontogeny of the teleost reproductive complex. The third section concludes with a discussion as to how ontogenetic processes may inform our understanding of the evolution of gonad morphological diversity among hermaphroditic gobiid fishes. It is hoped that this approach may offer new avenues of investigation into ontogenetic processes that are instrumental in the development of labile sexual patterns among vertebrates.

DEVELOPMENTAL PATTERNS OF GAMETOGENIC TISSUES IN HERMAPHRODITIC GOBIES

Among gonochore goby species, in which all individuals produce only one gamete type, either ova or sperm, the reproductive complex follows the basic teleost pattern. Paired gonadal lobes are united posteriorly and are continuous with an oviduct in females, and a sperm duct in males. The gobiid oviduct is typically a short, simple, tubular structure that conveys ova to the outside of the body. The gobiid sperm duct, however, is rarely simple. In the majority of goby taxa, expanded regions of the sperm duct form lobular secretory structures that vary extensively in their morphology across the family (Miller 1984; Fishelson 1991). Because of their association with the sperm duct, these structures have been termed sperm duct glands (Miller 1984).

In hermaphroditic goby species, either most or all individuals pass through an immature ovariform phase followed by an adult, ova-producing (i.e., female-functioning) phase (Cole & Robertson 1988; Cole & Shapiro 1992; Munday et al. 1998). If there is no preformed spermatogenic tissue present and the gonad is strictly ovariform, the oogenic tissue has one of two fates following sex change. It may completely disappear, resulting in the formation of a *secondary testis* (i.e., develops secondarily from an ovary); or, alternatively, healthy oocytes may persist during and following the development of spermatogenic tissue, and the gonad becomes a *secondary ovotestis* (i.e., develops secondarily from an initial single-sex gonad).

If the ovariform gonad of both the juvenile and initial ova-producing phases includes early-stage spermatogenic tissue, the gonad develops directly as an ovotestis. In hermaphroditic gobies, a shift to male function following an initial ova-producing phase is typically accompanied by a concomitant regression, but not disappearance, of ova-producing tissue. Therefore, in this ontogenetic pattern the direct development of an ovotestis that is subsequently retained throughout life results in a *persistent ovotestis*.

In addition to gonadal transformation, sex change in hermaphroditic gobies includes the development of secretory structures that appear to fulfill the function of sperm duct glands of gonochore species. Accessory secretory structures associated with male function in hermaphroditic gobiids may arise from one of two locations. In most species, precursive tissue masses (pAGS) which are associated with the wall of the ovariform gonad differentiate to form gonad-derived secretory structures (i.e., accessory gonadal structures, or AGS) (Cole 1988, 1990). However, in a small number of hermaphroditic species, secretory structures arise directly from the gonoduct as gonoduct-derived structures (i.e., accessory gonoduct structures, or AGdS), much like the sperm duct glands of gonochore gobies (Cole & Hoese 2001; Cole 2009)

From the anatomical diversity described above, the sequence of morphologies exhibited during immature, ova-producing and sperm-producing phases among hermaphroditic goby species has the potential to vary considerably. These differences can be expressed in terms of both the distribution of gamete-producing tissue(s) and the extent to which they co-occur. Applying these two criteria, hermaphroditic gobies exhibit a variety of developmental patterns of sequential morphological expression, which show a high degree of taxon specificity, as described by Cole earlier in this volume. Reproductive morphology patterns for hermaphroditic gobies fall into a number of different patterns, which include: (i) the interim ovotestis pattern of the Coryphopterus group, including the genera *Coryphopterus*, *Rhinogobiops*, *Lophogobius*, and *Fusigobius*; (ii) the modified interim ovotestis pattern of *Elacatinus multifasciatus*; (iii) the persistent integrated ovotestis pattern of *Eviota*; (iv) the regionalized ovotestis of *Lythrypnus*; (v) the persistent partitioned ovotestis of *Trimma* and *Bryaninops*, including the specialized persistent partitioned ovotestis pattern of *Priolepis hipoliti* and *P. eugenia,* which includes an interim functionally bisexual ovotestis; and (vi) the tripartite secondary ovotestis of *Gobiodon*. Most of these patterns are illustrated in detail in Chapter 5 of this volume (Figures 5.11, 5.12). Figure 6.1 illustrates ova-producing and sperm-producing gonad morphology for five of the six currently recognized gonad developmental patterns.

In the majority of reproductive morphology patterns listed above, the AGS arise from tissues associated with the gonadal wall and become fully differentiated and active at the onset of male function. In the sixth pattern shown by *Gobiodon*, however, accessory secretory structures are associated with the gonoduct rather than the gonadal wall. Although similar in position with the sperm duct glands male gonochore gobies *sensu* Miller (1984), the gonoduct-derived AGdS of *Gobiodon* differ in their expression. Unlike sperm duct glands, AGdS occur as fully differentiated structures in all individuals, including immatures and ova-producing adults and therefore are a persistent feature of the reproductive complex throughout all phases of sexual expression.

The diversity of morphology patterns associated with the reproductive complex of hermaphroditic gobies reflects the intersection of ontogeny, morphogenesis, and evolution. The question is, can variable patterns of morphology and development provide informative models for how reproductive structures and gene expression can be modified to generate complex functional morphologies?

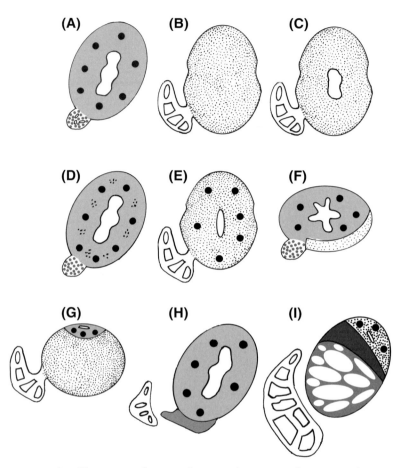

FIGURE 6.1. Illustration of ova-producing and sperm-producing gonad morphologies for four gonad developmental patterns exhibited by hermaphroditic
gobies. (A) Coryphopterus group ovary and (B) secondary testis of interim
ovotestis pattern. (C) *Elacatinus multifasciatus* secondary testis of modified interim ovotestis pattern. (D) *Eviota* ova-producing ovotestis and (E) sperm-
producing ovotestis of persistent integrated ovotestis pattern. (F) *Trimma* and
Bryaninops ova-producing ovotestis and (G) sperm-producing ovotestis of persistent partitioned ovotestis pattern. (H) *Gobiodon* ovary and (I) secondary
ovotestis of tripartite secondary ovotestis pattern. Light gray background represents oocyte-bearing tissue. Large black dots are oocytes. Open circle clusters
represent pAGS tissue. Light stipple (F) is early-stage, non-active spermatogenic tissue. Darker stipple is male-active spermatogenic tissue. Medium gray
background (H) represents undifferentiated region of future storage lobules.
White ovals (I) are storage lobules surrounded by supporting tissue (medium
gray); dark gray (I) is stromal tissue.

EARLY REPRODUCTIVE DEVELOPMENT IN TELEOSTS
Primordial Germ Cell Origins

The teleost reproductive complex initially develops as a composite of two cell sources: primordial germ cells (PGCs) arising from the germ cell line, and somatic cells. The germ cell line appears early in teleost development and the arrival of PGCs at the site of the future gonad is closely timed to that of the earliest stage of gonadal development, the gonadal anlagen. In the zebrafish, it has been shown that the origin of the germ cell line is predetermined by an aggregation of substances within the developing ovum, which are present long before ovum maturation and fertilization (Yoon et al. 1997; Braat et al. 1999; Pelegri & Schulte-Merker 1999; Howley and Ho 2000; Knaut et al. 2000; Pelegri 2003). These substances consist in part of an electron-dense material called "nuage" (Andre & Rouiller 1957) and the aggregate is referred to as "germ plasm" (e.g., Eddy 1975; Selman et al. 1993; Rongo et al. 1997). The presence of mRNA for the gene *vasa* (an RNA-binding protein), which is associated with germ plasm, is detectable in the early stages of oocyte development (stages I and II) in the maternal gonad and therefore predates fertilization (Yoon et al. 1997; Howley and Ho 2000; Pelegri 2003). Following fertilization, daughter cells that acquire germ plasm during cleavage events go on to form the germ cell line.

Vasa and *vasa*-like proteins associated with the germ cell line have been described for a number of teleosts including the Nile tilapia, *Oreochromis niloticus* (Kobayashi et al. 2000), rainbow trout, *Oncorhynchus mykiss* (Yoshizaki et al. 2000); Gibel carp, *Carassius auratus gibelio* (Xu et al. 2005); and most recently, bluefin tuna, *Thunnus orientalis* (Nagasawa et al. 2009). Early ontogeny studies of *vasa* expression associated with germ plasm have not yet been carried out for any goby species as of yet. However, a similar "nuage-like" aggregation of subcellular, electron-dense structures has been described for cells tentatively identified as PGCs in the goby *Leucopsarion petersii*, also known as shiro-uo, or the ice-goby (Miyake et al. 2006). Removal of this nuage-like substance during early ontogeny results in a significant decline in PGC numbers later in development (Miyake et al. 2006) suggesting that the origin of the germ cell line in gobiid taxa is similar to that of other examined teleosts.

In zebrafish, transcripts of the gene *vas* (zebrafish *vasa* homologue) provide a specific marker for germ line cells. Following fertilization, *vasa*-specific mRNA signal exhibits a predictable localization pattern (Yoon et al. 1997; Braat et al. 1999; Knaut et al. 2000). The signal is first expressed along the first and second cleavage furrows, and then becomes localized into four aggregates at each end of the two cleavage furrows. At the 32-cell stage, the aggregates ingress into four cells. During subsequent cleavage events, the four *vasa*-expressing cells undergo asymmetric segregation such that only one of the two daughter cells retains the *vasa*-positive germ plasm. In this way, the germ plasm remains restricted to four cells up until the 1-k cell (i.e., late blastula) stage (Knaut et al. 2000). At this point, *vasa*-positive cells undergo symmetric segregation such that both daughter cells now acquire a *vasa* signal, an event that marks the start of proliferation of germ plasm-containing cells and the consequent formation of a germ cell line. Thus, in the developing zebrafish embryo, two cell lines are formed as early as the blastula stage. One cell line consists of cells having no germ plasm, from which the somatic cell line arises. The other line includes cells that retain and often enrich the germ plasm and subsequently gives rise to the germ cell line and future PGCs.

In other teleosts, *vasa*-positive expression associated with the development of the germ cell line shows similar patterns across several teleost families and orders. In the loach, *Misgurnus anguillicaudatus* (F. Cobitidae, Cypriniformes) anywhere from four to eight *vasa* mRNA aggregates develop prior to the 32-cell stage. Here, symmetrical segregation within the *vasa*-positive cell line that marks the beginning of cell proliferation within the germ cell line also occurs at the 1 k-cell (i.e., late blastula) stage (Fujimoto et al. 2006). In goldfish, *Carassius auratus* (F. Cyprinidae, Cypriniformes) a total of eight *vasa*-specific mRNA signals develop initially and the development of the germ cell line, signaled by the onset of symmetric segregation, is initiated at the 256-cell (i.e., early blastula) stage (Otani et al. 2002). In medaka (F. Adrianichthyidae, Beloniformes), PGCs identified through cell-specific expression of the *vasa*-like gene, *olvas*, initiate symmetric segregation slightly later, in the late gastrula (stage 16)(Shinomiya et al. 2000).

The few studies of *vasa* expression in gobiid fishes show a similar pattern. In the ice-goby, *L. petersii*, the pattern of appearance and localization of *vasa* mRNA is similar to that of the goldfish in that eight *vasa* mRNA aggregates form in early cleavage by the 16-cell stage (Miyake et al. 2006). In another goby, *Gymnogobius urotaenia*, also known as ukigori, anywhere from 4 to 8 *vasa* aggregates develop by the 16-cell stage (Saito et al. 2004). In both the ice-goby and ukigori, the initiation of symmetric segregation signaled by an increase in the number of *vasa*-positive cells also occurs after the late blastula, 512-cell stage.

In summary, there are broad similarities in PGC formation across numerous examined teleost taxa, including two goby species, ice-goby, and ukigori. These similarities include: the location and timing of appearance of *vasa* signal early in cleavage; the number of cells that express *vasa* mRNA prior to the initiation of cell division and establishment of the germ cell line; and the timing of symmetric segregation during the blastula stage that initiates the development and proliferation of the germ cell line.

Primordial Germ Cell Migration

Studies of a number of teleost taxa, including killifish, *Fundulus heteroclitus* (Richards & Thompson 1921); medaka (Hamaguchi 1982; Shinomiya et al. 2000;

Kurokawa et al. 2006); Celebes medaka, *Oryzias celebensis* (Hamaguchi 1983); and zebrafish (Weidinger et al. 1999) have demonstrated that *vasa*-positive cells, initially located outside the embryo proper, move to the interior of the developing embryo and migrate to the target site of the future gonad. The pathway followed by PGCs within the embryo starts along the ventral coelomic wall and ends at a dorsally located position, just below the developing mesonephric ducts. During their migration, PGCs become encircled by a small number of cells derived from the lateral plate mesoderm, or LPM (Hamaguchi 1982; Kobayashi et al. 2004).

PGC tracking in a number of teleost species has revealed a combination of shared and differing behaviors of germ cell movement and subsequent development of the gonadal anlagen. In medaka, PGCs become incorporated into the outer somatic layer of the LPM and then move dorsolaterally to arrive at the location of the presumptive gonad (Gamo 1961; Kurokawa et al. 2006). Here, the proliferation of dorsal coelomic peritoneal cells adjacent to PGCs rapidly increases their numbers, resulting in the formation of the gonadal anlagen (Hamaguchi 1982; Kobayashi et al. 2004). In zebrafish, PGCs migrate into the embryo interior during early gastrulation and become clustered bilaterally in the dorsoanterior trunk region (Weidinger et al. 1999; Weidinger et al. 2002; Reichman-Fried et al. 2004). This is followed by directed movement until they reach the site of the future gonad (Doitsidou et al. 2002). In rainbow trout (F. Salmonidae, Salmoniformes) PGCs have been tracked moving from the ventral mesentery first along the inner splanchnic layer of the LPM on either side of the gut mesentery, then along the dorsal mesentery and laterally to the left, or right, along the roof of the coelom (Moore 1937).

In two gobies, ukigori and ice-goby, PGCs are reported to move dorsally along the inner splanchnic layer of the LPM to the dorsal region of the gut, then to the site of the future gonadal anlage (Saito et al. 2002, 2004). In the round goby, *Neogobius melanostomus*, PGC migration along the LPM has been described as occurring along both the inner splanchnic and outer somatic layers (Moiseyeva 1983). Thus, gobies exhibit PGC migration patterns seen in other teleosts.

Early Gonad Development

PGC migration stops upon reaching a dorsal position adjacent to the mesonephric ducts. Here, PGCs, their surrounding LPM-derived somatic cells and adjacent, proliferating peritoneal cells of the dorsal coelom form the gonadal anlagen. The first conspicuous morphological indication of the formation of the gonadal anlagen is the development of a longitudinal ridge of cells, referred to as the peritoneal gonadal, or genital ridge, which extends along the length of the dorsal coelom on either side of the midline. The timing of formation of the peritoneal ridge varies among teleosts such that in some species, the peritoneal ridge forms prior to the arrival of PGCs, while in others it appears to develop in response to the arrival of PGCs.

In rainbow trout, after PGCs reach a position below the mesonephric ducts, they become surrounded by adjacent peritoneal epithelial cells, which proliferate to form the gonad anlagen (Moore 1937; Takashima et al. 1980). Subsequently, PGCs migrate anteriorly such that the developing gonadal anlagen extends nearly the length of the abdominal cavity (Moore 1937). A similar pattern of PGC migration and arrival at the site of the future gonad prior to peritoneal cell proliferation has been reported for a number of species, including: brown bullhead, *Ameiurus nebulosus* (Bachmann 1914); killifish, *Fundulus heteroclitus* (Richards & Thompson 1921); goldfish (Stromsten 1931); medaka (Hamaguchi 1982); rosy barb, *Puntius conchonius* (Timmermans & Taverne 1983); and carp, *Cyprinus carpio* (Parmentier & Timmermans 1985).

Information is less precise for gobies. In ukigori, brief descriptions provided by Saito et al. (2002, 2004) imply that the formation of a peritoneal ridge predates the arrival of the PGCs. In the round goby, however, a brief description of the site of the future gonad implies that the arrival of the PGCs may precede the proliferation of adjacent peritoneal cells to form a ridge (Moiseyeva 1983).

Gonadogenesis among teleosts starts with the initial formation of anlagen, which characterizes the *gonadal anlagen* stage. The term "gonadal anlage" refers to the structure that includes PGCs already surrounded by enveloping LPM cells plus additional surrounding, but non-enveloping, somatic (i.e., peritoneal) cells (Parmentier & Timmermans 1985). With subsequent cell proliferation the (usually two) gonadal anlagen expand into the coelomic cavity, being separated from the body cavity only by the portion of the peritoneum which envelops each anlage. This peritoneal layer also usually forms the suspensory mesentery—the mesorchium of the testis or the mesovarium of the ovary—which suspends each anlage from the dorsal coelomic wall. In the subsequent *indifferent gonad* stage, the gonad is still insufficiently differentiated morphologically to be identified as either an ovary or a testis. This stage is characterized by the mitotic increase of PGCs, the transformation of PGCs to gonial cells that cannot yet be distinguished as either oogonia or spermatogonia, and the proliferation of somatic cells of the former gonadal anlagen. The next stage—that of a *sexually differentiated* gonad—applies to gonads exhibiting identifiable male or female sex cells, distinctive ovarian or testiform morphology, or both. The final gonadal stage is that of a *mature* gonad and is identified by the presence of mature gametes (i.e., ova, sperm, or both). These stages of gonadal development can be applied to all teleosts, including hermaphroditic gobies.

Evidence of Predetermined Sex in Some Gonochore Teleosts

In some teleosts, gonial cells are insufficiently distinctive to identify as either oogonia or spermatogonia until later in development. However, predetermined sex may still be evident by differences in mitotic division rates of germ cells within the indifferent gonad. Medaka, for example, exhibit no obvious ultrastructural differences between male and female gonial cells (Satoh 1974). However, gonia among genetic female medaka mitotically increase to four times their original number between the time they become incorporated into the gonad anlagen and hatching, while only doubling in number among genetic males (Hamaguchi 1982; Kobayashi et al. 2004). As a result, medaka having relatively fewer gonial cells prior to hatching are male, while similar-aged embryos having relatively more gonial cells are female (Satoh & Egami 1972). Similar findings of an initial female-specific increase in PGCs have also been reported in the threespine stickleback, *Gasterosteus aculeatus* (Lewis et al. 2008), rainbow trout, *Oncorhynchus mykiss* (Lebrun et al. 1982) and Mozambique tilapia, *Oreochromis mossambicus* (Nakamura et al. 1998). These examples reflect the establishment of sexual determination, albeit morphologically cryptic, well before sex-specific characteristics of gonia or the gonad become evident. In such instances, it should be noted, the term "indifferent gonad" serves primarily as a point of reference for the absence of visible characteristics of gonad ontogeny, rather than an actual absence of sex determination.

In addition to differences in rates of germ cell division, females and males frequently exhibit different time lines in sex cell differentiation. Among female medaka, gonial cells differentiate into primary oocytes at about five days post-hatch, thereby transforming the indifferent gonad into a recognizable ovary. Among males, however, mitotic proliferation of gonial cells does not occur until around 45 days post-hatch. The differentiation of spermatogonia into primary spermatocytes occurs even later, well after the differentiation of interstitial (i.e., Leydig) cells and their subsequent sex steroid production (Satoh 1974; Yoshikawa & Oguri 1979). In female pejerrey, *Odontesthes bonariensis*, the proliferation of both gonial cells and gonadal somatic cells among females begins 49 to 63 days post-hatch compared to 98 days among males (Strüssmann et al. 1996). Exhibiting a similar pattern, the number of sex cells in Celebes medaka is greater in females at hatching, and primary oocytes differentiate in females before primary spermatocytes do so in males (Hamaguchi 1983). In the goldfish, mitotic division of germ cells starts around 6 to 7 weeks post-fertilization regardless of genetic sex. However, subsequent meiosis of primary oocytes begins at approximately 16 weeks post-fertilization while secondary spermatocyte formation following meiosis does not begin until 20 weeks post-fertilization (Parmentier & Timmermans 1985; van Winkoop et al. 1992).

Gonad differentiation characterized by sex-specific gonad morphology frequently shows a similar sex-specific gap in timing. In female channel catfish, *Ictalurus punctatus*, the initiation of ovarian cavity formation occurs at around 14 days post-hatch, prior to PGC differentiation into gonia, while testis-typical lobule formation among males does not take place until about 96 days post-hatch (Patiño et al. 1996). In the cichlid, *Cichlasoma dimerus*, ovarian differentiation marked by the development of the ovarian cavity and the concomitant differentiation of gonia into primary oocytes occurs at about 40 days post-hatch (Meijide et al. 2005). In contrast, the earliest sign of testis differentiation among males occurs at 65 days post-hatch, as signaled by the appearance of presumptive Sertoli cells surrounding individual gonial cells. Primary spermatocyte differentiation occurs even later, 70 days post-hatch, followed by the differentiation of seminiferous lobules at about 100 days post-hatch (Meijide et al. 2005). In the closely related Nile tilapia, histological sex differentiation is characterized in females by the onset of oocyte meiosis, which is simultaneous with the development of the ovarian cavity at around day 28 post-fertilization. Among males, the onset of first meiotic prophase in future spermatocytes occurs at about 55 days post-fertilization, resulting in a time lag of about three weeks between female and male time lines (Kanamori et al. 1985; Nakamura & Nagahama 1989; Nakamura et al. 1998; D'Cotta et al. 2001a). In a slight deviation from the general trend, early ovarian lamellar architecture and testis lobular structure in rainbow trout are both evident at 100 days. However, primary oocytes in females are first recognizable at 67 days post-fertilization while primary spermatocytes among males are not distinguishable until 168 days post-fertilization (Takashima et al. 1980).

In zebrafish, testis morphogenesis follows a sequential pathway. In all individuals, an ovariform gonad develops first, in tandem with the initial differentiation of primary oocytes, at around 16 to 24 days post-fertilization. Among genetic females, the ovarian components persist throughout life. Among genetic males, oocyte development is followed by degeneration, which occurs anywhere from 30 to 40 days post-fertilization and is then followed by the development of testis features (Takahashi 1977; Maack & Segner 2003; XG Wang et al. 2007). This ovariform to testiform pattern of early gonad ontogeny in otherwise gonochoric species has also been reported for carp, *Cyprinus carpio* (Davies & Takashima 1980); Sumatra barb, *Barbus tetrazona* (Takahashi & Shimizu 1983); red sea bream, *Pagrus major* (Matsuyama et al. 1988); mosquitofish, *Gambusia affinis* (Koya et al. 2003);Mekong ricefish, *Oryzias mekongensis*; and up to 50 percent of genetic males in the Luzon ricefish, *O. luzonensis* (Otake et al. 2008).

Thus, the overall trend among examined teleosts is one of differentiation of sex-specific features of the female reproductive complex prior to that of the male. Among gobiids, there is little information available on sex-specific rates of gonial development, timing of gonial differentiation according to sex, or for sex differentiation processes in general. In the round goby (*Neogobius melanostomus*), Moiseyeva (1984) reported that the first indication of sex differentiation among females was oocyte differentiation at 20 days post-hatch, followed by ovarian morphogenesis by 30 days post-hatch. Among males, features of testis morphology were first evident at 40 to 45 days post-hatch but spermatocytes did not appear until close to a year after hatching, thereby repeating the generalized teleost pattern of a time lag in male sex differentiation.

Formation of the Gonadal Ducts

The other component of the teleost reproductive complex in addition to the gonad is the gonoduct, for which developmental details among teleosts are scant. In fish taxa that have paired gonadal lobes, the lobes either unite posteriorly and empty into a common gonoduct, or each empties into a gonoduct that then unites to form a single duct that terminates at the distal orifice at the tip of a genital papilla. In some species, gonoducts of one or both sexes may be reduced or absent. Male yellow perch, *Perca flavescens*, have a short sperm duct which links with the posteriorly united testis lobes. In females, however, there is no oviduct. An ovarial sac enclosing the single ovary is fused posteriorly with the body wall. Shortly before ovulation, a protrusion develops at the junction of the ovarian sac and abdominal wall. As ovulation approaches, the overlying tissue of the protrusion thins, then ruptures, thereby providing an exit point for the ova from the ovarian lumen to the external environment (Parker 1942). In the Japanese sea bass, *Lateolabrax japonicus*, the gonoducts of both sexes are very short and in the female, the oviduct opens to the outside only at the time of ovulation (Hayashi 1969). In a somewhat different ontogenetic pattern, the gonadal ducts of the silver-stripe round herring, *Spratelloides gracilis*, form early in development but are incomplete, being unconnected to the gonad until maturity (Hatakeyama et al. 2005).

The origin(s) of gonoducts have been variously ascribed to the dorsal peritoneum, the suspensory mesentery of the gonad, a posterior continuation of the surrounding peritoneum of the gonad which becomes narrowed down to form a trough or tube, an extension of a somatic cell mass that develops in the posterior gonad, or some combination thereof. In coho salmon, *Oncorhynchus kisutch*, the oviduct forms anteriorly from both a continuation of the portion of the peritoneum forming an ovarian covering and from the suspensory mesovarium. A left and a right anterior oviduct leading from the ovarian lobes each extend posteriorly for a short distance, and then merge to form a single channel. More posteriorly, folding of the dorsal mesentery forms the posterior oviduct. However, unlike the female reproductive complex of most teleosts, the oviduct remains open dorsally due to incomplete folding and fusion and forms a trough rather than a true duct (Kendall 1921).

In the swordtail, *Xiphophorus helleri* (F. Poeciliidae, Cyprinodontiformes), elements of anterior oviduct development are reminiscent of that found in coho

(Essenberg 1923). Early in development, the left and right gonadal anlagen fuse, first along the ventral margin, then subsequently along the dorsal margins, to form a single ovary and ovarian cavity. The associated oviduct develops from two points of origin. While the gonad is still incompletely fused dorsally, cell proliferation takes place at the caudal end of the ovary to form a cell mass. This cell mass then widens and develops a ventral groove, mimicking the cross-sectional V shape of the still dorsally open ovary. Therefore, for a brief period, the developing swordtail oviduct is open dorsally, like that of coho. Subsequently, the groove becomes covered by a portion of the suspensory mesovarium lying directly above to become completely enclosed, thereby forming a lumen at the anterior end of the developing oviduct. In the posterior region of the abdominal cavity, a concomitant proliferation of peritoneal cells develops along the dorsal midline, which results in the formation of a solid cord of cells. The cell cord develops forward along the dorsal wall of the abdominal cavity and fuses with the anterior portion of the oviduct arising from the posterior gonad. Subsequently, first a slit, then a lumen forms within the proliferating peritoneal cell cord that becomes continuous with the lumen of the anterior oviduct.

The extra-gonadal portion of the swordtail sperm duct develops in much the same manner as that of the oviduct, with the initial formation of anterior and posterior gonoduct primordia. Anteriorly, the proliferation of peritoneal cells at the caudal end of the testis effectively becomes an extension of the internal (i.e., efferent) sperm duct system such that the efferent sperm duct lumen within the testis becomes continuous with that of the extra-gonadal sperm duct. Thus, in the development of both the female and male swordtail reproductive complex, the gonoduct forms from the proliferative activity of cells both from the caudal portion of the gonad and from the peritoneum of the posterior region of the abdominal cavity.

In female medaka, the oviduct also develops from two different sources (Suzuki & Shibata 2004) but the process differs from that of the swordtail. The anterior portion of the oviduct arises from somatic cell proliferation from the posterior end of the ovary. This cell mass grows caudally to the posterior extent of the bladder. Subsequent cavitation results in the formation of a lumen, transforming the cell mass into a duct that is continuous with the lumen of the ovary. However, at this stage the caudal portion of the anterior oviduct is blind-ended. In contrast, the posterior portion of the oviduct arises from mesenchymal cells surrounding the urethra. The portion of the mesenchymal cell layer lying ventral to the urethra becomes thickened and forms a cortical tissue layer (so termed in Suzuki & Shibata [2004]). A proliferation of anterior cells of the cortical layer extends forward until it almost reaches the posterior end of the anterior oviduct. Subsequently, two bilateral cavities develop along the length of the cortical layer.

At this point, the anterior and posterior portions of the oviduct are still separate from one another. With the approach of first maturity, the anterior oviduct and cortical layer extension meet and fuse. The partition separating the right and left cavities of the cortical layer disappears, leaving behind a common lumen, which becomes continuous with the anterior oviduct lumen. At the terminus of the urogenital papilla, the distal end of the posterior oviduct becomes open to the outside of the body, thereby providing an exit for ova to the outside environment. Thus, the final medaka oviduct arises from an anterior portion that is derived from gonadal somatic cells and from a posterior portion that arises from urethra-associated mesenchymal tissue in the form of a cortical layer.

In male medaka, the sperm duct also develops from two sources (Suzuki & Shibata 2004). Information provided on its development is not as detailed as that of the oviduct, but appears to share a number of similarities with the latter. The anterior portion of the sperm duct arises from a cell mass forming from the proliferation of somatic cells in the posterior end of the testis, then subsequently develops a lumen by cavitation, and extends caudally towards the ventral region of the bladder. The posterior portion arises in the ventral area of urethra-associated mesenchyme and extends caudally almost until it reaches the urinary pore. Here, the posterior sperm duct and urethra merge and the inner lumen epithelia of the two structures become confluent. The subsequent fusion of the anterior and posterior portions results in the formation of a single, continuous sperm duct that transmits sperm from the efferent duct system within the testis to a urogenital sinus before exiting the body.

The development of the oviduct in the guppy, *Poecilia reticulata*, follows a somewhat different ontogeny than that of medaka. In the guppy, oviduct development involves both the dorsal mesentery and dorsal peritoneum located just posterior to the fused ovarian lobe (Anteunis 1959). Two lateral folds of the dorsal mesentery develop, extend in opposite directions away from the midline, then each deflect upward towards the dorsal peritoneum. The fusion of the two recurved folds, either with each other or with the dorsal peritoneum, and the subsequent disappearance of the internal partition formed by the enclosed portion of the dorsal mesentery, results in the formation of an undivided duct that is continuous with the posterior portion of the ovary.

The two sperm ducts in male guppies have a slightly different developmental pattern. The sperm ducts originate as somatic cell masses arising from the posterior end of the two testis lobes (Takahashi & Iwasaki 1973). These become extended as solid, bilateral ridges that form the sperm duct anlagen. As they approach the urogenital sinus, the two ridges become fused to form a single ridge. Subsequently, an internal slit develops along the length of the fused and separate portions of the ridge, and becomes lined with epithelial cells to form a proper sperm duct. The lumen of the anterior paired sperm ducts becomes continuous with the main lumen of the efferent duct system within the testis while the posterior, unified portion remains closed at its terminus until later in development.

Finally, in female three-spined stickleback, *Gasterosteus aculeatus*, bilateral, sterile genital ridges form along the dorsal coelomic wall posterior to the ovary. Medial and lateral cellular extensions grow outward from the ridge, become extended towards each other, and then fuse to form a tubular structure. The tubular structure subsequently becomes continuous with the ovarian lobe and its lumen and becomes the oviduct (Shimizu & Takahashi 1980).

In summary, the origins of the gonoducts show considerable variation across fish taxa. They may develop from peritoneal epithelium that variously surrounds the gonad, forms the gonad suspensory mesentery and dorsal mesentery, or lines the dorsal region of the abdominal cavity. Their formation may result from somatic cell proliferation from the posterior gonad, posterior tubular extensions of gonadal serosa, peritoneal folds, dorsal mesentery folds, cavitation of a solid cord of cells, or by some combination of these. As a result, there is considerable scope for the development of new features and anatomical structures, a possibility that may have been instrumental in the development of the considerable diversity of gonoduct-associated accessory structures documented across fish taxa in general, and gobies in particular.

Ontogeny Summary

The earliest stage of teleost gonad development, the gonadal anlagen, is composed of PGCs or PGC-derived gonial cells, enveloping LPM cells and dorsal coelom peritoneal cells. The anlagen develop following the formation of germ line cells, the subsequent migration of PGCs to a location below the mesonephric ducts and their integration into genital (i.e., peritoneal) ridges extending along the dorsal coelom. Subsequently, the anlagen increase in size through somatic (peritoneal) cell proliferation to form an indifferent gonad. The indifferent gonad is characterized by germ cells and associated support cells and tissues, and an absence of morphological features indicative of future sex. With the differentiation of germ cells, or sex-specific gonadal features, or both, the gonad enters the sexually differentiated stage that persists until gamete production, at which time it becomes a mature gonad. In contrast, the gonoduct is characterized by the absence of identifiable germ cells or any directly associated gametogenic tissue. It may form from peritoneally derived cells, mesenchymal cells, somatic cells of the gonad, or some combination thereof. Typically, the junction of the gonoduct and gonad comprises an expanded region with no associated gametogenic tissue, which in hermaphroditic gobies is termed the common genital sinus (Cole & Robertson 1988; Cole 1990).

Based on the above information, the ontogeny of the teleost reproductive complex can be viewed in one of two ways. The entire complex may form as a construct of separate, independently arising components consisting of the gonad proper (germ line cells and peritoneal derivatives) and the gonoduct (only peritoneal or other derivatives) that are linked by a non-germinal, transition region (i.e., the common genital sinus of gobies). Alternatively, the reproductive complex represents a continuous peritoneally derived structure that is characterized by: the localized presence of germ cells, support cells, and tissues forming the initial anlagen and subsequent gonad proper; and the remaining non-gametogenic sections making up the common genital sinus and gonoduct.

Ontogenetic information related to the development of the reproductive complex is still extremely limited for gobiid species, especially for early anlage and indifferent developmental stages and for gonocyte differentiation. However, the broad similarity of developmental patterns across teleosts, coupled with similar features of early germ cell-line formation and PGC migration exhibited by shiro-uo, ukigori and round goby, suggest that general features of the development of the teleost reproductive complex may be predictive of gobies. A comparative examination of diverse morphologies of the reproductive complex among hermaphroditic gobies may therefore suggest possible modifications of ontogenetic processes that result in labile sexual expression and reveal the underlying morphogenic nature of the hermaphroditic reproductive complex as a whole

REPRODUCTIVE COMPLEXITY IN HERMAPHRODITIC GOBY TAXA FROM AN ONTOGENETIC PERSPECTIVE

Origins of Morphology Patterns of the Hermaphroditic Gonad

To date, all reported hermaphroditic gobiid taxa are found within the Gobiidae subfamily, Gobiinae, *sensu* Pezold (1993). Functional hermaphroditism has been demonstrated or inferred from gonad morphology in fourteen genera. These include the following:

Bryaninops (Fishelson 1989)

Coryphopterus (Robertson & Justines 1982; Cole 1983; Cole & Shapiro 1990)

Eviota (Cole 1990)

Fusigobius (Cole 1990)

Gobiodon (Cole 1990; Nakashima et al. 1995, 1996; Munday et al. 1998; Cole & Hoese 2001)

Elacatinus (formerly *Gobiosoma*)(Robertson & Justines 1982)

Lophogobius (Cole 1990)

Luposicya (Fishelson 1989)

Lythrypnus (St. Mary 1993, 1994)

Paragobiodon (Lassig 1977; Cole 1990; Kuwamura et al. 1994; Nakashima et al. 1995)

Pleurosicya (Fishelson 1989)

Priolepis (Cole 1990; Sunobe & Nakazono 1999)

Rhinogobiops (previously *Coryphopterus*)(Cole 1983)

Trimma (Fishelson 1989; Cole 1990; Sunobe & Nakazono 1990, 1993)

Four hermaphroditic genera making up the Coryphopterus group *sensu* Cole (see Chapter 5 of this volume) include *Coryphopterus, Rhinogobiops, Lophogobius,* and *Fusigobius,* all of which exhibit an interim ovotestis development pattern. In this pattern, the production of sex-specific gametes occurs during temporally disassociated phases of ova and sperm production, and the ovotestis phase is both transient and afunctional (see Cole, this volume, for more detailed description). *Elacatinus multifasciatus,* another hermaphroditic goby species, exhibits a "modified" interim ovotestis pattern which is identical to that of the Coryphopterus group excepting the retention of a gonadal lumen. The interim and modified interim ovotestis developmental patterns appear to represent the simplest gonad ontogeny pattern among hermaphroditic gobiids. Based on what we now know of sex determination and differentiation in other teleosts, a shift from gonochorism to simple sequential hermaphroditism in gobiids may have been achieved by the modification of two ontogenetic processes. The first involves temporal shifts in gene expression associated with sexual differentiation, which change the timing of upregulation and downregulation of various aspects of gonadogenesis. The second is the development and retention of gonial stem cells, or bipotential gonia, that are capable of differentiating along either an oogenic or a spermatogenic pathway.

Among examined gonochore teleosts, gametocytes and gonad morphology typically differentiate earlier in females than in conspecific males. As described in the previous section, in a number of species genetic males first develop an ovariform gonad, and sometimes oocytes, before spermatocyte differentiation and testis development (Takahashi 1977; Davies & Takashima 1980; Takahashi & Shimizu 1983; Matsuyama et al. 1988; Koya et al. 2003; Maack & Segner 2003; XG Wang et al. 2007; Otake et al. 2008). From here, it would only take a short, additional developmental step for the initial feminized phase of gonad development that occurs in all individuals to become extended past first maturity such that all individuals function first as an ova-producing female. Secondary male function associated with sperm production would follow with the delayed development of male sex cells and a testiform gonad. By means of such a heterochronic shift, a gonochore sexual pattern might be transformed into a protogynous sequence of sexual function.

In unidirectional hermaphroditism as expressed in the Coryphopterus group and *Elacatinus multifasciatus,* the first appearance of male sex cells is delayed until after a period of ova-production and is accompanied by the disappearance of all ovarian tissues. A mechanism for delayed masculinization of the reproductive complex in the interim ovotestis pattern can be found in events associated with the regulation of normal sexual differentiation in teleosts. Cytochrome P450 aromatase (referred to here as aromatase) is a steroidogenic enzyme that acts to catalyze androgens (mostly testosterone) to oestrogens (mostly oestradiol-17, or E_2) and plays a feminizing role in sexual ontogeny. In a number of gonochore teleosts, normal ovarian development among females has been shown to be accompanied by elevated levels of aromatase and E_2 while normal testis development is associated with a decrease in both E_2 and aromatase (Guiguen et al. 1999; Kitano et al. 1999; D'Cotta et al. 2001a, b; Uchida et al. 2004), demonstrating a direct relationship for aromatase production, E_2 levels and feminization of the gonad. Moreover, aromatase and E_2 appear to be essential for ovarian development. Treatment during early gonad differentiation with the non-steroidal aromatase inhibitor, fadrozole, has been shown to result in the masculinization of genetic females in a number of fish species including zebrafish (Fenske & Segner 2004), golden rabbitfish, *Siganus guttatus* (Komatsu et al. 2006) and European sea bass, *Dicentrarchus labrax* (Navarro-Martín et al. 2009).

Inhibition of aromatase production can also be induced by maintenance at higher than normal water temperatures. Among genetically female Nile tilapia, *Oreochromis niloticus*, elevated levels of aromatase enzyme activity and E_2 are normally associated with a shift from an indifferent gonad to a differentiated ovariform gonad. However, when maintained under high temperature regimes, reduced aromatase expression is followed by masculinization (D'Cotta et al. 2001a, b). In female Atlantic salmon, *Salmo salar*, warmer maintenance temperatures are associated with an inhibition of aromatase activity, a decrease in plasma E_2 levels and an increase in testosterone levels (Watts et al. 2004). And in Japanese flounder, *Paralichthys olivaceus*, rearing genetically female larvae at high water temperatures causes a suppression of aromatase gene expression and the conversion of genetic females into phenotypic males (Yoshinaga et al. 2004). Thus, in gonochore teleost development, the presence of aromatase has a feminizing influence through its mediation of E_2 production, and its absence has a masculinizing influence.

A similar feminizing role for E_2 has been demonstrated in a number of hermaphroditic teleosts in which elevated E_2 levels are characteristic of the ovarian phase while decreased levels are associated with the testiform phase (see Piferrer & Guiguen 2008 for a review). In two hermaphroditic goby species, *Rhinogobius* (previously *Coryphopterus*) *nicholsii* and *Gobiodon erythrospilus,* the inhibition of aromatase activity by fadrozole triggers adult female-to-male sex change (Kroon & Liley 2000; Kroon et al. 2005), implicating a role for declining E_2 in normal sex change events. The subsequent application of E_2 to male-phase adult *Gobiodon erythrospilus*, a serially hermaphroditic species that can shift between female and male adult function, results in a shift back to female function (Kroon et al. 2005). In another serial hermaphrodite, *Gobiodon histrio*, whole-body concentrations of testosterone, 11-ketotestosterone, and E_2 suggest that the upregulation or down-regulation of the aromatase/testosterone-to-E_2 conversion pathway is a probable candidate for mediating serial sex change in this species (Kroon et al. 2003). Based on the above findings, the delayed masculinization characteristic of the interim ovotestis pattern of the Coryphopterus group may be mediated by a prolonged period of aromatase upregulation. The brief appearance of an afunctional ovotestis presumably reflects an overlap between declining levels of aromatase and E_2 associated with oogenic regression, and the concomitant upregulation of testis development.

Genes involved in the morphogenesis of ovariform and testiform features of the teleost gonad are becoming better known and patterns of gene expression are starting to emerge. *FOXl2* expression is increasingly associated with ovarian morphogenesis in teleosts through the initiation and regulation of aromatase transcription. In female medaka, somatic cells directly surrounding germ cells exhibit *FOXl2* expression at the time of ovarian differentiation (Nakamoto et al. 2006). Similar findings have been made for the Japanese flounder (Yamaguchi et al. 2007) and Nile tilapia (D-S Wang et al. 2004, 2007; Ijiri et al. 2008). In the Luzon medaka, *O. luzonensis*, the expression of *FOXl2* in the somatic cells of the indifferent gonad is found in all genetic females and in almost half of all genetic males. In the latter, the downregulation of *FOXl2* directly precedes testis development and is thought to reflect the upregulation of one or more testis-determining genes that act to inhibit further ovarian differentiation (Nakamoto et al. 2009). A close association between *FOXl2* and teleost ovarian development is also indicated by the findings of over-expression of *FOXl2* in the developing ovary, but not the testis, in the Southern catfish, *Silurus meridionalis* (Liu et al. 2007), Nile tilapia (D-S Wang et al. 2007) and rainbow trout (Baron et al. 2004). Among hermaphroditic species, the downregulation of *FOXl2* is associated with ovarian degeneration during female-to-male sex change in the protogynous honeycomb grouper, *Epinephelus merra* (Wu et al. 2008), while *FOXl2* upregulation is associated with testicular regression during male-to-female sex change in the protandrous black porgy, *Acanthopagrus schlegeli* (Alam et al. 2008). And in genetic female rainbow trout that were artificially masculinized by the application of either an aromatase inhibitor or active androgens, ovarian regression was accompanied by reduced *FOXl2* expression (Vizziano et al. 2007). Therefore, *FOXl2* and the cells that express it are likely key to ovarian development in both gonochoric and hermaphroditic fish species, and a decline in *FOXl2* expression is likely necessary for the development of either a secondary testis or secondary ovotestis.

The gene *SOX9* has also been implicated in early sexual development in a number of teleosts. In medaka of both sexes, *SOX9b* is expressed in enveloping cells that surround PGCs. These enveloping cells differentiate into granulosa cells in females and Sertoli cells in males. As such, *SOX9b*-expressing cells are intimately associated with germ cells and are precursors of both male and female gonocyte support cells (Nakamura et al. 2008). *SOX9a2* is another gene within the *SOX9* group, which in medaka is expressed in germ-cell enveloping cells in both males and females prior to differentiation. However, its expression persists only in genetic males and appears to play a role in testis morphogenesis, specifically in seminiferous lobule formation (Nakamoto et al. 2005).

If the first step towards functional, female-first hermaphroditism in gobies involved a heterochronic shift in gonial and gonad differentiation, then this shift was likely mediated at least in part by the delayed expression of testis-determining gene(s). In medaka, *DMY* (also known as *Dmrt1bY*) has been identified as a testis-determining gene for this species (Matsuda et al. 2002; Nanda et al. 2002) In genetic males, its expression is found in somatic cells directly surrounding germ cells (i.e., presumptive Sertoli cells), is initiated at the end of the indifferent gonad phase, just prior to testis differentiation, and is involved in testis-specific PGC proliferation and testis differentiation (i.e., gonad morphogenesis)(Kobayashi et al. 2004). Spermatogonial differentiation in this species, however, is regulated by another gene, *Dmrt1* (Kobayashi et al. 2004). Similarly, in the protogynous orange-spotted grouper, *Epinephelus coioides*, the expression of an intronless version of *Dmrt1* is associated with male-phase germ cells and appears to play a central role in stimulating spermatogenesis (Xia et al. 2007). The role of *Dmrt1* in testis differentiation has been either implicated or demonstrated in a number of other teleosts including southern catfish (Liu et al. 2007), Nile tilapia (Injiri et al. 2008; Kobayashi & Nagahama 2009), and pejerrey, *Odontesthes bonariensis* (Fernandino et al. 2008). Further evidence of its importance in the development of male function is evident in several examples of ovary-testis transitioning events. *Dmrt1* upregulation occurs in genetically female, experimentally masculinized, rainbow trout at the time of testis development (Vizziano et al. 2008). Conversely, reduced *Dmrt1* expression characterizes male-to-female sex change in two protandrous fish species, gilthead bream, *Sparus auratus* (Liarte et al. 2007) and black porgy (Shin et al. 2009).

Thus, *DMY/Dmrt1bY* and *Dmrt1* play important roles in testis development across a wide range of teleost taxa. However, similarities of expression and proposed function across several taxa do not necessarily indicate automatic conservation within taxa. *DMY/Dmrt1bY* expression, which is present in both medaka and a congener, *O. curvinotus* (Matsuda et al. 2003), is absent in two other congeners, *O. mekongensis* (Otake et al. 2008) and *O. celebensis* (Kondo et al. 2003). Therefore, even closely related taxa may have different testis-determining regulatory genes. So far, little is known regarding gene(s) involved in the regulation of testis development and spermatogenesis in gobiids. In one study of the hermaphroditic blue-banded goby, *Lythrypnus dalli*, *Dmrt1* was found to undergo rapid upregulation and exhibit a two-fold increase shortly after the initiation of a transition from ova to sperm production (Rogers 2007). Whether *Dmrt1* is widely expressed across gobiid taxa, and what precise role it plays in testis development, remain to be discovered. Certainly, the identification of testis-determining gene(s) and their regulation properties will be central to the reconstruction of the evolutionary steps that have led to the development of functional hermaphroditism among the Gobiidae.

A second ontogenetic modification likely associated with a shift from gonochorism to functional hermaphroditism, particularly when involving the secondary development of spermatogenic tissue, entails the successful maintenance of a reservoir of gonial cells capable of developing into either oocytes or spermatocytes. There are a variety of potential sources for new oogonia and spermatogonia. In the mouse, embryonic stem cells have been shown to produce both male and female gametes under experimental conditions (Geijsen et al. 2003; Nayernia et al. 2006; Kerkis et al. 2007). In the rainbow trout, spermatogonia have been shown to be both developmentally plastic and sexually bipotent. Green fluorescent protein (GFP)-labeled spermatogonia taken from adult rainbow trout testes and transplanted into the undifferentiated genital ridges of conspecific fry subsequently differentiated into spermatocytes in the recipient gonad of genetic males and into oocytes in genetic females (Okutsu et al. 2006). Recent findings have demonstrated that there are different types of spermatogonia in several fish species. Genetic male zebrafish have several types of spermatogonia including: undifferentiated type A (two types), differentiated type A, type B (early), and type B (late). When undifferentiated type A spermatogonia divide, one daughter cell remains an undifferentiated type A spermatogonium, while the other becomes a differentiated type A spermatogonium. The latter gives rise to type B spermatogonia, which in turn give rise to spermatocytes (Leal et al. 2009). Based on the findings of Okutsu et al. (2006) regarding the lability of rainbow trout spermatogonia, it appears that spermatogonia with stem cell properties may also have the ability to dedifferentiate and give rise to oogonia. In hermaphroditic teleosts in which male sex cells are not evident prior to the development of male function (i.e., interim ovotestis and secondarily derived ovotestis patterns in gobiids), the source of newly-developed spermatogonia may be either embryonic stem cells or undifferentiated type A spermatogonia which are present in the gonad during the initial ovariform phase. Among serial hermaphrodites with alternating ova and sperm production, oocytes may arise from retained oogonia, embryonic stem cells, or type A spermatogonia.

The combination of extended feminization (likely mediated by prolonged aromatase upregulation and E_2 production) and the maintenance of a source of bipotential gonia or embryonic stem cells that are capable of differentiating into either oogonia or spermatogonia is sufficient to generate all known gobiid hermaphroditic sexual patterns. Hermaphroditic members of the Coryphopterus group (i.e., *Coryphopterus, Rhinogobiops, Lophogobius,* and *Fusigobius*) all express an interim ovotestis pattern in which initial ovariform development is either widespread or universal, and future male potential is not evident within the gametogenic tissue prior to sex change. The subsequent, unidirectional shift to a male phase is accompanied by the loss of ovarian features and female gametocytes such that healthy, late-stage male and female sex cells never co-occur. A similar developmental pattern is found during early gonad ontogeny in zebrafish in which an ovariform gonad develops in all individuals regardless of genetic sex. The initial development of oocytes within the zebrafish indifferent gonad which transforms it into an ovariform gonad is associated with an enhanced expression of *vasa*-dependent green fluorescent protein (EGFP)(XG Wang et al. 2007). Subsequently, among genetic males there is a drop in *vasa*-dependent EGFP signal concurrent with the first appearance of spermatocytes, which transforms the ovariform gonad into a testiform gonad (XG Wang et al. 2007). Thus, zebrafish exhibit the same ontogenetic pattern of interim ovotestis development seen in the Coryphopterus group, with the exception that the latter extend the ovariform phase in most or all individuals beyond first maturity, thereby generating a functional hermaphroditic sexual pattern. In both zebrafish and hermaphroditic members of the Coryphopterus group, a shift from ova to sperm production likely involves the downregulation of aromatase pathways such as *FOXl2* that maintain ovariform expression while upregulating pathways that initiate testiform development and expression (i.e., testis-determining genes such as *DMY, Dmrt1* and *SOX9* variants)(Kobayashi et al. 2004; Yoshinaga et al. 2004; Guiguen et al. 2009).

In the interim ovotestis pattern, all ovarian tissues disappear following testis development. Whether female expression among secondary male protogynous hermaphrodites is completely lost, or only suppressed following the development of male expression, is unknown. In two protogynous wrasses including the three-spot, *Halichoeres trimaculatus,* and the Chinese wrasse, *H. tenuispinis,* direct-developing males can be induced with estrogen treatment to secondarily become female, indicating a female potential among males that normally never function as females (Kojima et al. 2008; Miyake et al. 2008). In rainbow trout, testicular germ cells removed from an adult male and transplanted into the indifferent gonad of genetic males and females differentiate into viable spermatogonia and oogonia, respectively (Okutsu et al. 2006). Taken together, these findings suggest that some form of stem cell or bipotential sex cell may be retained in the gonad of both gonochoric and hermaphroditic species.

In the interim and modified interim ovotestis developmental pattern that is expressed by the Coryphopterus group and *Elacatinus multifasciatus,* the initial gonadal phase appears entirely ovarian. This is followed by the transient presence of an ovotestis until ovarian tissue is completely replaced by testis tissue. The tissue of the secondary testis here arises apparently *de novo* within the field of regressing ovarian tissue (Cole 1983; Cole & Robertson 1988; Cole & Shapiro 1990, 1992). Following the degeneration of oocytes, their supporting cells, and ovarian-specific tissues that marks the end of the transient ovotestis phase, the gonadal tissue of the secondary phase appears entirely spermatogenic. If gonadogenesis regulation in gobiids is similar to that of other teleosts, this likely reflects an end to *FOXl2*, or *FOXl2*-like, upregulation and the initiation of *DMY, Dmrt*, or similar testis-determining gene upregulation. There have been no reports of secondary males naturally reversing sex and returning to a female-active phase within the Coryphopterus group or *E. multifasciatus*, and efforts to reverse the sex-change process by manipulation of the social environment among some *Coryphopterus* species have been unsuccessful (Cole & Robertson 1988; Cole unpublished data). This failure suggests that either oogonial or bipotential gonial cells do not persist past sex change, alternative sources of oogonia are not available; upregulation of aromatase production is blocked, testis-determining genes cannot be downregulated, or that some combination of these is the case.

Members of the hermaphroditic genus *Eviota* exhibit a different developmental pattern in which all individuals develop an ovotestis early in ontogeny and retain it throughout life. Within the ovotestis, early-stage oogenic and spermatogenic tissues and their associated gonocytes are intermingled in close proximity to one another, resulting in a persistent integrated ovotestis pattern. In this type of gonad, aromatase/E_2 and testis-determining genes have to be jointly upregulated, at least to some extent, to simultaneously sustain gametocytes and support cells of both ova and sperm-generating tissues. Consequently, in *Eviota*, these two regulatory pathways are unlikely to have inhibitory effects on one another, as suggested by Nakamoto et al. (2009) for the Luzon rice fish, unless the inhibition is incomplete. The close proximity of intermingled ova-fated and sperm-fated gonocytes suggests that sex-specific regulatory cues for gametocyte differentiation and support tissue proliferation are highly localized and in close proximity to target cells. Germ cell enveloping cells meet the proximity requirement, and as future granulosa and Sertoli cells, their expression of *FOXl2* and *DMY/Dmrt1*, respectively argues for their central regulatory role in functional shifts of sexual expression in hermaphroditic fishes, including gobies.

Among the hermaphroditic genera *Trimma, Bryaninops*, and *Priolepis*, the gonad also develops initially as an ovotestis. However, early-stage oogenic and spermatogenic cells and tissues are physically separated from one another by a connective tissue boundary, resulting in a persistent partitioned ovotestis. In most instances, when one gonadal tissue is active it occupies much of the gonad while the other is reduced to a small area consisting of early-stage gonocytes and their support cells. The shared persistent ovotestis features of *Trimma*, *Bryaninops*, and *Priolepis* suggest shared patterns of regulatory gene expression.

Given the close proximity of healthy oocytes and spermatocytes in the *Eviota* pattern, it seems unlikely that partitioning is requisite for ovotestis development in *Trimma*, *Bryaninops*, and *Priolepis*. Consequently, the function of partitioning is unclear. Its presence, however, may have facilitated the apparent addition of an interim functional bisexual stage to the gonad developmental sequence in *Priolepis hipoliti* and *P. eugenius*. In these two species, mature, healthy oocytes and free spermatozoa co-occur for an undetermined period of time (Cole 1990, this volume). Consequently, a putative functional bisexual phase exists in addition to the strictly ova-producing and sperm-producing phases in *Priolepis hipoliti* and *P. eugenius*, which is novel among currently known sexual patterns of hermaphroditic gobiid taxa.

Gobiodon, which has a secondary ovotestis, initially mimics the interim ovotestis pattern of the Coryphopterus group. However, ovigerous tissue is retained during the sperm-producing phase, resulting in the formation of a secondary ovotestis. In this case, the testis-determining gene(s) may act to downregulate aromatase activity and E_2 production, leading to the development of a secondary ovotestis. Consequently, the Gobiodon developmental pattern of secondary ovotestis formation may reflect a combination of sequential and serial gene expression patterns associated with several other hermaphroditic goby developmental patterns. For example, the initial, strictly ovariform gonad development that persists through the juvenile and early maturation phases is similar to the interim ovotestis pattern of the sequentially protogynous Coryphopterus group. In *Gobiodon*, however, when testicular tissue develops after a period of ova production, oogenic tissue does not disappear. Instead, early-stage oocytes and supporting cells persist scattered throughout the testicular tissue, much like the integrated ovotestis of *Eviota*, which in *Gobiodon* results in a secondary ovotestis. Subsequently, the alternation of ova-producing and sperm-producing phases reflects shifts between the upregulation and downregulation of gene expression associated with these two gonadal phases. The serial expression of male and female function that is characteristic of *Gobiodon* likely involves maintaining a balance between the activities of testis-determining and aromatase-upregulating genes, which results in the generation of two functional gonad states in an alternating fashion. Such possibilities as those described above remains to be tested.

Origins of Accessory Structures

The most conspicuous and consistent synapomorphy of the reproductive complex of gobiid fishes is the presence among males of accessory secretory structures (i.e., sperm duct glands). These structures, described as "expanded leaflike hyaline appendages" are formed in most gobiids by the elaboration of the sperm duct wall (Miller 1984, 1992). In the majority of known hermaphroditic goby taxa however, the reproductive complex of male-active fishes does not have associated sperm duct glands *sensu* Miller (1984). Instead, lobulated secretory structures having a similar appearance and apparent function as sperm duct glands arise from small tissue masses associated with the ovariform gonadal wall during a transition to male function (Cole 1988; Cole & Robertson 1988; Cole & Shapiro 1992). In these instances, the resulting secretory lobules are referred to as accessory gonadal structures or AGS, and the tissue masses they arise from are referred to as precursive accessory gonadal structures, or pAGS. The pAGS are typically present throughout both the immature and female-active phases prior to sex change.

Neither the sperm duct glands of gonochore gobies nor the AGS of hermaphroditic gobies normally exhibit male gametogenic tissue or gametocytes. In spite of this, there are several sources of evidence that suggest a close ontogenetic affinity between the gametogenic regions of the gonad and the non-gametogenic regions consisting of specialized tissues and structures of the male gobiid reproductive complex. Testicular tissue, sperm duct glands, secretory and storage AGS, and AGdS are anatomically similar in that all are lobulated structures. Testicular tissues are organized into seminiferous lobules that extend along the longitudinal axis of the gonad. As discussed in Cole (this volume), both sperm duct glands of gonochore goby taxa and the AGdS found in *Gobiodon* are made up of longitudinally oriented secretory lobules that are associated with the gonoduct. The secretory AGS of hermaphroditic gobiids, which arise from gonad-associated pAGS, appear similar to both sperm duct glands and AGdS in their lobular structure and cytology. And the storage AGS of *Gobiodon,* which arise from the posterior region of the gonad proper, consist of longitudinally oriented lobules that are lined with a thin squamous epithelium.

In addition to shared lobular architecture, testicular tissue and the AGS of hermaphroditic gobiids also have a positional affinity. pAGS, which are an ovariform feature of all gonad developmental patterns of hermaphroditic gobies excepting that of *Gobiodon*, develop from the ventro-caudal region of each gonadal lobe just anterior to the common genital sinus. In the partitioned ovotestis of *Trimma, Priolepis,* and *Bryaninops*, the narrow region occupied by spermatogenic tissue during the juvenile and ova-producing phases also comprises the ventro-lateral margin of the ovotestis lobe (Cole 1990, this volume). And in the regionalized ovotestis of *Lythrypnus,* one of two centers of spermatogenic tissue originates in the same ventrolateral location (Cole, this volume). Collectively, these findings suggest that the ventro-lateral portion of the gonadal lobe in hermaphroditic gobies is a reactive site for the future development of testis-associated tissues and structures.

Lastly, there may be a close affinity between gametogenic and AGS regions of the male reproductive complex of hermaphroditic gobies based on ontogenetic origins and ontogenetic potential. In the gonochore goby, *Bathygobius soporator*, the surgical removal of the testis lobes from adult males resulted in the ectopic development of spermatogenic seminiferous lobules originating from the sperm duct glands (Tavolga 1955). These newly formed testis lobes arising from the body of the sperm duct gland contained all stages of male sex cells, including mature sperm. They were also functionally associated with the sperm duct gland, as evidenced by the presence of mature sperm within sperm duct gland lobule lumina. Thus, under certain conditions, sperm duct gland tissue of *B. soporator* appears competent to differentiate into functional spermatogenic tissue.

The source of spermatogonia found in the regenerated testis tissues of *B. soporator* is unknown. Tavolga (1955) thought that the removal of testis tissue may have been incomplete and that residual spermatogonial cells subsequently migrated into the AGS to induce the formation of spermatogenic support tissues. However, subsequent findings on the lability of germ cells associated with the teleost gonad suggest a possible alternative explanation. When 85 juvenile kokanee salmon (*Onchorhynchus nerka*) were gonadectomized and followed over a period of nine years, 95% of 59 males and 38% of 26 females had regenerated fully functional gametogenic tissue including mature ova among females and spermatozoa among males (Robertson 1961). Similarly, in an experimental group of grass carp, *Ctenopharyngodon idella*, ten months after the removal of the gonads and surrounding mesenteries in their entirety, all but one male and one female of the 30 surviving fish showed complete gonad regeneration. Most were producing either mature ova or releasing seminal fluid, the latter presumed to be laden with sperm (Underwood et al. 1986).

In the case of experimental male *B. soporator* described by Tavolga (1955), bipotential gonial cells or embryonic stem cells capable of differentiating into spermatogonia may reside within the AGS, the posterior region of the gonoduct, or both. Under appropriate conditions, these cells can be upregulated and provide the source of cells and tissues required for gonad regeneration. In the reproductive complex of hermaphroditic gobies, the ventrolateral portion of the gonadal lobes may have similar developmental capabilities. The retention of bipotential or embryonic stem cells in this region following initial gonad differentiation provides numerous options for differing gonad ontogeny patterns. Depending on clade-specific patterns of gene expression associated with gonad development, either pAGS, spermatogenic tissues, or in the case of the persistent regionalized and partitioned ovotestis, both can be formed in this region. A test of this possibility would be to remove all gonadal tissues, excepting undifferentiated pAGS, from the reproductive complex of pAGS-expressing hermaphroditic gobies. If the pAGS region contains bipotential or gonial stem cells, any regeneration of gonadal tissue would arise from the pAGS tissue. Depending on whether the regeneration occurred in an aromatase/E_2, or a *DMY/Dmrt1*, dominated environment, the newly formed gonad would presumably be either ovariform, or testiform, respectively.

In the past, it has been presumed that sperm duct glands of gonochore gobies and secretory AGS of hermaphroditic gobiids are not homologous (Cole & Robertson 1988). However, the findings of Tavolga (1955), Robertson (1961) and Underwood et al. (1986), in combination with the morphological similarities of spermatogenic tissue, sperm duct glands, AGS (both secretory and storage) and AGdS of hermaphroditic gobies suggest a much closer relationship. This has two interesting implications. The first is that the morphogenesis of lobulated structures of the reproductive complex may all be regulated by the same gene (or genes) being expressed in different regions of the reproductive complex. These include testis lobules, secretory AGS and AGdS lobules, and storage AGS lobules. The second is that most, if not all, regions of the reproductive complex are competent to form lobulated structures. If so, differences in the location of, or ontogenetic timing for, the development of the various lobulated structures that distinguish different gonad ontogeny patterns may simply reflect differences in temporal and regional patterns of gene expression.

In gonochore goby taxa, the upregulation of sperm duct gland development is localized to a region of the gonoduct and only occurs in males. In *Gobiodon*, the upregulation of AGdS formation is also localized to a region of the gonoduct but takes place early in ontogeny in all individuals, presumably in the presence of elevated aromatase/E_2 levels associated with ovariform gonad development. In other hermaphroditic goby taxa, the localization of pAGS and subsequent AGS development to the region of the posterior, ventrolateral portion of the gonadal lobes suggests a similar localization of either the presence of, or competence to respond to, associated developmental induction cues.

SUMMARY

Although little is known regarding the early development of germ cells and the gonadal anlagen among gobiids, what is known conforms to that of other teleosts. Consequently, general features of teleost gonadogenesis and development of the reproductive complex may be representative of similar processes occurring among gobiids. A number of morphological modifications have likely been instrumental in the evolution of various hermaphroditic patterns among gobies. Simple unidirectional hermaphroditism as expressed by the Coryphopterus group and *E. multifasciatus* may have involved the least number of ontogenetic alterations and been accomplished by a simple heterochronic shift in gonochore gonadogenesis events. As a consequence, instead of regulatory pathways becoming canalized along one of two mutually exclusive pathways, ovariform gonad development and gonia differentiation associated with higher levels of aromatase and E_2 occurs either in most, or all, individuals and persists into adulthood. Subsequently, the disappearance of oocytes, ova-producing tissues and ovariform features occurs in tandem with testiform gonad development and spermatocyte differentiation. In this manner, a pre-existing male pattern of early ovariform ontogeny followed by a shift to testiform development, as found in zebrafish and some other fish species, becomes universalized to all individuals, and produces a protogynous, unidirectional hermaphroditic sexual pattern.

Among goby taxa having a persistent ovotestis, the upregulation of both oocyte and spermatocyte development early in gonad differentiation is maintained throughout life. In the subsequent serial expression of male and female function, alternation of oogenic and spermatogenic function may be mediated through respective regulatory oscillations (i.e., possible alternation of *FOXl2*/gonadal aromatase and *DMRT1/SOX9* expression [Kobayashi et al. 2004]). In the case of the persistent ovotestis pattern, partial downregulation resulting in the regression of somatic tissue of one gonad morphology co-occurs with the upregulation of morphogenesis of the other gonadal tissue, and both sex-specific germ cell lines are retained.

FOXl2 has been implicated in ovarian differentiation of *Oryzias luzonensis* (Nakamoto et al. 2009) while the upregulation of *DMY* in medaka (*Oryzias latipes*) leads to testis differentiation (Kobayashi et al. 2004). In the protogynous hermaphrodite, *Epinephelus coioides*, *SOX3* protein is found within differentiating PGCs, oogonia and multiple developmental stages of oocytes of ovarian tissue, suggesting that *SOX3* is responsible, at least in part, for oogenesis (Yao et al. 2007). However, among males, *SOX3* expression is found only in the Sertoli cells of testicular tissue, suggesting a different role in testis differentiation (Yao et al. 2007). In medaka, *Dmrt1* may also be an important regulator of spermatogenesis (Kobayashi et al. 2004). Consequently, in hermaphroditic gobiids, sex cell and gonadal differentiation respectively are likely controlled by differing regulatory pathways. Whether these regulators act in concert or in opposition in hermaphroditic gobies ultimately determines the taxon-specific pattern of reproductive morphogenesis.

The pattern of secondary ovotestis development in *Gobiodon* appears intermediate to that of the Coryphopterus group and of *Eviota*. It differs from the persistent ovotestis pattern in its sole expression of ovariform development throughout the juvenile and early adult stages (i.e., a Coryphopterus group trait). During this period, the pattern of upregulation is likely similar to, or identical with, that of the Coryphopterus group. However, at the point where upregulation of male reproductive tissues coincides with the downregulation and extinction of female reproductive tissues in the Coryphopterus group pattern, female reproductive somatic tissues of the Gobiodon pattern simply regress to small groups of oocytes and support cells distributed throughout the ovotestis, much like the integrated ovotestis of *Eviota*. The oocytes, which are early-stage and relatively few in number during the sperm-producing phase, can subsequently be prompted to resume differentiation. Thus, the underlying regulatory processes governing gonad expression and function in *Gobiodon* shifts to that of the persistent integrated ovotestis pattern in which male and female adult function may be serially expressed.

In a number of *Gobiodon* species, it has been shown that the oocytes and surrounding cells within an ovotestis can be upregulated for a return to ova production (Nakashima et al. 1996; Munday et al. 1998; Cole & Hoese 2001). This form of serial or bidirectional hermaphroditism has been demonstrated in a number of other gobiid genera maintaining an ovotestis, including *Lythrypnus* (St. Mary 1994, 1996, 2000), *Trimma* (Sunobe & Nakazono 1993; Shiobara 2000; Manabe et al. 2007, 2008) and *Paragobiodon echinocephalus* (Kuwamura et al. 1994). Serial hermaphroditism has also been suggested, based on the presence of a persistent ovotestis, for *Priolepis cincta* (Sunobe & Nakazono 1999) and *Bryaninops yongei* (Munday et al. 2002).

The diversity of gonad and gonad-associated morphology described herein suggests that there is considerable ontogenetic lability in the reproductive complex of hermaphroditic gobies. A repeated structural pattern found in the male-active reproductive complex is one of lobulation. The teleost testis is lobule-based with seminiferous lobules (i.e., seminiferous tubules consisting of developing gametes and support tissues) typically making up the majority of the gonad. In hermaphroditic goby taxa, the close anatomical association between the seminiferous lobules and pAGS-derived secretory lobules, as well as their shared architecture, invites speculation of shared ontogenetic origins. Such speculation is strengthened by the findings following gonadectomy in one goby species, *B. soporator*. In gonadectomized males, the sperm duct gland was the site of testis regeneration and the subsequent production of mature, viable sperm. In *Gobiodon*, a portion of the gonad develops into a highly lobulated, non-gametogenic, secretion storage region following a shift to male function. These examples suggest that there is both considerable ontogenetic lability, and broadly distributed competency, of the reproductive complex of hermaphroditic gobies to develop into a diversity of tissues and structures serving diverse reproductive functions.

Among hermaphroditic gobies, the diversity and distinctiveness of different patterns of reproductive ontogeny described here suggest that modifications in ontogenetic processes have taken several different directions across hermaphroditic goby clades. For example, secretory structures secondarily arise from the gonad in some hermaphroditic taxa and from the gonoduct in others. The presence of an ovotestis can be transitory, persistent, or secondary. And the distribution of gametogenic tissues within the ovotestis may be intermingled, regionalized, or completely separated. All of these variations tend to be distributed along clade-specific lines. Based on known developmental processes associated with teleost gonad and germ cell differentiation, small changes in germ cell behavior and ontogenetic processes appear to be sufficient to explain the variety of developmental patterns of gonad ontogeny, despite the extensive morphological diversity among hermaphroditic gobies. In addition, the number of ontogenetic alterations required for the evolution of both hermaphroditic function, and for the considerable morphological diversity of the reproductive complex found among hermaphroditic gobiids, may be relatively small.

ACKNOWLEDGMENTS

Portions of this chapter benefitted greatly from conversations with A.C. Gill, D.W. Greenfield, A.S. Harold, D.F. Hoese, H.K. Larson, P.L. Munday, L.R. Parenti, F. Pezold, C.E. Thacker, J.L. Van Tassel and R. Winterbottom and from insightful review comments from J. Burns and J. Godwin. Specimens were generously provided by D.W. Greenfield, D.F. Hoese, R.C. Langston, H.K. Larson, D.R. Robertson and R. Winterbottom. Additional assistance and support in the field were kindly provided by: L. Orsak, J. Masey (at Christensen Research Institute, Papua New Guinea); A. Hoggett, J. Leis, S. Reader, L.Vail (at Lizard Island Research Station, Australia); and L. Bell, P. Colin, Y. Sadovy (at Coral Reef Research Foundation, Palau). Portions of the research reported on in this chapter were supported by grants from the Smithsonian Institution through the National Museum of Natural History's Caribbean Coral Reef Ecosystems Program (CCRE Contribution no. 889), a Curatorial Fellowship from the Australian Museum and from funds provided by the University of Hawaii at Mānoa.

REFERENCES

Alam MA, Kobayashi Y, Horiguchi R, Hirai T, Nakamura M (2008). Molecular cloning and quantitative expression of sexually dimorphic markers *Dmrt1* and *Foxl2* during female-to-male sex change in *Epinephelus merra*. General and Comparative Endocrinology 157: 75–85.

Andre J, Rouiller C (1957). L'ultrastructure de la membrane nucleaire des ovocytes de l'araignee (*Tegeneraria domestica* Clark). In: Sjostrand F, Rhodin J (eds.), *Proceedings of the European Conference on Electron Microscopy, Stockholm, 1956*. Academic Press, New York, pp. 162–164.

Anteunis A (1959). Recherches sur la structure et la développement de l'ovaire et l'oviducte chez *Lebistes reticulatus* (Téléostéen). Archives de Biologie (Liege) 70: 783–809.

Bachman FM (1914). The migration of the germ cells in *Amiurus nubulosus*. Biological Bulletin 26: 351–366.

Baron D, Cocquet J, Xia X, Fellous M, Guiguen Y, Veitia R (2004). An evolutionary and functional analysis of *FoxL2* in rainbow trout gonad differentiation. Journal of Molecular Endocrinology 33: 705–715.

Braat AK, Speksnijder JE, Zivkovic D (1999). Germ line development in fishes. International Journal of Developmental Biology 43: 745–760.

Cole KS (1983). Protogynous hermaphroditism in a temperate zone territorial marine goby, *Coryphopterus nicholsii*. Copeia 1983: 809–812.

Cole KS (1988). Predicting the potential for sex change on the basis of ovarian structure in gobiid fishes. Copeia 1988: 1082–1086.

Cole KS (1990). Patterns of gonad structure in hermaphroditic gobies (Teleostei: Gobiidae). Environmental Biology of Fishes 28: 125–142.

Cole KS (2009). Modifications of the reproductive complex and implications for the reproductive biology of *Gobiodon oculolineatus* (Teleostei: Gobiidae). Environmental Biology of Fishes 84: 261–273.

Cole KS, Hoese DF (2001). Gonad morphology, colony demography and evidence for hermaphroditism in *Gobiodon okinawae* (Teleostei, Gobiidae). Environmental Biology of Fishes 61: 161–173.

Cole KS, Robertson DR (1988). Protogyny in a Caribbean reef goby, *Coryphopterus personatus*: gonad ontogeny and social influences on sex change. Bulletin of Marine Sciences 42: 317–333.

Cole KS, Shapiro DY (1990). Gonad structure and hermaphroditism in the gobiid genus *Coryphopterus* (Teleostei: Gobiidae). Copeia 1990: 996–1003.

Cole KS, Shapiro DY (1992). Gonadal structure and population characteristics of the protogynous goby *Coryphopterus glaucofraenum*. Marine Biology 113: 1–9.

Davies PR, Takashima F (1980). Sex differentiation in common carp, *Cyprinus carpio*. Journal of the Tokyo University of Fisheries 66: 191–199.

D'Cotta H, Fostier A, Guiguen Y, Govoroun M, Baroiller J-F (2001a). Aromatase plays a key role during normal and temperature-induced sex differentiation of tilapia *Oreochromis niloticus*. Molecular Reproduction and Development 59: 265–276.

D'Cotta H, Fostier A, Guiguen Y, Govoroun M, Baroiller J-F (2001b). Search for genes involved in the temperature-induced gonadal sex differentiation in the tilapia, *Oreochromis niloticus*. Journal of Experimental Zoology 290: 574–585.

Doitsidou M, Reichman-Fried M, Stebler J, Köprunner M, Dörries J, Meyer D, Esguerra CV, Leung T, Raz E (2002). Guidance of PGC migration by the chemokine SDF-1. Cell 111: 647–659.

Eddy E (1975). Germ plasm and the differentiation of the germ cell line. International Review of Cytology 43: 229–280.

Essenberg JM (1923). Sex-differentiation in the viviparous teleost *Xiphophorus helleri* Heckel. Biological Bulletin 45: 46–97.

Fenske M, Segner H (2004). Aromatase modulation alters gonadal differentiation in developing zebrafish (*Danio rerio*). Aquatic Toxicology 67: 105–126.

Fernandino JI, Hattori RS, Shinoda T, Kimura H, Strobl-Mazzulla PH, Strüssmann CA, Somoza GM (2008). Dimorphoic expression of *dmrt1* and *cyp19a1* (ovarian aromatase) during early gonadal development in pejerrey, *Odontesthes bonariensis*. Sexual Development 2: 316–324.

Fishelson L (1989). Bisexuality and pedogenesis in gobies (Gobiidae: Teleostei) and other fish, or: why so many little fish in tropical seas? Senckenbergiana Maritima 20: 147–160.

Fishelson L (1991). Comparative cytology and morphology of seminal vesicles in male gobiid fishes. Japanese Journal of Ichthyology 38: 17–30.

Fujimoto T, Kataoka T, Sakao S, Saito T, Yamaha E, Arai K (2006). Developmental stages and germ cell lineage of the loach (*Misgurnus anguillicaudatus*). Zoological Science 23: 977–989.

Gamo H (1961). On the origin of germ cells and the formation of gonad primordia in the medaka, *Oryzias latipes*. Japanese Journal of Zoology 13: 101–115.

Geijsen N, Horoschak M, Kim K, Gribnau J, Eggan K, Daley GQ (2003). Derivation of embryonic germ cells and male gametes from embryonic stem cells. Nature 427: 148–154.

Guiguen Y, Baroiller J-F, Ricordel M-J, Iseki K, McMeel OM, Martin SAM, Fostier A (1999). Involvement of estrogens in the process of sex differentiation in two fish species: the rainbow trout (*Oncorhynchus mykiss*) and a tilapia (*Oreochromis niloticus*). Molecular Reproduction and Development 54: 154–162.

Guiguen Y, Fostier A, Piferrer F, Chang C-F (2009). Ovarian aromatase and estrogens: a pivotal role for gonadal sex differentiation and sex change in fish. General and Comparative Endocrinology 165: 351–558.

Hamaguchi S (1982). A light- and electron-microscopic study on the migration of primordial germ cells in the teleost, *Oryzias latipes*. Cell and Tissue Research 227: 139–151.

Hamaguchi S (1983). Asymmetrical development of the gonads in the embryos and fry of the fish, *Oryzias celebensis*. Development, Growth and Differentiation 25: 553–561.

Hatakeyama R, Shirafuji N, Nishimura D, Kawamura T, Watanabe Y (2005). Gonadal development in early life stages of *Spratelloides gracilis*. Fisheries Science 71: 1201–1208.

Hayashi I (1969). Some observations on the reproductive duct of the Japanese sea bass, *Lateolabrax japonicus* (Cuvier and Valenciennes). Japanese Journal of Ichthyology 16: 68–73.

Howley C, Ho RK (2000). mRNA localization patterns in zebrafish oocytes. Mechanisms of Development 92: 305–309.

Ijiri S, Keneko H, Kobayashi T, Wang D-S, Sakai F, Paul-Prasanth B, Nakamura M, Nagahama Y (2008). Sexual dimorphic expression of genes in gonads during early differentiation of a teleost fish, the Nile tilapia *Oreochromis niloticus*. Biology of Reproduction 78: 333–341.

Kanamori A, Nagahama Y, Egami N (1985). Development of the tissue architecture in the gonads of the medaka, *Oryzias latipes*. Zoological Science 2: 695–706.

Kendall WC (1921). Peritoneal membranes, ovaries, and oviducts of salmonoid fishes and their significance in fish-cultural practices. Bulletin of the United States Bureau of Fisheries 37: 183–208.

Kerkis A, Fonseca SAS, Serafim RC, Lavagnolli TMC, Abdelmassih S, Abdelmassih R, Kerkis I (2007). *In vitro* differentiation of male mouse embryonic stem cells into both presumptive sperm cells and oocytes. Cloning and Stem Cells 9: 535–548.

Kitano T, Takamune K, Kobayashi T, Nagahama Y, Abe S-I (1999). Suppression of P450 aromatase gene expression in sex-reversed males produced by rearing genetically female larvae at a high water temperature during a period of sex differentiation in the Japanese flounder (*Paralichthys olivaceus*). Journal of Molecular Endocrinology 23: 167–176.

Knaut H, Pelegri F, Bohmann K, Schwarz H, Nusslein-Volhard C (2000). Zebrafish *vasa* RNA but not its protein is a component of the germ plasm and segregates asymmetrically before germline specification. Journal of Cell Biology 149: 875–888.

Kobayashi T, Nagahama Y (2009). Molecular aspects of gonadal differentiation in the teleost fish, the Nile tilapia. Sexual Development 3: 109–117.

Kobayashi T, Kajiura-Kobayashi H, Nagahama Y (2000). Differential expression of *vasa* homologue gene in the germ cells during oogenesis and spermatogenesis in a teleost fish, tilapia, *Oreochromis niloticus*. Mechanisms of Development 99: 139–142.

Kobayashi T, Matsuda M, Kajiura-Kobayashi H, Suzuki A, Saito N, Nakamoto M, Shibata N, Nagahama Y (2004). Two DM domain genes, DMY and DMRT1, involved in testicular differentiation and development in the medaka, *Oryzias latipes*. Developmental Dynamics 231: 518–526.

Kojima Y, Bhandari RK, Kobayashi Y, Nakamura M (2008). Sex change of adult initial-phase male wrasse, *Halichoeres trimaculatus* by estradiol-17beta treatment. General and Comparative Endocrinology 156: 628–632.

Komatsu T, Nakamura S, Nakamura M (2006). Masculinization of female golden rabbitfish *Siganus guttatus* using an aromatase inhibitor treatment during sex differentiation. Comparative Biochemistry and Physiology Part C 143: 402–409.

Kondo M, Nanda I, Hornung U, Asakawa S, Shimizu N, Mitani H, Schmid M, Shima A, Schart M (2003). Absence of the candidate male sex-determining gene dmrt1b(Y) of medaka from other fish species. Current Biology 13: 416–420.

Koya Y, Fujita A, Niki F, Ishihara E, Miyama H (2003). Sex differentiation and pubertal development of gonads in the viviparous mosquitofish, *Gambusia affinis*. Zoological Science 20: 1231–1242.

Kroon FJ, Liley NR (2000). The role of steroid hormones in protogynous sex change in the blackeye goby, *Coryphopterus nicholsii*. General and Comparative Endocrinology 118: 273–283.

Kroon FJ, Munday PL, Pankhurst NW (2003). Steroid hormone levels and bi-directional sex change in the coral dwelling goby *Gobiodon histrio* (Teleostei: Gobiidae). Journal of Fish Biology 62: 153–167.

Kroon FJ, Munday PL, Westcott DA, Hobbs J-P, Liley NR (2005). Aromatase pathway mediates sex change in each direction. Proceedings of the Royal Society of London Series B 272: 1399–1405.

Kurokawa H, Aoki Y, Nakamura S, Ebe Y, Kobayashi D, Tanaka M (2006). Time-lapse analysis reveals different modes of primordial germ cell migration in the medaka *Oryzias latipes*. Development, Growth and Differentiation 48: 209–221.

Kuwamura T, Nakashima Y, Yogo Y (1994). Sex change in either direction by growth-rate advantage in the monogamous coral goby, *Paragobiodon echinocephalus*. Behavioral Ecology 5: 434–438.

Lassig BR (1977). Socioecological strategies adopted by obligate coral-dwelling fishes. In: Taylor DL (ed.), Proceedings of the Third International Coral Reef Symposium. Vol. 1, Biology. Rosenstiel School of Marine and Atmospheric Science, Miami, Florida, pp. 565–570.

Leal MC, Cardoso ER, Nóbrega RH, Batlouni SR, Bogerd J, França LR, Schulz RW (2009). Histological and stereological evaluation of zebrafish (*Danio rerio*) spermatogenesis with an emphasis on spermatogonial generations. Biology of Reproduction 81: 177–187.

Lebrun C, Billard R, Jalabert B (1982). Changes in the number of germ cells in the gonads of the rainbow trout (*Salmo gairdneri*) during the first 10 post-hatching weeks. Reproduction Nutrition Développement 22: 405–412.

Lewis ZR, McClellan MC, Postlethwait JH, Cresko WA, Kaplan RH (2008). Female-specific increase in primordial germ cells marks sex differentiation in threespine stickleback (*Gasterosteus aculeatus*). Journal of Morphology 269: 909–921.

Liarte S, Chaves-Pozo E, García-Alcazar A, Mulero V, Meseguer J, García-Ayala A (2007). Testicular involution prior to sex change in gilthead seabream is characterized by a decrease in DMRT1 gene expression and by massive leukocyte infiltration. Reproductive Biology and Endocrinology 5: 20.

Liu Z, Wu F, Jiao B, Zhang X, Hu C, Huang B, Zhou L, Huang X, Wang Z, Zhang Y, Nagamaha Y, Cheng CHK, Wang D (2007). Molecular cloning of doublesex and mab-3-related transcription factor 1, forkhead transcription factor gene 2, and two types of cytochrome P450 aromatase in Southern catfish and their possible roles in sex differentiation. Journal of Endocrinology 194: 223–241.

Maack G, Segner H (2003). Morphological development of the gonads in zebrafish. Journal of Fish Biology 62: 895–906.

Manabe H, Ishimura M, Shinomiya A, Sunobe T (2007). Field evidence for bi-directional sex change in the polygynous gobiid fish *Trimma okinawae*. Journal of Fish Biology 70: 600–609.

Manabe H, Matsuoka M, Goto K, Dewa S, Shinomiya A, Sakurai M, Sunobe T (2008). Bi-directional sex change in the gobiid fish *Trimma* sp.: does size-advantage exist? Behaviour 145: 99–113.

Matsuda M, Nagahama Y, Shinomiya A, Sato T, Matsuda C, Kobayashi T, Morrey CE, Shibata N, Asakawa S, Shimizu N, Hori H, Hamaguchi S, Sakaizumi M (2002). DMY is a Y-specific DM-domain gene required for male development in the medaka fish. Nature 417: 559–563.

Matsuda M, Sato T, Toyazaki Y, Nagahama Y, Hamaguchi S, Sakaizumi M (2003). *Oryzias curvinotus* has DMY, a gene that is required for male development in the medaka, *O. latipes*. Zoological Science 20: 159–161.

Matsuyama M, Torres Lara R, Matsuura S (1988). Juvenile bisexuality in the red sea bream, *Pagrus major*. Environmental Biology of Fishes 21: 27–36.

Meijide FJ, Lo Nostro FL, Guerrero GA (2005). Gonadal development and sex differentiation in the cichlid fish *Cichlasoma dimerus* (Teleostei, perciformes): a light- and electron-microscopic study. Journal of Morphology 264: 191–210.

Miller PJ (1984). The tokology of gobioid fishes. In: Potts GW, Wootton RJ (eds.), *Fish Reproduction: Strategies and Tactics*. Academic Press, London, pp. 119–153.

Miller PJ (1992). The sperm duct gland: a visceral synapomorphy for gobioid fishes. Copeia 1992: 253–256.

Miyake A, Saito T, Kashiwagi T, Ando D, Yamamoto A, Suzuki T, Nakatsuji N, Nakatsuji T (2006). Cloning and pattern of expression of the shiro-au *vasa* gene during embryogenesis and its roles in PGC development. International Journal of Developmental Biology 50: 619–625.

Miyake Y, Fukui Y, Kuniyoshi H, Sakai Y, Hashimoto H (2008). Examination of the ability of gonadal sex change in primary males of the diandric wrasses *Halichoeres poecilopterus* and *Halichoeres tenuispinis*: estrogen implantation experiments. Zoological Science 25: 220–224.

Moiseyeva YB (1983). The development of the gonads of the round goby, *Neogobius melanostomus* (Gobidae) during the embryonic period. Journal of Ichthyology 23: 64–74.

Moore GA (1937). The germ cells of the trout (*Salmo irideus* Gibbons). Transactions of the American Microscopical Society 56: 105–112.

Munday PL, Caley MJ, Jones GP (1998). Bi-direction sex change in a coral-dwelling goby. Behavioral Ecology and Sociobiology 43: 371–377.

Munday PL, Pierce SJ, Jones GP, Larson HK (2002). Habitat use, social organization and reproductive biology of seawhip goby *Bryaninops yongei*. Marine and Freshwater Research 53: 769–775.

Nagasawa K, Takeuchi Y, Miwa M, Higuchi K, Morita T, Mitsuboshi T, Miyaki K, Kadomura K, Yoshizaki G (2009). cDNA cloning and expression analysis of a vasa-like gene in Pacific bluefin tuna *Thunnus orientalis*. Fisheries Science 75: 71–79.

Nakamoto M, Suzuki A, Matsuda M, Nagahama Y, Shibata N (2005). Testicular type *Sox9* is not involved in sex determination but might be in the development of testicular structures in the medaka, *Oryzias latipes*. Biochemical and Biophysical Research Communications 333: 729–736.

Nakamoto M, Matsuda M, Wang DS, Nagahama Y, Shibata N. (2006). Molecular cloning and analysis of gonadal expression of *Foxl2* in the medaka, *Oryzias latipes*. Biochemical and Biophysical Research Communications 344: 353–361.

Nakamoto M, Muramatsu S, Yoshida S, Matsuda M, Nagahama Y, Shibata N (2009). Gonadal sex differentiation and expression of *Sox9a2*, *Dmrt1*, and *Foxl2* in *Oryzias luzonensis*. Genesis 47: 289–299.

Nakamura M, Nagahama Y (1989). Differentiation and development of Leydig cells, and changes of testosterone levels during testicular differentiation in tilapia *Oreochromis niloticus*. Fish Physiology and Biochemistry 7: 211–219.

Nakamura M, Kobayashi T, Chang X, Nagahama Y (1998). Gonadal sex differentiation in teleost fish. Journal of Experimental Zoology 281: 362–372.

Nakamura S, Aoki Y, Saito D, Kuroki Y, Fujiyama A, Naruse K, Tanaka M (2008). *Sox9b/sox9a2*-EGFP transgenic medaka reveals the morphological reorganization of the gonads and a common precursor of both the female and male supporting cells. Molecular Reproduction and Development 75: 472–476.

Nakashima Y, Kuwamura T, Yogo Y (1995). Why be a both-ways sex changer? Ethology 101: 301–307.

Nakashima Y, Kuwamura T, Yogo Y (1996). Both-ways sex change in monogamous gobies, *Gobiodon* spp. Environmental Biology of Fishes 46: 281–288.

Nanda I, Kondo M, Hornung U, Asakawa S, Winkler C, Shimizu A, Shan Z, Haaf T, Shimizu N, Shima A, Schmid M, Schartl M (2002). A duplicated copy of DMRT1 in the sex-determining region of the Y chromosome of the medaka, *Oryzias latipes*. Proceedings of the National Academy of Sciences 99: 11778–11783.

Navarro-Martín L, Blázquez M, Piferrer F (2009). Masculinization of the European sea bass (*Dicentrarchus labrax*) by treatment with an androgen or aromatase inhibitor involves different gene expression and has distinct lasting effects on maturation. General and Comparative Endocrinology 160: 3–11.

Nayernia K, Nolte J, Michelmann HW, Lee JH, Rathsack K, Drusenheimer N, Dev A, Wulf G, Ehrmann IE, Elliott DJ, Okpanyi V, Zechner U, Haaf T, Meinhardt A, Engel W (2006). In vitro-differentiated embryonic stem cells give rise to male gametes that can generate offspring mice. Developmental Cell 11: 125–132.

Okutsu T, Suzuki K, Takeuchi Y, Takeuchi T, Yoshizaki G (2006). Testicular germ cells can colonize sexually undifferentiated embryonic gonad and produce functional eggs in fish. Proceedings of the National Academy of Sciences 103: 2725–2729.

Otake H, Shinomiya A, Kawaguchi A, Hamaguchi S, Sakaizumi M (2008). The medaka sex-determining gene DMY acquired a novel temporal expression pattern after duplication of DMRT1. Genesis 46: 719–723.

Otani S, Maegawa S, Inoue K, Arai K, Yamaha E (2002). The germ cell lineage identified by vas-mRNA during the embryogenesis in goldfish. Zoological Science 19: 519–526.

Parker JB (1942). Some observations on the reproductive system of the yellow perch (Perca flavescens). Copeia 1942: 223–226.

Parmentier HK, Timmermans LPM (1985). The differentiation of germ cells and gonads during development of carp (Cyprinus carpio L.). A study with anti-carp sperm monoclonal antibodies. Journal of Embryology and Experimental Morphology 90: 13–32.

Patiño R, Davis KB, Schoore JE, Uguz C, Strüssmann CA, Parker NC, Simco BA, Goudie CA (1996). Sex differentiation of channel catfish gonads: normal development and effects of temperature. Journal of Experimental Zoology 276: 209–218.

Pelegri F (2003). Maternal factors in zebrafish development. Developmental Dynamics 228: 535–554.

Pelegri F, Schulte-Merker S (1999). A gynogenesis-based screen for maternal-effect genes in the zebrafish, Danio rerio. In: Detrich W, Zon LI, Westerfield M (eds.), The Zebrafish: Genetics and Genomics, vol. 60. Academic Press, San Diego, pp. 1–20.

Pezold F (1993). Evidence for monophyletic Gobiinae. Copeia 1993: 634–643.

Piferrer F, Guiguen Y (2008). Fish gonadogenesis. Part II. Molecular biology and genomics of sex differentiation. Reviews in Fisheries Science 16(S1): 33–53.

Reichman-Fried M, Minina S, Raz E (2004). Autonomous modes of behavior in primordial germ cell migration. Developmental Cell 6: 589–596.

Richards A, Thompson JT (1921). The migration of the primary sex-cells of Fundulus heteroclitus. The Biological Bulletin 40: 325–348.

Robertson DR, Justines G (1982). Protogynous hermaphroditism and gonochorism in four Caribbean reef gobies. Environmental Biology of Fishes 7: 137–142.

Robertson OH (1961). Prolongation of the life span of Kokanee salmon (Oncorhynchus nerka Kennerlyi) by castration before beginning of gonad development. Proceedings of the National Academy of Sciences 47: 609–621.

Rogers EW (2007). Sexual plasticity in a marine goby (Lythrypnus dalli): social, endocrine, and genetic influences on functional sex. Ph.D. dissertation, Georgia State University, Atlanta.

Rongo C, Broihier HT, Moore L, Van Doren M, Forbes A, Lehmann R (1997). Germ plasm assembly and germ cell migration in Drosophila. Cold Spring Harbor Symposia on Quantitative Biology 62: 1–11.

Saito T, Otani S, Nagai T, Nakatsuji T, Arai K, Yamaha E (2002). Germ cell lineage from a single blastomere at 8-cell stage in shiro-uo (ice goby). Zoological Science 19: 1027–1032.

Saito T, Otani S, Fujimoto T, Suzuki T, Nakatsuji T, Arai K, Yamaha E (2004). The germ line lineage in ukigori, *Gymnogobius* species (Teleostei: Gobiidae) during embryonic development. The International Journal of Developmental Biology 48: 1079.

Satoh N (1974). An ultrastructural study of sex differentiation in the teleost, *Oryzias latipes*. Journal of Embryology and Experimental Morphology 32: 195–215.

Satoh N, Egami N (1972). Sex differentiation of germ cells in the teleost, *Oryzias latipes*, during normal embryonic development. Journal of Embryology and Experimental Morphology 28: 385–395.

Selman S, Wallace RA, Sarka A, Qi X (1993). Stages of oocyte development in the zebrafish, *Brachydanio rerio*. Journal of Morphology 218: 203–224.

Shimizu M, Takahashi H (1980). Process of sex differentiation of the gonad and gonoduct of the three-spined stickleback, *Gasterosteus aculeatus* L. Bulletin of the Faculty of Fisheries, Hokkaido University 31: 137–148.

Shin HS, An KW, Park MS, Jeong MH, Choi CY (2009). Quantitative mRNA expression of *sox3* and DMRT1 during sex reversal, and expression profiles after GnRHa administration in black porgy, *Acanthopagrus schlegeli*. Comparative Biochemistry and Physiology B 154: 150–156.

Shinomiya A, Tanaka M, Kobayashi T, Nagahama Y, Hamaguchi S (2000). The *vasa*-like gene, *olvas*, identifies the migration path of primordial germ cells during embryonic body formation stage in the medaka, *Oryzias latipes*. Development, Growth and Differentiation 42: 317–326.

Shiobara Y (2000). Reproduction and hermaphroditism of the gobiid fish, from Suruga Bay, central Japan. Science Reports of the Museum, Tokai University 2: 19–30.

St. Mary CM (1993). Novel sexual patterns in two simultaneous hermaphroditic gobies, *Lythrypnus dalli* and *Lythrypnus zebra*. Copeia 1993: 1062–1072.

St. Mary CM (1994). Sex allocation in the simultaneous hermaphrodite, the bluebanded goby (*Lythrypnus dalli*): the effects of body size and behavioral gender and the consequences for reproduction. Behavioral Ecology 5: 304–313.

St. Mary CM (1996). Sex allocation in a simultaneous hermaphrodite, the zebra goby *Lythrypnus zebra*: insights gained through a comparison with its sympatric congener, *Lythrypnus dalli*. Environmental Biology of Fishes 45: 177–190.

St. Mary CM (2000). Sex allocation in *Lythrypnus* (Gobiidae): variations on a hermaphroditic theme. Environmental Biology of Fishes 58: 321–333.

Sunobe T, Nakazono A (1990). Polygynous mating system of *Trimma okinawae* (Pisces: Gobiidae) at Kagoshima, Japan with a note on sex change. Ethology 84: 133–143.

Sunobe T, Nakazono A (1993). Sex change in both directions by alteration of social dominance in *Trimma okinawae* (Pisces: Gobiidae). Ethology 94: 339–345.

Sunobe T, Nakazono A (1999). Mating system and hermaphroditism in the gobiid fish, *Priolepis cincta*, at Kagoshima, Japan. Ichthyological Research 46: 103–105.

Suzuki A, Shibata N (2004). Developmental process of genital ducts in the medaka, *Oryzias latipes*. Zoological Science 21: 397–406.

Stromsten FA (1931). The development of the gonads in the goldfish *Carassius auratus* L. Iowa Studies in Natural History 13: 3–45.

Strüssmann CA, Takashima F, Toda K (1996). Sex differentiation and hormonal feminization in pejerrey *Odontesthes bonariensis*. Aquaculture 139: 31–45.

Takahashi H (1977). Juvenile hermaphroditism in the zebrafish, *Brachydanio rerio*. Bulletin of the Faculty of Fisheries, Hokkaido University 28: 57–65.

Takahashi H, Iwasaki Y (1973). The occurrence of histochemical activity of 3-beta-hydroxysteroid dehydrogenase in the developing testes of *Poecilia reticulata*. Development, Growth and Differentiation 15: 241–253.

Takahashi H, Shimizu M (1983). Juvenile intersexuality in a cyprinid fish, the Sumatra barb, *Barbus tetrazona tetrazona*. Bulletin of the Faculty of Fisheries, Hokkaido University 34: 69–78.

Takashima F, Patiño R, Nomura M (1980). Histological studies on the sex differentiation in rainbow trout. Bulletin of the Japanese Society of Scientific Fisheries 46: 1317–1322.

Tavolga W N (1955). Effects of gonadectomy and hypophysectomy on prespawning behavior in males of the gobiid fish, *Bathygobius soporator*. Physiological Zoology 28: 218–233.

Timmermans LPM, Taverne N (1983). Origin and differentiation of primordial germ cells (PGCs) in the rosy barbs, *Barbus conchonius*, (Cyprinidae, Teleostei). Acta Morphologica Neerlando-Scandinavica 21: 182.

Uchida D, Yamashita M, Kitano T, Iguchi T (2004). An aromatase inhibitor or high water temperature induce oocyte apoptosis and depletion of P450 aromatase activity in the gonads of genetic female zebrafish during sex reversal. Comparative Biochemistry and Physiology A 137: 11–20.

Underwood JL, Hestand II RS, Thompson BZ (1986). Gonad regeneration in grass carp following bilateral gonadectomy. The Progressive Fish-Culturist 48: 54–56.

van Winkoop A, Booms GHR, Dulos GJ, Timmermans LPM (1992). Ultrastructural changes in primordial germ cells during early gonadal development of the common carp (*Cyprinus carpio* L., teleostei). Cell and Tissue Research 267: 337–346.

Vizziano D, Randuineau G, Baron D, Cauty C, Guiguen Y (2007). Characterization of early molecular sex differentiation in rainbow trout, *Oncorhynchus mykiss*. Developmental Dynamics 236: 2198–2206.

Vizziano D, Baron D, Randuineau G, Mahè S, Cauty C (2008). Rainbow trout gonadal masculinization induced by inhibition of estrogen synthesis is more physiological than masculinization induced by androgen supplementation. Biology of Reproduction 78: 939–946.

Wang D, Kobayashi T, Zhou L, Nagahama Y (2004). Molecular cloning and gene expression of *FOXl2* in the Nile tilapia, *Oreochromis niloticus*. Biochemical and Biophysical Research Communications 320: 83–89.

Wang D-S, Kobayashi T, Zhou L-Y, Paul-Prasanth B, Ijiri S, Sakai F, Okubo K, Morohashi K-i, Nagahama Y (2007). *Foxl2* up-regulates aromatase gene transcription in a female-specific manner by binding to the promoter as well as interacting with Ad4 binding protein/steroidogenic factor 1. Molecular Endocrinology 21: 712–725.

Wang XG, Bartfait R, Sleptsova-Freidrich I, Orban L (2007). The timing and extent of "juvenile ovary" phase are highly variable during zebrafish testis differentiation. Journal of Fish Biology 70 (SA): 33–44.

Watts M, Pankhurst NW, King HR (2004). Maintenance of Atlantic salmon (*Salmo salar*) at elevated temperature inhibits cytochrome P450 aromatase activity in isolated ovarian follicles. General and Comparative Endocrinology 135: 381–3990.

Weidinger G, Wolke U, Köprunner M, Raz E (1999). Identification of tissues and patterning events required for distinct steps in early migration of zebrafish primordial germ cells. Development 126: 5295–5307.

Weidinger G, Wolke U, Köprunner, M, Thisse C, Thisse C, Raz E (2002). Regulation of zebrafish primordial germ cell migration by attraction towards an intermediate target. Development 129: 25–36.

Wu G-C, Tomy S, Nakamura M, Chang C-F (2008). Dual roles of *cyp19a1a* in gonadal sex differentiation and development in the protandrous black porgy, *Acanthopagrus schlegeli*. Biology of Reproduction 79: 1111–1120.

Xia W, Zhou L, Yao B, Li C-J, Gui J-F (2007). Differential and spermatogenic cell-specific expression of DMRT1 during sex reversal in protogynous hermaphroditic groupers. Molecular and Cellular Endocrinology 263: 156–172.

Xu H, Gui J, Hong Y (2005). Differential expression of vasa RNA and protein during spermatogenesis and oogenesis in the Gibel carp (*Carassius auratus gibelio*), a bisexually and gynogenetically reproducing vertebrate. Developmental Dynamics 233: 872–882.

Yamaguchi T, Yamaguchi S, Hirai T, Kitano T (2007). Follicle-stimulating hormone signaling and Foxl2 are involved in transcriptional regulation of aromatase gene during gonadal sex differentiation in Japanese flounder, *Paralichthys olivaceous*. Biochemical and Biophysical Research Communications 359: 935–940.

Yao B, Zhou L, Wang Y, Xia W, Gui J-F (2007). Differential expression and dynamic changes of SOX3 during gametogenesis and sex reversal in protogynous hermaphroditic fish. Journal of Experimental Zoology 307A: 207–219.

Yoon C, Kawakami K, Hopkins N (1997). Zebrafish *vasa* homologue RNA is localized to the cleavage planes of 2- and 4-cell-stage embryos and is expressed in the primordial germ cells. Development 124: 3157–3166.

Yoshikawa H, Oguri M (1979). Gonadal sex differentiation in the medaka, *Oryzias latipes*, with special regard to the gradient of the differentiation of the testes. Bulletin of the Japanese Society of Scientific Fisheries 45: 1115–1121.

Yoshinaga N, Shiraishi E, Yamamoto T, Iguchi T, Abe S, Kitano T (2004). Sexually dimorphic expression of a teleost homologue of Müllerian inhibiting substance during gonadal sex differentiation in Japanese flounder, *Paralichthys olivaceus*. Biochemical and Biophysical Research Communications 322: 508–513.

Yoshizaki G, Sakatani S, Tominaga H, Takeuchi T (2000). Cloning and characterization of a *vasa*-like gene in rainbow trout and its expression in the germ cell lineage. Molecular Reproduction and Development 55: 364–371.

7

Fertilization in Marine Fishes

A Review of the Data and Its Application to Conservation Biology

Christopher W. Petersen and Carlotta Mazzoldi

Fertilization in fishes appears to be a relatively simple phenomenon to discuss. Individuals live in an aquatic medium, so sperm and eggs can be broadcast into the environment with spawning partners releasing gametes synchronously in close proximity. This suggests that fertilization success (the proportion of eggs fertilized) will be high in marine fishes, that ejaculates do not need special modifications, and that male ejaculates with sperm and seminal fluid similar to blood plasma would be adequate to allow for successful fertilization. In the early seminal papers on social and mating system evolution, sperm production costs were seen as trivial, and it was hypothesized that enough sperm would be produced by a single male to fertilize all of a female's clutch (e.g., Trivers 1972, p. 138).

This simplified scenario ignores a significant arena in which tradeoffs in reproductive allocation may play out. By viewing sperm production, morphology, and behavior as elements of allocation to reproduction in males to maximize lifetime reproductive success, with explicit tradeoffs among the ways males can partition their resources, a much clearer picture of fertilization and the evolution of reproductive traits in fishes emerges.

There are several places where tradeoffs may occur in fish fertilization strategies. Although maximizing the number of spawns a male takes part in might be seen as maximizing reproductive success, individuals also maximize reproductive success by reducing the number of spawns they take part in but increasing their proportional fertilization in those spawns by excluding other males. This appears to have led to a range of allocation patterns to testicular tissue and sperm production, both interspecifically and intraspecifically, with some males defending females or spawning sites while maintaining low levels of sperm production, and some males having high sperm production associated with chronic sperm competition (Stockley et al. 1997; Petersen & Warner 1998; Taborsky 1998). Another area for potential tradeoffs is the composition of the ejaculate of each spawn; males release not just sperm but other substances, some of which affect sperm performance. Additionally, individual sperm may vary in either the quantity of provisioned energy or their morphology and behavior, where characteristics like sperm swimming speed or longevity might be expected to trade off against one another (Snook 2005).

Given the diversity of fish reproductive biology, we expect that these patterns might play out differently for species with different types of fertilization (internal vs. external) and different types of eggs (demersal vs. pelagic). In addition, the degree of sperm competition should also affect tradeoffs in reproductive allocation.

In this chapter, we examine fertilization mode and fertilization success from the perspective of behavioral and evolutionary ecology. We start with an overview of the dynamics of fertilization in fishes, and then discuss how fertilization success has been measured in the field. We then summarize data on marine fishes with external fertilization (pelagic and demersal egg species) and internal fertilization. Following that overview, we examine the evolution of male ejaculates and examine how different ecological circumstances may select for different allocation strategies. Finally, we ask how this information interacts with and informs applied areas such as conservation biology and exploited species, such as Atlantic cod and large sex-changing tropical species such as groupers. This chapter will not include several very interesting and important areas of the field of fertilization; in particular, we do not discuss the cellular mechanisms of fertilization (e.g., Amanze & Iyengar 1990; Yanagimachi et al. 1992; Munehara et al. 1997; and for review on external fertilizers Cosson et al. 2008a, b). We also restrict our analysis to bony fishes.

FERTILIZATION DYNAMICS IN MARINE FISHES

In the behavioral ecology and evolutionary biology literature, the first authors to suggest that fertilization might be less than 100% in fishes were Nakatsuru & Kramer (1982). In this study, they showed that when male lemon tetras (*Hyphessobrycon pulchripinnis*, Characinidae) were serially exposed to receptive females, fertilization success fell off rather quickly so that after 40 spawning acts fertilization was near zero. Females discriminated against males that had recently spawned suggesting that females did encounter this situation in the field, and had developed behaviors that would increase their fertilization success.

The result poses a question: If sperm are cheap to produce, why do males not always produce enough sperm to fertilize all of a female's clutch? To understand how this quandary is resolved, an informative starting point is a fertilization curve for an externally fertilizing fish (Figure 7.1A). As the number of sperm released in a spawn increases, fertilization increases, but with diminishing returns as number of sperm released increases (Levitan & Petersen 1995; Warner et al. 1995). The marginal return in fitness to the male, the amount of fertilization gained for each incremental increase in sperm production, decreases to near zero as most of the eggs are fertilized (Figure 7.1B). Importantly, as long as the marginal return is above zero, not all of the eggs have been fertilized. Even though sperm are cheap, at some point the returns on increased male gametic investment may be less than fitness returns for alternative allocations of this energy.

Unlike males, female fitness from a spawning event may always be maximized when all of their eggs are fertilized. But even here, potential exceptions exist since females may do better by mating with males that produce higher quality offspring even if there is a fertilization cost, or mating events that give higher fertilization may have higher mortality risks for females. Thus, in many cases fertilization is probably best seen as a form of sexual conflict, in which the male's interests and the female's interests may not completely coincide.

Some work in the past has suggested that there is a threshold number of sperm needed to effect fertilization. We believe that this misconception has probably arisen due to the use of plotting log fertilization curves (Figure 7.2). This type of plot is a common way to graphically illustrate the effect of sperm number on fertilization success (e.g., Ginzburg 1968; marine invertebrate literature). This sharp increase after a range of x-values with very little increase gives the impression of a critical sperm concentration. This idea is embedded in the models of Shapiro & Giraldeau (1996). If we replot Figure 7.2, but instead plot number of sperm released on a linear scale we get Figure 7.1A. There is no threshold of sperm needed to begin fertilization, just as there is no guaranteed upper level of sperm release that will always fertilize all of the eggs in a spawning event.

These figures assume that the main factor affecting fertilization success of a batch of eggs is the concentration of sperm around the eggs at the time of spawning. In aquaculture, when determining how to maximize fertilization, researchers often refer to the ratio of sperm to eggs (e.g., Suquet et al. 1995; Rurangwa et al. 1998), thus explicitly including egg concentration into fertilization dynamics. In the previous examples, however, we have been assuming that fertilization success depended exclusively on the number of sperm, and was independent of egg concentration (or number). What is the cause of this difference? For egg number to affect the fertilization success of a spawn, eggs have to be at such a high concentration relative to sperm that by having an egg nearby being fertilized by one sperm (or attracting multiple sperm), it must be reducing the probability of fertilization for nearby eggs. The inclusion of egg number in models of fertilization rate implies that eggs are strongly competing for sperm, and that increasing egg concentration without increasing sperm number would decrease fertilization success.

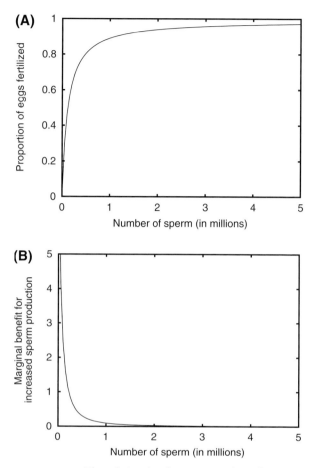

FIGURE 7.1. The relationship between number of
sperm released and (A) the proportion of eggs fertilized
and (B) the marginal benefit for producing more
sperm at different sperm release levels. The fertilization
curve used in this calculation is proportion of eggs
fertilized = (0.994 × number of sperm released in millions) /
(0.117 + number of sperm released in millions). It is the curve
used to fit data for bluehead wrasse (*Thalassoma bifasciatum*) field data. Equation first used in Warner et al. 1995.

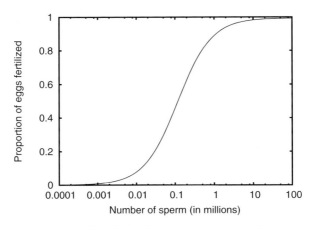

FIGURE 7.2. The relationship between number of sperm released and the proportion of eggs fertilized, plotted with a log-scale x-axis. The equation used is the same as in Figure 7.1.

The inclusion of egg concentrations in fertilization equations is probably warranted when egg concentrations are very high, as will occur in a bowl of eggs being fertilized by sperm, but is probably not important at the concentrations of sperm and eggs seen in natural environments. The exclusion of egg concentration from field estimates of fertilization success is supported by statistical analysis of field data in the best-studied example (Table 1 in Warner et al. 1995). In addition, Kiflawi (2000) modeled both types of scenarios in his detailed theoretical treatment of fertilization curves, and found that the data did not strongly support the results predicted from fertilization dynamics dependent on egg concentration. All of this suggests that sperm concentration will be the most important causal factor affecting fertilization success in nature.

METHODS FOR MEASURING FERTILIZATION SUCCESS IN THE FIELD

Measuring fertilization success in a marine fish species that spawns pelagic eggs is conceptually an easy task. Imagine that all eggs float post spawning, that spawning is predictable in time and in space, that fertilized and unfertilized eggs exhibit similar behavior in the water column, and that collecting does not damage eggs or interfere with fertilization. It would also be convenient if all eggs could easily be identified to species when multiple species synchronize spawning spatially and temporally. Then, by putting a plankton net downcurrent of a spawning aggregation, you could estimate fertilization for a species without too much difficulty. In fact, many of these conditions are probably met for some species, and some investigators have used similar techniques to estimate fertilization success (Colin 1983; Howell et al. 1991). The fact that sperm in at least the one tropical reef fish where it has been measured have a very short span of time that they are capable of fertilizing eggs (<1 minute) reduces many of the potential biases of taking eggs from the environment soon after spawning (*T. bifasciatum*, Petersen et al. 1992).

Unfortunately, nets, especially more durable nitex netting used in plankton nets, damage eggs collected immediately after spawning in a wide range of species (Markle & Waiwood 1985; Petersen et al. 1992; Marconato et al. 1997; Petersen pers. obs.). One alternative to nitex nets that has been used is softer hand held nylon aquarium nets (Petersen et al. 1992) although these nets have also been shown to create damage to eggs, biasing estimates of fertilization success (Marconato et al. 1997).

The methodology that appears to produce the least amount of bias in estimating fertilization success in spawns is collecting the eggs and the water around the spawn in a large plastic bag that is then taken to an above-water platform where samples can be taken for egg and sperm counts and for estimating fertilization. This technique was first developed by Andrea Marconato and has been used successfully in different species of small reef fishes (Shapiro et al. 1994; Marconato et al. 1995, 1997; Warner et al. 1995; Marconato & Shapiro 1996; Petersen et al. 2001). Collecting eggs in a bag after the spawn creates two logistical challenges. First, the diver needs to be relatively close to the spawning event to successfully sample eggs before they disperse in the water column, so only species that allow relatively close approach by divers can be collected with good reliability. Second, the bag needs to be taken to a boat, where a sample of eggs (and potentially sperm) can be taken. This requires more labor than the net technique. The potential for bags to produce biases in measurements of fertilization success was studied extensively by Marconato et al. (1997). They concluded that the bag used immediately after spawning caused a slight decrease in estimates of fertilization success (0.2% to 3.8% reported in Marconato et al. 1997).

In species with demersal eggs, eggs do not disperse and fertilization can be measured hours after fertilization when eggs have hardened and are less vulnerable to handling. Although few studies report fertilization success for marine fishes with demersal eggs, fertilization success appears to be very close to 100% (Petersen & Warner 2002). One possible bias in the estimation of fertilization rate hours after egg deposition comes from parental cannibalism. Cannibalism by the parental male has been recorded in several demersal spawners with male egg care, and part of this cannibalism is performed on unfertilized eggs (Manica 2002). If such cannibalism occurs in a short time after fertilization, fertilization rate would be overestimated.

MODES OF FERTILIZATION

Among marine fishes, patterns and modes of spawning and fertilization are characterized by an amazing degree of diversity; however, we have found it convenient to divide our summary and discussion of fertilization into three categories, fertilization of pelagic eggs, fertilization of demersal eggs, and internal fertilization. Much of the work is focused on a few well-studied species, including the Caribbean bluehead wrasse, *Thalassoma bifasciatum*, the gobies of the Mediterranean and the sculpins (Cottidae) of the Pacific.

Fertilization Success in Species with Pelagic Eggs

For tropical fishes with pelagic eggs, there have been no published accounts of fertilization success in the field since the data were last reviewed by Petersen & Warner (2002). Several of these papers have concluded that sperm limitation exists because there is a strong relationship between either direct measurements of sperm released in a spawn and fertilization success (measured as successful development of embryos), or a relationship between fertilization success and a feature of the mating system believed to correlate with sperm release patterns, such as number of males releasing sperm or the spawning rate of a male. Here we focus on data from bluehead wrasse, *Thalassoma bifasciatum*, and then briefly compare those results with other small tropical species.

Bluehead Wrasse *(Thalassoma bifasciatum).* The bluehead wrasse is one of the most common shallow coral reef fishes in the Caribbean. Fertilization success of this fish in the wild has been studied more than any other marine fish. Bluehead wrasse spawn daily on coral reefs at predictable times, allowing collections of gametes and embryos from individual spawns. In a series of papers, Marconato, Shapiro, Warner, and Petersen have reported on a variety of details of fertilization in populations from Puerto Rico and St. Croix (Petersen et al. 1992; Shapiro et al. 1994; Warner et al. 1995; Marconato et al. 1997; Petersen et al. 2001).

Fertilization success of individual spawning events in bluehead wrasse is high, with most spawns well over 90% and an average of 95.3% (Petersen et al. 2001). The early reports of fertilization averaging 75% by Petersen et al. (1992) appear to be due to damage caused by nets causing reduced fertilization success. However, even with these high average levels of fertilization success, these studies consistently show that sperm limitation does cause a small reduction in fertilization success in some spawns.

These papers have shown that several factors predictably affect fertilization success in bluehead wrasse, and virtually all of them can be related to the concentration of sperm around eggs at fertilization. Fertilization success is positively related to the number of sperm released in a spawn, either when comparing pair spawns (single male spawning with a single female) to group spawns (single female and multiple males spawning in a single spawning rush)(Marconato et al. 1997) or when just examining the patterns within pair spawns (Warner et al. 1995; Petersen et al. 2001). The difference in average fertilization success for pair-spawning males can be considerable, with average male fertilization success ranging from 86 to 99% in St. Croix (Warner et al. 1995; Marconato et al. 1997; Petersen et al. 2001).

The lowest fertilization rates occur for pair spawns involving males with a high daily mating frequency who release less sperm per spawn than males with lower rates of spawning (Warner et al. 1995). Group spawns, which have much higher levels of sperm released per spawn (approximately 50x)(Shapiro et al. 1994), and much lower variation in fertilization success than pair spawns (Marconato et al. 1997), have as high or higher fertilization success than pair spawns on the same reef, and typically average 98% to 99% (Marconato et al. 1997). Within a reef, spawns involving pair-spawning males with the highest mating success (number of matings per day) have an average fertilization success as much as 12% below the average for group spawns on the same reef (Marconato et al. 1997).

At least one physical factor, water flow at a site, also appears to affect fertilization success in a spawn. Sites with higher level of water flow have lower fertilization success for the same amount of sperm released by a male, which would be expected if flow caused gametes to disperse faster (Petersen et al. 1992, 2001).

In addition to these very intuitive results, Petersen et al. (2001) also found that fertilization effectiveness, the fertilization success for a given number of sperm released per spawn, was negatively correlated with female size, and also negatively correlated with the mating success of the male after the number of sperm released per spawn was statistically removed. Since larger females tend to release more eggs per spawn, this might suggest that egg numbers affect fertilization success, but Warner et al. (1995) found that fertilization success was independent of number of eggs released in a spawn. If larger females create more turbulence in their spawns, gametes could be dispersed over a wider area, reducing the sperm concentration and subsequent fertilization success of the spawn, but this interpretation is speculative. The negative correlation between male mating success and fertilization success, controlling for the number of sperm released, could result from males with higher mating success either not being as well positioned with the female during spawning or having sperm of lower quality, but neither of these hypotheses has been tested.

In addition to the descriptive studies, this research group conducted two field experiments to determine how sperm release responded to changes in energy budget and mating success. In the first experiment, Warner et al. (1995) experimentally enhanced the energy budgets of territorial, terminal-phase male bluehead wrasse by feeding them daily for two weeks. They found that individuals increased mating success, but did not increase sperm output per spawn. In a second experiment, Petersen et al. (2001) experimentally changed mating success of individual males by either removing females from a patch reef or removing competing males. When a male's mating success was changed, his fertilization success decreased with increasing mating success, mirroring the natural population data.

Although fertilization success varies in predictable ways, and is less than 100% due to sperm limitation, there is no evidence in *T. bifasciatum* that females preferentially spawn in situations that provide higher fertilization success. None of the patterns that could be used to predict higher fertilization success for specific types of spawns, individual males, or individual sites, appear to be used by females to secure higher fertilization success (Warner et al. 1995; Petersen et al. 2001). The evolution of female choice in this species may be influenced by several selective forces, including not just fertilization success but many factors that influence lifetime reproductive success including mortality rate associated with spawning, male quality, and the energetic or time constraints of getting to a spawning site.

Other Species of Pelagic Spawners.   Although fertilization success has only been measured in the field for a few other species with pelagic eggs, the results from these studies suggest that the picture that emerges from bluehead wrasse may be a common one. In comparisons among spawns in another wrasse, the slippery dick, *Halichoeres bivattatus*, increased fertilization success was found in spawns with streakers (additional males that quickly join pair spawners, releasing sperm in competition with the male-role pair-spawning individual)(Petersen 1991). The addition of streakers presumably increases the total sperm released in a spawn compared to pair spawns. Fertilization success decreased with rougher sea conditions and increased water velocity (Petersen 1991). In the razorfish *Xyrichthys novacula*, Marconato et al. (1995) found high fertilization success (96% to 98%) and higher sperm release per spawn in two males with lower mating success, while a third male that had recently enlarged his harem had much lower sperm released per spawn and lower (87%) fertilization success. In the bucktooth parrotfish, *Sparisoma radians*, Marconato & Shapiro (1996) found fertilization success positively correlated with the amount of sperm released in a pair spawn, and more sperm released in spawns with larger females. The median fertilization success reported in this study was 93%.

One study that did not find differences in fertilization success with changes in the number of sperm presumably released in a spawn is the study by Kiflawi et al. (1998) for the group-spawning surgeonfish *Acanthurus nigrofuscus*. These investigators obtained very high fertilization success measurements, averaging 98% to 99%, for both group spawns (single spawns with multiple males) and mass spawning (multiple group spawns in an areas over a very short period, leaving large and visually persistent gamete clouds). In contrast to previous studies, they also found no effect of ambient current speed on fertilization success. Because all of the spawns in this species involved high levels of sperm competition, the total amounts of sperm released were probably quite high, and at high levels of sperm release, we expect consistently high levels of fertilization success, with little difference over large differences in sperm release (Figure 7.1A)(Petersen & Warner 2002).

Fertilization Success in Species with Demersal Eggs

In pelagic spawners, fertilization is affected by the rapid dispersal and dilution of sperm and eggs in the environment, and consequently, highly synchronous gamete release appears to be a prerequisite in order to achieve high fertilization rates. However, in demersal spawners eggs are attached to a substrate and not dispersed, providing the possibility of asynchronous or extended gamete release. Demersal spawners may scatter their eggs on the open substrate or release a defined clutch, often inside a nest. While in the first case sperm may be rapidly dispersed as in pelagic spawners, reproduction within a nest makes the oviposition site highly predictable and sperm may be retained for a long period inside the nest. This mode of reproduction is exhibited by dozens of different families, often with male parental care (Breder & Rosen 1966; Blumer 1979; Baylis 1981; Gross & Shine 1981; Thresher 1984), and with males occupying and/or building the nest and attracting females.

In one such family, the Gobiidae, asynchronous gamete release has been demonstrated, with males starting sperm release before females begin egg deposition. Such asynchronous gamete release is achieved by releasing sperm embedded in a mucous-enriched seminal fluid on the ceiling of the nest. During the spawning act, males turn upside down, intermittently rubbing the urogenital papilla and releasing sperm on the roof of the nest site (Marconato et al. 1996; Ota et al. 1996). The mucous component of these trails keeps sperm inactive, and sperm activate only when in direct contact with seawater. The mucins of the trails slowly dissolve, liberating active sperm into the seawater for hours (Scaggiante et al. 1999; Rasotto & Mazzoldi 2002). A single trail may release sperm for up to about 30 hours, while sperm in the water may be active for an average of 80 minutes, as observed in the grass goby *Zosterisessor ophiocephalus* (Scaggiante et al. 1999). Sperm reach eggs via the surrounding water, and eggs do not need to be released directly on sperm trails (Scaggiante et al. 1999; Giulianini et al. 2001). In these species females take a very long time (up to several hours) to lay an entire clutch one egg at a time (Mazzoldi et al. 2000). The intermittent gamete release that characterizes the laying of sperm trails frees males from spending all the spawning time releasing sperm and possibly reduces sperm production costs, allowing males to save energy (Mazzoldi et al. 2005).

There are many other families that deposit eggs cared for by males in a nest (Breder & Rosen 1966; Blumer 1979, 1982; Thresher 1984). In families such as the Blenniidae, it has been observed that males rub the urogenital region on the nest ceiling releasing sperm before (Giacomello pers. comm.) and during egg deposition (Giacomello et al. 2006). In some species, males actually spend part of the time outside the entrance while females are laying eggs (Kraak 1996). This fertilization style might be quite common among demersal spawners using enclosed nests.

Eggs of other demersal spawners, such as the cottid *Hemilepidotus gilberti,* are laid in a mass enclosed in the viscous ovarian fluid. In this species, egg deposition takes around one hour, and males start emitting sperm a few minutes after the beginning of oviposition, intermittently quivering the posterior trunk (Hayakawa & Munehara 1996). Fertilization occurs while the eggs are still embedded in the ovarian fluid. After several hours, the ovarian fluid dissolves and the demersal egg mass develops on the bottom (Hayakawa & Munehara 1998).

In mouthbrooding fish, such as cardinalfish, females release eggs, bound in a mass by threads, which are immediately taken in the mouth by males. The partners maintain a close proximity during courtship, turning in a circle side by side. Although Garnaud (1950) hypothesized internal fertilization for *Apogon imberbis,* external fertilization appears more likely for this species (Mazzoldi et al. 2008) as reported for all the other Apogonidae. Fertilization has been hypothesized to occur before the male turns to grab the eggs (Kuwamura 1983; Vagelli 1999). Males often wrap their anal fin around the female's abdomen (Thresher 1984) but given that the entire mating process takes two seconds (Vagelli 1999), males could also gulp sperm together with eggs, and actual fertilization might occur or continue within the male buccal cavity. If this is the case, fertilization would take place in a restricted volume without sperm competition, and consequently few sperm would be needed. An indication of low sperm production is suggested by the low values of the gonadosomatic index (GSI, the proportion of body weight devoted to gonads) reported for cardinalfish males (Okuda 2001). Jawfish, another group of marine mouthbrooders, also appears to have a low male GSI in both monogamous and polygynous species (Hess 1993; Marconato & Rasotto 1993; Hess pers. comm.).

Fertilization within a restricted volume also occurs in some pipefish and seahorses in which males have a brood pouch. In those species, eggs are released inside the male pouch where fertilization takes place (Watanabe et al. 2000; Foster & Vincent 2004). The confined volume where fertilization occurs allows the production of extremely small amounts of sperm, enabling the male to allocate more energy to embryo development (Watanabe et al. 2000; Van Look et al. 2007). In the seahorse *Hippocampus kuda* and the pipefish *Nerophis ophidion,* which does not have a brood pouch, sperm may be activated by ovarian fluid (Ah-King et al. 2006; Van Look et al. 2007).

Although a smaller amount of sperm is required for fertilization in demersal versus pelagic spawners, nonetheless, sperm limitation may occur in these species. The first fish species in which sperm limitation was reported is the freshwater demersal spawner, lemon tetra (Nakatsuru & Kramer 1982). Many species with demersal eggs exhibit high degrees of polygyny, which are correlated with sperm limitation in pelagic species. However, we expect that lower water flow around demersal oviposition sites will increase the probability of contact between sperm and eggs and enhance fertilization success for similar initial sperm concentrations, reducing sperm limitation for most demersal species. Sperm limitation is least likely in species in which males care for just one egg clutch at time, like some mouthbrooders and syngnathids (Kuwamura 1985; Okuda 2001; Foster & Vincent 2004; Kolm & Berglund 2004). In these species, males can devote all of their sperm during an entire brood cycle to just one predictable fertilization event.

In addition, we think it will be much less likely to find sperm depletion in species spawning in enclosed nests such as burrows, cavities, or shells, in which the exchange of water is limited. In addition to limited sperm dispersal, the nest surface often constitutes a limitation to the number of egg clutches a male may receive (e.g., Mazzoldi et al. 2002; Giacomello & Rasotto 2005). For a group like gobies, these factors combined with prolonged sperm release from sperm trails and the predictable and limited degree of polygyny makes sperm depletion unlikely.

For demersal spawners that lay eggs on open substrate, sperm limitation is more likely to occur, but only in some circumstances. If eggs are laid on a definite substrate, as is the case for most pomacentrid species in which males defend a substrate for egg deposition and perform parental care (Thresher 1984), a physical limitation on the number of egg clutches may still occur. In these species, we expect adaptations in the seminal fluid to reduce sperm dispersal, and we do not expect the occurrence of sperm depletion. The limited accounts for groups like pomacentrids suggest fertilization of clutches at or near 100% with little or no sperm limitation (Petersen & Warner 2002). Sperm limitation is most likely in species that have the potential for extended breeding cycles, with individual males spawning with multiple females and no mating constraints due to limited oviposition sites. In demersal spawners, this is most likely in species like the lemon tetra that scatter eggs on the open substrate and exhibit a mating system in which male mating success is not limited by nest size and paternal care is absent.

Fertilization in Internal Fertilizers

Internal fertilization has evolved independently in fishes a large number of times (Wourms 1981; Goodwin et al. 2002). Internal fertilization is less common than external fertilization in marine bony fish, occurring only in a few families spread over several orders: Atheriniformes, Beloniformes, Cyprinodontiformes, Ophidiiformes, Perciformes, and Scorpaeniformes. This mode of fertilization may be associated with viviparity, ovoviviparity, or oviparity. Usually, but not always, internal insemination implies the presence of a copulatory organ in males, devoted to sperm transfer, that is often constituted by a modification of the urogenital papilla to an intromittent organ (Breder & Rosen 1966). Internal fertilization is also associated with a major change in parental care patterns in fishes, with female care predominating whereas male care predominates in parental species with external fertilization (Gross & Shine 1981; Clutton-Brock 1991). With internal fertilization comes a number of changes in fertilization dynamics, including the physical environment at the time of fertilization and the degree or dynamics of sperm competition.

True internal fertilization implies the fusion of gamete nuclei within the female genital tract. However in the family Cottidae a peculiar form of internal insemination occurs, called internal gametic association (IGA). In several species belonging to this family, sperm are transferred to the female oviduct, then remain within the ovarian lumen where they associate with ovulated eggs, with one or more subsequently entering the micropyle (i.e. gametic association). However, penetration of the sperm into the ooplasm and fusion between gamete nuclei occur only after exposure to seawater when ova are released to the external environment (Munehara et al. 1989, 1991; see also Muñoz herein). An unusual plasticity in fertilization mode exists in this family, with species having both types of fertilization: the typical external fertilization and internal gametic association (Munehara 1988; Petersen et al. 2005). In *Alcichthys alcicornis*, a species with internal gametic association (Munehara et al. 1989), males copulate after females lay eggs. This behavior implies that the first egg batch of the season must be fertilized externally by sperm leaking from the females during and after copulation (Munehara 1988). In two other species, *Artedius fenestralis* and *A. lateralis*, although the most common fertilization mode is external, in some cases experimentally extruded clutches of eggs developed into embryos without the external addition of sperm (Petersen et al. 2005), implying internal insemination. While in species with internal gametic association the urogenital papilla is usually modified into a copulatory organ (Morris 1952, 1956; Ragland & Fisher 1987), in the two *Artedius* species mentioned above the urogenital papillae are quite small (Petersen et al. 2005). Although it has not been investigated how sperm are transferred to females in these two species, the possibility to have internal insemination without an obvious copulatory organ has been clearly found in the sculpins *Blepsias cirrhosus* (Munehara et al. 1991) and *Hemitripterus villosus* (Munehara 1996). In this last species, it has been observed that females actually evert the genital duct, producing a jellylike material in which sperm become embedded and which is then drawn back into the female reproductive tract (Munehara 1996).

While fertilization success has not been well investigated in internal fertilizers, it appears to be quite high, and some data exist suggesting long sperm lifetime within female genital tracts. In the cottid *Clinocottus analis*, females may store sperm for two months at least (Hubbs 1966), while in another scorpaeniform species, *Helicolenus dactylopterus dactylopterus*, sperm may be retained for 10 months; after that they are activated by a change in pH, from acid to basic, of the surrounding ovarian fluid (Munõz et al. 2002). In this last species, sperm are stored in crypts in the ovarian lumen that secrete polysaccharides that may provide nutrients for sperm (Munõz et al. 2002).

Postcopulatory sexual selection, consisting of sperm competition (Parker 1970, 1998) and/or cryptic female choice, occurring when female bias sperm use in favor of a certain male (Eberhard 1996), may occur in species with either internal fertilization or internal gametic association. In addition, copulations may be cooperative or coercive, as in the case of the well-studied poeciliids, a family of freshwater fish with internal fertilization and different degrees of coercive, sneaky copulations with respect to cooperative copulations (Constantz 1989; Houde 1997). In cooperative copulations of the guppy *Poecilia reticulata*, males may transfer larger amounts of sperm than during sneaky copulations (Pilastro & Bisazza 1999), suggesting that females may control the number of transferred sperm in at least some spawns and consequently exert cryptic mate choice (Pilastro et al. 2004). However, in several other poeciliid species, a mating advantage has been either shown or implied for small males in sneaky copulations (Bisazza & Pilastro 1997; Pilastro et al. 1997). The complex picture emerging from the well-studied poeciliids suggests that several aspects of postcopulatory sexual selection in marine internal fertilizers deserve greater attention.

EVOLUTION OF MALE EJACULATES

The male ejaculate includes sperm and seminal fluid. At both the interspecific and intraspecific levels, males may differ in total sperm production, sperm concentration, and sperm traits. Sperm traits analyzed in the light of their role in fertilization include morphology, in particular size, velocity, motility and longevity (lifespan), and adenosine triphosphate (ATP) content (reviewed by Snook 2005). Most of the work on variation in seminal fluid and sperm characteristics in fishes has been done on freshwater or anadromous (i.e., marine species that migrate to freshwater to spawn) fishes. Here we give an overview of data on sperm production, male ejaculates, and sperm characteristics in marine fishes, also including some examples from freshwater species when appropriate, with an emphasis on the role of life-history tradeoffs in shaping the high degree of variation in male ejaculates in fishes.

Sperm Production and Sperm Competition

In terms of sperm number and concentration, Stockley et al. (1997), Petersen and Warner (1998), and Taborsky (1998) reviewed male gonadal investment in fishes and found a strong, positive interspecific correlation between the degree of relative testis size, the standing crop of sperm, the volume of seminal fluid, and the estimated degree of sperm competition based on behavioral studies. Many authors have also shown that within species with multiple male mating tactics, males that experience higher levels of sperm competition have higher male allocation as measured by the GSI (e.g., Petersen & Warner 1998). To cope with the high level of sperm competition, males adopting alternative male mating tactics release larger amounts of sperm than do conspecifics that are exposed to lower levels of sperm competition, as predicted by theoretical models (Parker 1990), and this may be achieved by producing ejaculates of larger volumes or with more concentrated sperm.

In species with multiple male mating tactics, large territorial males often spend more time aggressively defending females or mating sites from potential competitors, and have reduced sperm output compared with non-territorial conspecifics. Sperm limitation appears to be more extreme in spawns of these pair-spawning, territorial males (e.g., Marconato et al. 1997), but females do not discriminate against them in favor of group spawns in species like bluehead wrasse. Even when supplemented with food, bluehead wrasse territorial males increase mating success, not fertilization success, apparently by devoting more time to male-male aggression and territorial defense (Warner et al. 1995). Similarly, when spawning in the presence of sneakers in the laboratory, neither territorial grass gobies nor black gobies (*Gobius niger*) increase their sperm release levels, but instead respond with increased aggression towards potential sperm competitors (Scaggiante et al. 2005). However, sneakers in both goby species do respond to differences in perceived sperm competition by changing the amount of sperm they release in a spawn. As the number of sneakers increases they first increase, then decrease the amount of sperm released per spawn (Pilastro et al. 2002), in agreement with the theoretical models for changes in sperm allocation with changes in the intensity of sperm competition (Parker et al. 1996).

In species not subjected to sperm competition and in which few sperm are needed to fertilize eggs, economy in sperm production may be obtained through a low rate of sperm maturation. In semi-cystic spermatogenesis, cysts open in the lumen of the testicular lobules before the completion of sperm maturation, releasing spermatids that complete their maturation along the reproductive duct system (Lahnsteiner et al. 1990; Manni & Rasotto 1997). This form of spermatogenesis causes an asynchronous maturation of the spermatids and, consequently, a low number of sperm are simultaneously mature. Semi-cystic spermatogenesis has been observed in species with demersal eggs having either a monogamous mating system (Rasotto et al. 1992; Marconato & Rasotto 1993; Mazzoldi 2001; Fishelson et al. 2006), or an enclosed nest with low numbers of eggs (Giacomello et al. 2008).

Sperm Production versus Production of Other Seminal Substances

The significance of sperm concentration in the seminal fluid of fishes is not well understood. It is not clear what the benefits are for having sperm concentrated in seminal fluid versus releasing a larger seminal volume to get the same number of sperm released. Interestingly, there is an emerging intraspecific pattern among teleosts of higher sperm concentration in the ejaculate of individuals that have higher sperm numbers. In three species, bluehead wrasse (Schärer & Robertson, 1999) and both black and grass gobies (Mazzoldi et al. 2000; Rasotto & Mazzoldi 2002), males that engage in chronic sperm competition had higher sperm concentrations in their ejaculates compared to males who experienced low levels of sperm competition. Schärer & Robertson (1999) proposed three nonexclusive hypotheses to explain this pattern: (1) that more concentrated sperm in individuals experiencing more intense sperm competition allows for the release of more sperm, faster; (2) milt may have different optimal characteristics (e.g., sperm concentration, viscosity) in pair versus group spawns; or (3) that by having more dilute sperm, males that release small amounts of sperm can have better control of the total amount of sperm released in a spawn. To date there have been no tests done to distinguish among these hypotheses in any fish species with this pattern.

Seminal fluid may include ions, proteins, polysaccharides, and other organic molecules. Seminal fluid may play a key role in shaping sperm performance, in particular prolonging sperm viability, and reducing sperm dispersal, as described in gobies (Marconato et al. 1996; Scaggiante et al. 1999), or in the rainbow trout, *Oncorynchus mykiss* (Lahnsteiner et al. 2004). In species with alternative mating tactics and accessory structures involved in seminal fluid production, intraspecific variability in their development has been found between territorial/parental males and sneaker males. In the studied species, parental males have well developed accessory structures while males that engage in alternative reproductive tactics (sneaking, streaking) have extremely reduced accessory structures (de Jonge et al. 1989; Ruchon et al. 1995; Scaggiante et al. 1999; Barni et al. 2001; Neat 2001; Rasotto & Mazzoldi 2002; Neat et al. 2003) and these males produce reduced amounts of mucins in their seminal fluid. For species that produce sperm trails, parental males produce trails that last for hours and slowly release sperm, while sneakers have trails that dissolve faster in seawater resulting in the release of a large amount of sperm that are immediately active (Scaggiante et al. 1999; Mazzoldi et al. 2000; Rasotto & Mazzoldi 2002). In the one study that performed a comparative analysis of these structures in gobies, however, the development of accessory structures in territorial males was positively correlated with the occurrence of polygynous mating systems, but not to the degree of sperm competition (Mazzoldi et al. 2005). The low number of species included in this latter study does not definitely rule out the possible additional influence of sperm competition, or of other factors, in the development of accessory structures.

Allocating energy away from sperm production and into the production of mucins may be differentially beneficial for territorial males in two ways. First, by allowing for a more constant release of sperm from a slowly dissolving sperm trail, males may be able to fertilize eggs with a lower total sperm production. Second, males may be freed to defend oviposition sites from potential sperm competitors while females are spawning, while still having sperm release taking place at the oviposition site from the sperm trail. For non-territorial males, the ubiquity of sperm competition probably puts a premium on sperm production, and the same accessory structures are reduced and/or used as sperm-storage organs (Scaggiante et al. 1999; Rasotto & Mazzoldi 2002).

In addition to differences in seminal fluid, the ejaculates of internal fertilizers can differ in how sperm are organized in the seminal fluid. In the transfer of sperm from male to female genital tracts sperm can be free in the seminal fluid or organized in encapsulated (spermatophores) or unencapsulated (spermatozeugmata) bundles (Grier 1981). The formation of sperm packets is hypothesized to improve sperm transfer to female genital tracts (Ginzburg 1968). Within bundles, sperm are specifically oriented with heads facing in the same direction, or either arranged around the periphery or the center of the bundle (Grier et al. 1978). Unpacked sperm have been observed in species, such as *Anableps anableps* (Anablepidae), which have a tubular gonopodium that can ensure an efficient transfer of sperm, even if they are not aggregated in bundles (Grier et al. 1981). This seminal fluid may also interact with the female genital tract and/or ejaculates from other males, opening the possibility for the rise of sexual conflict (Simmons 2001; Snook 2005).

The simple structure of the testis in most teleost species with pelagic eggs suggests that these species do not have as complex a seminal fluid as many species with demersal eggs.

Sperm Characteristics

Sperm morphology in fishes has been exhaustively reviewed in two very good books by Jamieson (1991, 2009). At the interspecific level, sperm morphology has been shown to be correlated with the mode of fertilization, with internal fertilizers having sperm with more elaborate, elongate heads (introsperm) than sperm from species with external fertilization that tend to have morphologically simple anacromsomal "aquasperm" (Jamieson 1991; Stockley et al. 1997). In some species, such as cardinalfish, batrachoids, and gobiesocids, sperm may be biflagellate; however the functional role of the presence of two flagella is not clear (Mattei 1988).

One prediction that follows from the level of gametic competition among males is that as the intensity of sperm competition increases in external fertilizers, there should be selection on sperm to increase speed, even if this incurs a cost in sperm longevity (Snook 2005). This idea is supported by studies such as those by Gage et al. (2004) in *Salmo salar* that show that the velocity of a male's sperm is the best predictor of his fertilization success in artificial, competitive spawnings. This has led several authors to look for a positive correlation between sperm competition intensity and sperm size, assuming that larger sperm will be faster swimmers, which appears to be the case in mammals (Gomendio & Roldan 1991). Although a review suggested that the relationship between sperm size and longevity or sperm swimming speed is not common in fishes (Snook 2005), a more recent paper found a strong interspecific relationship between sperm size and speed in a group of freshwater fishes (Fitzgerald et al. 2009). At the interspecific level, two studies have had opposite results, with sperm size negatively related with sperm competition intensity in a broad review of fishes (Stockley et al. 1997), and a positive relationship found among Lake Tanganyika cichlids (Fitzgerald et al. 2009).

At the intraspecific level, in most cases no differences in sperm size have been observed between males facing different levels of sperm competition (Gage et al. 1995; Schärer & Robertson 1999; Locatello et al. 2007), while in the freshwater sunfish *Lepomis macrochirus*, sneaker males had sperm with longer tails compared to sperm from males that defended nest sites (Burness et al. 2004).

Lifespan (the period of viability) of sperm released into saltwater appears to range from well under a minute in bluehead wrasse (Petersen et al. 1992) to days in herring (Ginzburg 1968). Some of this difference is probably due to the very different water temperatures experienced by marine species, with species in colder water having the greater potential for long-lived sperm.

The close juxtaposition of males and females during spawning and the synchrony of gamete release in most species should translate into a quickly diminishing probability for an individual sperm to fertilize an egg. As time from spawning increases, the number of eggs not fertilized from a spawn will decrease, and in a pelagically spawning species, the density of eggs will decrease as well. Given this, we might expect that sperm from species from demersal eggs will have higher longevity than sperm from pelagic species. Indeed, in some demersal spawners with an enclosed nest, prolonged egg deposition, and eggs that remain fertilizable for several hours, long sperm longevity has been recorded. For example, in the grass goby, *Zosterisessor ophiocephalus*, sperm released from trails can last for an average of 80 minutes (Scaggiante et al. 1999). In contrast, Petersen et al. (1992) found that artificial fertilization in the pelagic spawner *T. bifasciatum* was greatly reduced after sperm had spent just 15 seconds in seawater. In some species with pelagic eggs, like cod, sperm are relatively long lived and still have substantial fertilizing ability after one hour (Trippel & Morgan 1994). Species with pelagic eggs and spawning over an extended time in a restricted space may undergo selection for increased sperm longevity. This may help explain the long lifespan of sperm in species like herring, where spawning can occur over an extended time in an area. If fertilization is not simultaneous and immediate for all the eggs, sperm longevity can play an important role in the outcome of sperm competition (Ball & Parker 1996; Snook 2005).

In two species with alternative mating tactics, the grass goby, *Z. ophiocephalus*, and the black goby, *Gobius niger*, grass goby males performing different mating tactics did not differ in any of the studied sperm traits, while in the black goby, sneakers had faster sperm with higher viability and with more ATP content than territorial males (Locatello et al. 2007). These differences were not correlated with sperm size. Such differences between the two species may be due to the differences in nest type and sneaker behavior between the two species (Locatello et al. 2007). In the grass goby, the nest is an enclosed chamber where sneakers may hide and remain for extended periods (Mazzoldi et al. 2000), while in black gobies, the nest is more open, and sneakers are only able to enter for short time before being chased away (Mazzoldi & Rasotto 2002). In the grass goby, territorial males are exposed to prolonged sperm competition and may be forced to maintain as high a quality of sperm as sneakers in order to cope with the high level of sperm competition (Locatello et al. 2007). Differences in sperm quality, with sneakers having longer-lived sperm than guarding males, have also been found in the corkwing wrasse, *Symphodus melops*, another nesting species (Uglem et al. 2001), and in *Salmo salar* (Gage et al. 1995). In the freshwater species *Lepomis macrochirus*, however, sperm viability seems to trade off against speed, and sneakers have a higher proportion of motile sperm after activation, with faster initial swimming speeds but shorter longevity than parental males (Burness et al. 2004). In this species, it has been demonstrated that these differences imply a differential success in fertilization, with parental males having higher fertilization success late in the sperm activation cycle (Schulte-Hostedde & Burness 2005).

The presumed trade-off between sperm speed and longevity assumes a fixed investment per sperm, but at the level of the adult, allocation decisions may often involve tradeoffs between either quality versus number of sperm or investment in sperm production versus other allocations that could increase reproductive success. Thus, we may often find that there is either no correlation or a positive correlation among measures of sperm performance as might be expected when individual sperm differ in their initial energetic investment (Locatello et al. 2007). The most recent studies also suggest that measures of energetic investment in sperm such as ATP content may be a better metric of investment than morphological characteristics such as sperm size.

In another variation of sperm morphology and performance, in some species individual males produce two types of sperm, some of which are functionally able to fertilize (eusperm) and some that are morphologically different and cannot fertilize eggs (parasperm). These eusperm and parasperm have been observed in the sculpin *Hemilepidus gilberti*, an external fertilizer (Hayakawa et al. 2002b). In this species it has been demonstrated that parasperm may play two roles: they can reduce lateral dispersion of the ejaculate, increasing the distance that eusperm may travel (Hayakawa et al. 2002a); and they aggregate in lumps that might constitute a barrier for later arriving eusperm, such as those of sneaker males, thereby reducing the possibility of sperm competition (Hayakawa et al. 2002b).

A final consideration involves the role of ovarian fluid in fertilization. Besides cottid species, ovarian fluid has been demonstrated to influence sperm characteristics in terms of sperm longevity and velocity in species such as *Salmo trutta f. fario* (Lahnsteiner 2002), *Salvelinus alpinus* (Turner & Montgomerie 2002), *Gasterosteus aculeatus* (Elofsson et al. 2003), and *Gadus morhua* (Litvak & Trippel 1998). In *S. alpinus,* ovarian fluid from different females has differential effects on sperm, with ovarian fluid from some females enhancing sperm speed more than that of others. Even more interesting, sperm velocity was also found to vary depending on the individual male-female pair. This result opens the possibility, in species with sperm competition, for the occurrence of cryptic female choice, even in external fertilizers (Urbach et al. 2005).

HOW FERTILIZATION ECOLOGY INFORMS CONSERVATION BIOLOGY

Until very recently, fertilization success was not an explicit part of fisheries models, with the implicit assumption being that fertilization success was complete, or at least a constant, for a species. Among some commercially important species, the recent crashes of populations with very little subsequent recovery, and the documentation of populations with very low proportions of males, has heightened concerns about sperm limitation in exploited species and has led several authors to explicitly include fertilization success in their modeling of the population biology of exploited species. Two cases where the discussion of fertilization has been an explicit part of the conservation biology dialogue have been in Atlantic cod and in sex-changing fishes such as grouper. Both of these groups of fishes suffer from very high levels of exploitation, and both bring unique concerns for the effects of sperm limitation on population dynamics. Each of these two cases is reviewed in detail below.

Atlantic Cod (Gadus morhua)

In temperate marine fishes, the largest amount of work on fertilization ecology and reproductive behavior has been done by researchers on Atlantic cod (*Gadus morhua*). Atlantic cod, particularly off the Atlantic coast of Canada, have been severely depleted (Hutchings & Myers 1994; Myers et al. 1997), leading to severe restrictions of some fisheries in the Gulf of Maine, to complete closures of some fisheries in Atlantic Canada. The slow recovery of these species has led some to hypothesize that the decimation of these populations has led to an Allee effect or depensation, where populations at low densities suffer lower per capita growth rates, and that some of the failure of populations to rebound is caused by reduced fertilization due to reduced male numbers (Rowe and Hutchings 2003; Rowe et al. 2004). Although an interesting suggestion, the authors have made it clear that this is an untested hypothesis, with no empirical data to directly support depensation (Rowe & Hutchings 2003; Hutchings & Reynolds 2004).

Populations of cod have prolonged spawning periods of 6 to 12 weeks (Myers et al. 1993). Spawning in at least some populations has occurred at historically consistent locations, making these spawning aggregations vulnerable to overfishing (Ames 2004). Individuals appear to migrate to spawning sites where large numbers of males and females mate for several weeks (Hutchings et al. 1993; Morgan & Trippel 1996). Spawning occurs over the continental shelf, slightly off the bottom in 50 to 200m of water, where females tend to be found closer to the bottom than males. During the reproductive season, individual trawls typically have sex-biased catches, with a male-bias when individuals of both sexes are most reproductively active (Morgan & Trippel 1996). From these and other field collections, investigators have hypothesized that males form large spawning aggregations, while females visit these aggregations, spawn, and then return to their feeding grounds that in Canada appear to be in deeper and warmer water (Morgan & Trippel 1996).

Spawning in *G. morhua* has never been observed in the field. Despite this handicap, combining the field data with observations of spawning individuals in large tanks, several researchers have built a reasonable if slightly speculative scenario concerning spawning and fertilization ecology of *G. morhua* in the western Atlantic.

Descriptions of spawning from large tanks report very stereotyped pair-spawning behavior of individuals, where a male positions himself directly under a female and the two individuals swim slowly together while releasing gametes (Brawn 1961; Hutchings et al. 1999). However, Hutchings et al. (1999) did report satellite males joining in three spawning events in a large tank containing multiple males and females. Within observation tanks, males have been observed to act aggressively towards one other, and there appeared to be some size-assortative pairing in spawning partners (Bekkevold et al. 2002).

During the reproductive season, male Atlantic cod have relatively large testes (typically around 7% of their total body weight) that are similar in relative size to ovaries of females, implying a mating system with significant amounts of sperm competition (Cyr et al. 1998; Schwalme & Chouinard 1999; Rowe pers. comm.). Fertilization has been studied in a laboratory setting, and paternity of offspring in spawns has been determined using microsatellites as variable genetic markers. In artificial fertilizations, male size does not influence fertilization success (Rakitin et al. 1999). Fertilization in spawns in captivity averaged 91.5% in the one study where it was explicitly reported (Bekkevold et al. 2002), but fertilization was typically above 90% in spawns and only rarely below 80% in any of the reported spawns (see Figures 1 through 3 in Rowe et al. 2004). Based on genetic markers and using groups of eggs from a single female as a proxy for spawning, fertilization success increased when paternity increased from one to two males, but then appeared to level off for cases where more than two males successfully fertilized some of the eggs (Rakitin et al. 2001; Bekkevold et al. 2002; Rowe et al. 2004). In addition, in one of three studies the variance in fertilization success decreased with an increasing number of males in a spawning event. Thus, there may be a fertilization success increase with multiple males in spawns, although the increase appears to be less than 10% for eggs fertilized (Rowe et al. 2004). The arguments from several authors that there is an advantage in terms of genetic diversity to having multiple males in a spawn (cf. Rowe et al. 2004) appears to be minimized by the observation that these fish spawn multiple times in a season, almost undoubtedly with different partners.

In a more recent study, Rowe et al. (2007) showed that in large contained systems, mating success was positively correlated with partner size for both sexes by using genetic microsatellite markers to estimate the relative proportion of offspring produced by individuals in the tank. It was not clear if the increase in success for larger males was due to their higher fertilizing ability of clutches, or due to females releasing more eggs, or spawning more often, when paired with larger males.

Using these observations of fish in captivity to inform how fertilization success might change with increasing exploitation in this species is highly speculative. It is not clear at what point increasing the take of males either leads to sperm limitation directly, by limiting total sperm production in the population, or indirectly, by disturbing the social system in the field, or both. Although lab study results combined with population data in other species have shown the potential for substantial reductions in population growth rate due to decreases in fertilization success with fewer milt donors (Purchase et al. 2007), the application of these results to natural populations is not easily made, and other results using data from bluehead wrasse suggest that this effect will not be very strong under many circumstances (Petersen & Levitan 2001; Petersen & Warner 2002).

Sex Change, Spawning Aggregations, and Fertilization Success

One very simple question in fisheries is whether sex-changing species are more vulnerable to exploitation than gonochorisitc (separate-sexed) species. This is really a comparison of three different types of populations: protogynous (female-first) hermaphrodites, protandrous (male-first) hermaphrodites, and gonochoristic populations. Simultaneous hermaphroditism is rare in fishes and does not occur in any widely exploited fish species. The answer to this question depends on how we estimate the dynamics of fertilization and sex change within species.

Fisheries tend to differentially take larger, older individuals. The effects of fishing on egg production will be most strongly felt in protandrous species, followed by gonochores, followed by protogynous species. Thus, in the absence of any sperm limitation or behavioral disruption of spawning by a fishery, we would expect populations of protandrous species to be most affected by depletion due to overfishing.

Several authors have pointed out that any effect of differentially taking one sex in a fishery will be diluted by the rate at which individuals of the less-exploited sex change (Alonso & Mangel 2005; Molloy et al. 2007). Social control of sex change is known for many species of smaller reef fish (e.g., Robertson 1972; Shapiro 1980) and can occur in approximately two weeks in many of the species studied. In larger, exploited species, especially those that migrate to spawning aggregations and having limited spawning periods during the year, we expect that this compensation will be much slower. In groupers, limited evidence suggests that individuals change sex between spawning seasons (McGovern et al. 1998) or show little predictable temporal pattern for sex change with respect to spawning aggregation time (Shapiro 1987), so that any short-term changes in sex ratio on the spawning grounds would not be compensated for by immediate sex change. In addition, given that sex change may take several weeks, it might be disadvantageous for individuals to change sex in the middle of a short, intensive spawning season, with the cost of sex change including missing most of the reproductive season for a year. However, sex ratios or absolute numbers of males at spawning aggregations might influence the probability of sex change for the next season, which would increase the number of spawning males compared to a model of fixed size or age at sex change.

Using traditional population biology models, population growth is typically viewed as an exercise in modeling female growth, fecundity, and mortality in the absence of any paternal care (e.g., Wilson & Bossert 1971). If decreased male density or population size does not reduce the number or quality of successfully fertilized eggs, then any scenario that focuses fishing mortality on males will have less of an effect on population growth than scenarios where mortality is independent of sex or female-biased.

One concern about exploitation and fertilization success has been that in protogynous hermaphrodites, removal of the larger, older, and relatively rarer males will cause catastrophic reductions in fertilization success, so that protogynous species will be more vulnerable to exploitation and Allee effects (Smith 1982, as cited in Punt et al. 1993; Shapiro 1987; Petersen & Warner 2002; Alonso & Mangel 2004). In models where fertilization dynamics have been included, the impact on production of fertilized eggs for protogynous populations relative to gonochoristic populations depends on the relative intensity of fishing on males, the plasticity of sex change in the population, and on how fertilization is modeled with regard to either relative or absolute male numbers.

The most straightforward way to model fishing intensity would be to model effect of fishing mortality, F, on population structure and then apply a fertilization model to the population (Côté 2003). The effect on the male population will be particularly intense if fishing includes all size classes that include males (Alonso & Mangel 2004). Any model will be complicated by the size-selective nature of most fisheries, and will be sensitive to whether or not F should be a constant or size-class specific.

Modeling fertilization dynamics in exploited species to date has been a theoretical exercise. Several models use a relatively unrealistic assumption that fertilization success of eggs will be proportional either to the absolute or relative number of males (Bannerot et al. 1987; Huntsman & Schaaf 1994; Armsworth 2001; Côté 2003). More recent models have attempted to incorporate the asymptotic nature of the fertilization curve (Figure 7.1A)(Alonso & Mangel 2004; Heppel et al. 2006 for protogynous species; Molloy et al. 2007 for protandrous species). These latter models, although more realistic, still relate fertilization to total sperm production of the males in the population, not the amount of sperm released in a spawning event, and so miss most of the richness of the reproductive behavior in species in attempting to estimate the effects of sperm limitation on total fertilized egg production in populations. Some authors have picked two very different ways to model fertilization, hoping that their range of outcomes might include the true fertilization dynamics (Côté 2003; Heppel et al. 2006). All of the models to date show that under certain model parameters, sperm limitation can be a major factor influencing population recruitment and growth. However, until we can actually get accurate estimates of these fertilization parameters from the field, these results will be speculative.

In large species that have limited spawning times and migrate distances to spawn in large aggregations, the lack of some threshold number of males at a spawning site may cause females to abandon the site and search for a more populated locale to spawn. This problem is potentially more important than fertilization failure in these exploited species.

In protandrous (male-first) sequential hermaphrodites, several authors have concluded that populations run the risk of strong egg limitation when larger females are disproportionately taken in the fishery (Milton et al. 1998; Molloy et al. 2007). Sex change without quick replacement of taken females by smaller males leads to the largest declines in egg production in the population (Molloy et al. 2007), since female fecundity is directly correlated with body size. In protandrous species, fertilization success may decrease when populations are exploited, but the direct effects of exploitation on decreasing egg production are much stronger in limiting productivity of young in a population.

One problem with attempting to apply work on fertilization biology of fishes in the wild to conservation biology is that most conservation biology is focused on large, exploited species, while fertilization work has been focused on smaller, more tractable species that can be studied more easily (Vincent & Sadovy 1998). Generally, we can only guess how the results from these small species informs conservation biology of larger species.

The current evidence suggests that as the number or ratio of males in a local population decreases, sperm limitation will become more likely for the population, and to some extent it will compromise the embryo production of the population. Exploitation that reduces population size will have other potential negative impacts on populations tied to reproduction, such as the loss of traditional spawning sites. Until we know the real effect of these practices on fertilization success and successful recruitment, it is clearly in our best interest to take a precautionary approach to this problem and try to protect these populations both at spawning aggregations and at other places where impacts on population growth have the potential to severely deplete populations.

FERTILIZATION ECOLOGY IN FISHES AND MARINE INVERTEBRATES: IS THERE A DICHOTOMY?

There is currently substantial debate in the marine invertebrate literature on the importance of sperm limitation on the population dynamics and evolution of gametic and adult traits. Sedentary benthic marine invertebrates, ranging from attached species like corals, to slow moving species like sea urchins that have separate sexes and external fertilization, have the potential to be severely sperm limited, especially at low densities (e.g., Pennington 1985; Levitan & Petersen 1995). Alternatively, the limited data on externally fertilizing marine fishes suggests that sperm limitation makes a real but relatively small contribution to components of fitness in marine fishes (Levitan & Petersen 1995; Petersen & Warner 2002), probably due to the tight temporal and spatial synchrony of ova and sperm release (Petersen et al. 1992, 2001; Warner et al. 1995; Yund 2000).

One emerging view is that most animal taxa show evidence of sperm limitation in their evolutionary history, and that they have adaptations to reduce sperm limitation, including spawning synchronization and selection on gamete characteristics (Yund 2000). Invertebrate eggs have one additional problem that does not appear to be a major problem in bony fishes: polyspermy. At very high sperm concentrations, the percentage of eggs that successfully develop can actually decrease, due to developmental problems that result from polyspermy (Franke et al. 2002; Levitan et al. 2007). Thus in some invertebrate species, the probability of polyspermy in nature might be a stronger selective force for less permeable or fertilizable eggs compared to the selective force for ease of fertilization caused by sperm limitation.

Polyspermy does not appear to be a problem in the vast majority of bony fishes that have a single micropylar canal (Ginzburg 1968). This absence of polyspermy should push selection more towards ease of fertilization of fish eggs. This, along with behavioral synchrony of gamete release in time and space, should allow for high fertilization rates under most circumstances although there still may be predictable variation in fertilization in the wild that could be an evolutionary selective force.

Fertilization ecology and the potential for sperm limitation do appear to have different relative intensities between marine invertebrates and fishes, with sperm limitation more likely to be of greater concern in marine invertebrates. However, there will be variation in sperm limitation among species in both groups, and there is probably overlap in the levels of sperm limitation among these two groups, with some marine invertebrates having less sperm limitation than some fishes.

CONCLUSIONS: WHAT IS MISSING?

A large amount of research is being currently done on several aspects of fertilization, including the potential effects of limited fertilization in exploited species, methods to achieve high fertilization in aquaculture, and the evolution of gametic traits. In concluding, we would like to mention several areas that we believe are both currently understudied and would reward new research.

Measurements of Fertilization in Exploited Species

It is striking that despite extensive theoretical work by several research groups, there are no *in situ* measurements of fertilization success in the field for exploited species, especially for exploited tropical reef fishes such as snappers and groupers which are both currently overexploited, and have predictable spawning times and locations so that collecting eggs immediately post-spawning is possible.

Fertilization Dynamics in Group Spawners

Despite the growth of DNA fingerprinting techniques such as the use of hypervariable microsatellite loci to determine paternity in freshwater fish, with the exception of Syngnathids (seahorses and pipefish)(reviewed in Avise et al. 2002) there has been very little work done on distribution of paternity in spawns of marine fishes, and no work that we are aware of for group spawners. By tagging females and taking small fin clips at the same time, females can be genotyped, and it would be possible to determine the distribution of paternity in that female's spawn in a group-spawning situation. Microsatellites have been isolated for *T. bifasciatum* (Wooninck et al. 1998) but have not been applied to this question. Currently, we do not know the distribution of reproductive success among males, or even the number of males that actually contribute paternity to an individual spawn in any group-spawning species. Such information would help us to better understand both the degree of sperm competition and behavioral tactics within group-spawning species. For example, although Shapiro et al. (1994) report the amount of sperm released in groups spawns of the bluehead wrasse, it is unknown how many males released this sperm.

Male Reproductive Apparatus and Fertilization Dynamics

Species belonging to several different families have modifications of the male reproductive apparatus. Such modifications can be in the form of accessory structures such as sperm duct glands, seminal vesicles or blind pouches, as in Gobiidae (Miller 1984; Fishelson 1991), Blennioidea (Rasotto 1995; Richtarski & Patzner 2000), or Batrachoididae (Barni et al. 2001). They can also be in the form of specializations of different parts of the sperm transport system for sperm storage and secretion of components of the seminal fluid, as, for example, in Blenniidae (Lahnsteiner & Patzner 1990), Cottidae (Petersen et al. 2005), or Batrachoididae (Barni et al. 2001); maturation of sperm, as in Blennioidea (Rasotto 1995; Richtarski & Patzner 2000; Giacomello et al. 2008); or the packaging of sperm in internal fertilizers, as in the Scorpaenidae (Muñoz et al. 2002) or several freshwater species (Grier et al. 1978; Burns et al. 1995). In addition, modifications related to reproduction have been observed in the male urinary system (Haemulidae: Rasotto & Sadovy 1995). The components secreted by the different parts of the male reproductive apparatus play an important role in fertilization dynamics as well as in sperm competition mechanisms, as highlighted in gobies and blennies (Scaggiante et al. 1999; Mazzoldi et al. 2005; Giacomello et al. 2008). An extended investigation of the different families having some modification of the male reproductive apparatus in the context of the relationship between morphology and fertilization dynamics and/ or sperm competition represents an interesting field of study.

In concluding, our understanding of fertilization in marine organisms has come a long way since the time when both field ecologists and theoretical population biologists assumed that fertilization was absolute and invariant in natural populations. We now have both theory that incorporates variable fertilization into population modeling, and direct measurements of fertilization *in situ* for a number of marine fishes. In order to determine if our theoretical work is adequate to describe the real world, we need to both expand our collection of data on fishes and collect data that will directly inform theory. This will not be easy, but it is clearly the next step if we are to use our increasing understanding to help manage and conserve marine fishes in both temperate and tropical oceans.

ACKNOWLEDGMENTS

S. Rowe and P. Molloy were very helpful in answering multiple questions about their work, R. Warner helped with his knowledge of how to best frame the interesting questions, and H. Hess and M. Rasotto reviewed the manuscript and provided many helpful comments that increased the clarity of the chapter. C.P. was partially supported by a grant from the David Rockefeller Foundation while working on this chapter.

REFERENCES

Ah-King M, Elofsson H, Kvarnemo C, Rosenqvist G, Berglund A (2006). Why is there no sperm competition in a pipefish with externally brooding males? Insights from sperm activation and morphology. Journal of Fish Biology 68: 958–962.

Alonso SH, Mangel M (2004). The effects of size-selective fisheries on the stock dynamics of and sperm limitation in sex-changing fish. Fishery Bulletin 102: 1–13.

Alonso SH, Mangel M (2005). Sex-change rules, stock dynamics, and the performance of spawning-per-recruit measures in protogynous stocks. Fishery Bulletin 103: 229–245.

Amanze D, Iyengar A (1990). The micropyle: a sperm guidance system in teleost fertilization. Development 109: 495–500.

Ames EP (2004). Atlantic cod stock structure in the Gulf of Maine. Fisheries 29: 10–28.

Armsworth PR (2001). Effects of fishing on a protogynous hermaphrodite. Canadian Journal of Fisheries and Aquatic Sciences 58: 568–578.

Avise JC, Jones AG, Walker D, DeWoody JA, collaborators (2002). Genetic mating systems and reproductive natural histories of fishes: lessons for ecology and evolution. Annual Review of Genetics 36: 19–45.

Ball MA, Parker GA (1996). Sperm competition games: external fertilization and "adaptive" infertility. Journal of Theoretical Biology 180: 141–150.

Bannerot SP, Fox WW Jr, Powers JE (1987). Reproductive strategies and the management of groupers and snappers in the Gulf of Mexico and the Caribbean. In: Polovina JJ, Ralston S (eds.), *Tropical Snappers and Groupers: Biology and Fisheries Management*. Westview Press, Boulder, Colorado, pp. 561–606.

Barni A, Mazzoldi C, Rasotto MB (2001). Reproductive apparatus and male accessory structures in two batrachoid species (Teleostei, Batrachoididae). Journal of Fish Biology 58: 1557–1569.

Baylis JR (1981). The evolution of parental care in fishes, with reference to Darwin's rule of male sexual selection. Environmental Biology of Fishes 6: 223–251.

Bekkevold D, Hansen MM, Loeschcke V (2002). Male reproductive competition in spawning aggregations of cod (*Gadus morhua* L.). Molecular Ecology 11: 91–102.

Bisazza A, Pilastro A (1997). Small male mating advantage and reversed size dimorphism in poeciliid fishes. Journal of Fish Biology 50: 397–406.

Blumer LS (1979). Male parental care in the bony fishes. Quarterly Review of Biology 54: 149–161.

Blumer LS (1982). A bibliography and categorization of bony fishes exhibiting parental care. Zoological Journal of the Linnean Society (London) 76: 1–22.

Brawn VM (1961). Reproductive behavior of the cod (*Gadus callarias* L.). Behaviour 18: 177–198.

Breder CM, Rosen DE (1966). *Modes of Reproduction in Fishes*. Natural History Press, Garden City, New York.

Burness G, Casselman SJ, Schulte-Hostedde AI, Moyes CD, Montgomerie R (2004). Sperm swimming speed and energetics vary with sperm competition risk in bluegill (*Lepomis macrochirus*). Behavioral Ecology and Sociobiology 56: 65–70.

Burns JR, Weitzman SH, Grier HJ, Menezes NA (1995). Internal fertilization, testis and sperm morphology in glandulocaudinae fishes (Teleostei: Characidae: Glandulocaudinae). Journal of Morphology 224: 131–145.

Clutton-Brock TH (1991). *The Evolution of Parental Care*. Princeton University Press, Princeton, New Jersey.

Colin PL (1983). Spawning and larval development of the hogfish, *Lachnolaimus maximus*. (Pisces: Labridae). Fishery Bulletin 80: 853–862.

Constantz GD (1989). Reproductive biology of poeciliid fishes. In: Meffe GK, Snelson FF (eds.), *Ecology and Evolution of Livebearing Fishes (Poeciliidae)*. Prentice Hall, Englewood Cliffs, New Jersey, pp. 33–50.

Cosson J, Groison A-L, Suquet M, Fauvel C, Dreanno C, Billard R (2008a). Marine fish spermatozoa: racing ephemeral swimmers. Reproduction 136: 277–294.

Cosson J, Groison A-L, Suquet M, Fauvel C, Dreanno C, Billard R (2008b). Studying sperm motility in marine fish: an overview on the state of the art. Journal of Applied Ichthyology 24: 460–486.

Côté IM (2003). Knowledge of reproductive behavior contributes to conservation programs. In: Festa-Bianchet M, Apollonio M (eds.), *Animal Behavior and Wildlife Conservation*. Island Press, Washington, DC, pp. 77–92.

Cyr DG, Idler DR, Audet C, McLeese JM, Eales JG (1998). Effects of long-term temperature acclimation on thyroid hormone deiodinase function, plasma thyroid hormone levels, growth, and reproductive status of male Atlantic Cod, *Gadus morhu*a. General and Comparative Endocrinology 109: 24–36.

de Jonge J, Rutter AJH, Van den Hurk R (1989). Testis-testicular gland complex of two *Tripterygion* species (Blennioide, Teleostei): differences between territorial and non-territorial males. Journal of Fish Biology 35: 497–508.

Eberhard WG (1996). *Female Control: Sexual Selection by Cryptic Female Choice*. Princeton University Press, Princeton, New Jersey.

Elofsson H, Mcallister BG, Kime DE, Mayer I, Borg B (2003). Long lasting stickleback sperm: is ovarian fluid a key to success in fresh water? Journal of Fish Biology 63: 240–253.

Fishelson L (1991). Comparative cytology and morphology of seminal vesicles in male gobiid fishes. Japanese Journal of Ichthyology 38: 17–30.

Fishelson L, Delarea Y, Gon O (2006). Testis structure, spermatogenesis, spermatocytogenesis, and sperm structure in cardinal fish (Apogonidae, Perciformes). Anatomy and Embryology 211: 31–46.

Fitzgerald JL, Montgomerie R, Desjardins JK, Stiver KA, Kolm N, Balshine S (2009). Female promiscuity promotes the evolution of faster sperm in cichlid fishes. Proceedings National Academy Sciences of the United States of America 106: 1128–1132.

Foster SJ, Vincent ACJ (2004). Life history and ecology of seahorses: implications for conservation and management. Journal of Fish Biology 65: 1–61

Franke ES, Babcock RS, Styan CA (2002). Sexual conflict and polyspermy under sperm-limited conditions: in situ evidence from field simulations with the free-spawning marine echinoid *Evechinus chlorotincus*. American Naturalist 160: 485–496.

Gage MJG, Stocley P, Parker GA (1995). Effects of alternative male mating strategies on characteristics of sperm production in the Atlantic salmon (*Salmo salar*). Philosophical Transactions: Biological Sciences 350: 391–399

Gage MJG, Macfarlane CP, Yeates S, Ward RG, Searle JB, Parker GA (2004). Spermatozoa traits and sperm competition in Atlantic salmon: relative sperm velocity is the primary determinant of fertilization success. Current Biology 14: 44–47.

Garnaud J (1950). La reproduction et l'incubation branchiale chez *Apogon imberbis* G. et L. Bulletin de l'Institut Océanographique, Monaco 977: 1–10.

Giacomello E, Rasotto MB (2005). Sexual dimorphism and male mating success in the tentacled blenny, *Parablennius tentacularis* (Teleostei: Blenniidae). Marine Biology 147: 1221–1228.

Giacomello E, Marchini D, Rasotto MB (2006). A male sexually dimorphic trait provides antimicrobials to eggs in blenny fish. Biology Letters 2: 330–333.

Giacomello E, Neat FC, Rasotto MB (2008). Mechanisms enabling sperm economy in blenniid fishes. Behavioral Ecology and Sociobiology 62: 671–680.

Ginzburg AS (1968). Fertilization in fishes and the problem of polyspermy. Translated from Russian by Israel Program for Scientific Translations, Jerusalem 1972.

Giulianini PG, Ota O, Marchesan M, Ferrero EA (2001). Can goby spermatozoa pass through the filament adhesion apparatus of laid eggs? Journal of Fish Biology 58: 1750–1752.

Gomendio M, Roldan ERS (1991). Sperm competition influences sperm size in mammals. Proceedings of the Royal Society of London B 243: 181–185.

Goodwin NB, Dulvy NK, Reynolds JD (2002). Life-history correlates of the evolution of live bearing in fishes. Philosophical Transactions: Biological Sciences 357: 259–267

Grier HJ (1981). Cellular organization of the testis and spermatogenesis in fishes. American Zoologist 21: 345–357.

Grier HJ, Fitzsimons JM, Linton JR (1978). Structure and ultrastructure of the testis and sperm formation in Goodeid teleosts. Journal of Morphology 156: 419–438.

Grier HJ, Burns JR, Flores JA (1981). Testis structure in three species of teleosts with tubular gonopodia. Copeia 1981: 797–801.

Gross, MR, Shine R (1981). Parental care and mode of fertilization in ectothermic vertebrates. Evolution 35: 775–793.

Hayakawa Y, Munehara H (1996). Non-copulatory spawning and female participation during early egg care in a marine sculpin *Hemilepidotus gilberti*. Ichthyological Research 43: 73–78.

Hayakawa Y, Munehara H (1998). Fertilization environment of the non-copulating marine sculpin, *Hemilepidotus gilberti*. Environmental Biology of Fishes 52: 181–186.

Hayakawa Y, Akiyama R, Munehara H, Komaru A (2002a). Dimorphic sperm influence semen distribution in a non-copulatory sculpin *Hemilepidotus gilberti*. Environmental Biology of Fishes 65: 311–317.

Hayakawa Y, Munehara H, Komaru A (2002b). Obstructive role of the dimorphic sperm in a non-copulatory marine sculpin, *Hemilepidotus gilberti*, to prevent other males' eusperm from fertilization. Environmental Biology of Fishes 64: 419–427.

Heppel SS, Heppel SA, Coleman FC, Koenig CC (2006). Models to compare management options for a protogynous fish. Ecological Applications 16: 238–249.

Hess HC (1993). Male mouthbrooding in jawfishes (Opistognathidae): constraints on polygyny. Bulletin of Marine Science 52: 806–818.

Houde AE (1997). *Sex, Color, and Mate Choice in Guppies*. Princeton University Press, Princeton, New Jersey.

Howell BR, Child AR, Houghton RG (1991). Fertilization rate in a natural population of the common sole, *Solea solea* (L.). ICES Journal of Marine Science 48: 53–59.

Hubbs C (1966). Fertilization, initiation of cleavage, and developmental temperature tolerance of the cottid fish, *Clinocottus analis*. Copeia 1966: 29–42.

Huntsman GR, Schaaf WE (1994). Simulation of the impact of fishing on reproduction of a protogynous grouper, the graysby. North American Journal Fisheries Management 14: 41–52.

Hutchings JA, Myers RA (1994). What can we learn from the collapse of a renewable resource? Atlantic cod, *Gadus morhua*, of Newfoundland and Labrador. Canadian Journal of Fisheries and Aquatic Sciences 51: 2126–2146.

Hutchings JA, Reynolds JD (2004). Marine fish population collapses: consequences for recovery and extinction risk. Bioscience 54: 297–309.

Hutchings JA, Myers RA, Lilly GR (1993). Geographic variation in the spawning of Atlantic cod, *Gadus morhua*, in the northwest Atlantic. Canadian Journal of Fisheries and Aquatic Sciences 50: 2457–2467.

Hutchings JA, Bishop TD, McGregor-Shaw CR (1999). Spawning behaviour of Atlantic cod, *Gadus morhua*: evidence of mate competition and mate choice in a broadcast spawner. Canadian Journal of Fisheries and Aquatic Sciences 56: 97–104.

Jamieson BGM (1991). *Fish Evolution and Systematics: Evidence from Spermatozoa*. Cambridge University Press, Cambridge and New York.

Jamieson BGM, ed. (2009). *Reproductive Biology and Phylogeny of Fishes (Agnathans and Bony Fishes): Phylogeny, Reproductive System, Viviparity, Spermatozoa*. Science Publishers, Enfield, New Hampshire.

Kiflawi M (2000). Adaptive gamete allocation when fertilization is external and sperm competition is absent: Optimization models and evaluation using coral reef fish. Evolutionary Ecology Research 2: 1045–1066.

Kiflawi M, Mazeroll AI, Goulet D (1998). Does mass spawning enhance fertilization in coral reef fish? A case study of the brown surgeonfish. Marine Ecology Progress Series 172: 107–114.

Kolm N, Berglund A (2004). Sex-specific territorial behaviour in the Banggai cardinalfish, *Pterapogon kauderni*. Environmental Biology of Fishes 70: 375–379

Kraak SBM (1996). A quantitative description of the reproductive biology of the Mediterranean blenny *Aidablennius sphynx* (Teleostei, Blenniidae) in its natural habitat. Environmental Biology of Fishes 46: 329–342.

Kuwamura T (1983). Spawning behavior and timing of fertilization in the mouthbrooding cardinalfish *Apogon notatus*. Japanese Journal of Ichthyology 30: 61–71.

Kuwamura T (1985). Social and reproductive behavior of three mouthbrooding cardinalfishes, *Apogon doederleini*, *A. niger*, and *A. notatus*. Environmental Biology of Fishes 13: 17–24.

Lahnsteiner F (2002). The influence of ovarian fluid on the gamete physiology in the Salmonidae. Fish Physiology and Biochemistry 27: 49–59.

Lahnsteiner F, Patzner R (1990). The spermatic duct of blenniid fish (Teleostei, Blenniidae): fine structure, histochemistry and function. Zoomorphology 110: 63–73.

Lahnsteiner F, Richtarski U, Patzner RA (1990). Functions of the testicular gland in two blenniid fishes, *Salaria* (=*Blennius*) *pavo* and *Lipophrys* (=*Blennius*) *dalmatinus* (Blenniidae, Teleostei) as revealed by electron microscopy and enzyme histochemistry. Journal of Fish Biology 37: 85–97.

Lahnsteiner F, Mansour N, Berger B (2004). Seminal plasma proteins prolong the viability of rainbow trout (*Oncorynchus mykiss*) spermatozoa. Theriogenology 62: 801–808.

Levitan D, Petersen CW (1995). Sperm limitation in marine organisms. Trends in Ecology and Evolution 10: 228–231.

Levitan DR, Terhorst C.P, Fogarty ND (2007). The risk of polyspermy in three congeneric sea urchins and its implications for gametic incompatability and reproductive isolation. American Naturalist 61: 2007–2014.

Litvak MK, Trippel EA (1998). Sperm motility patterns of Atlantic cod (*Gadus morhua*) in relation to salinity: effects of ovarian fluid and egg presence. Canadian Journal of Fisheries and Aquatic Sciences 55: 1893–1898.

Locatello L, Pilastro A, Deana R, Zarpellon A, Rasotto MB (2007). Variation pattern of sperm quality traits in two gobies with alternative mating tactics. Functional Ecology 21: 975–981.

Manica A (2002). Filial cannibalism in teleost fish. Biological Review 77: 261–277.

Manni L, Rasotto MB (1997). Ultrastructure and histochemistry of the testicular efferent duct system and spermiogenesis in *Opistognathus whitehurstii* (Teleostei, Trachinoidei). Zoomorphology 117: 93–102.

Marconato A, Rasotto MB (1993). The reproductive biology of *Opistognathus whitehurstii* (Pisces, Opistognathidae). Biologia Marina Mediterranea, suppl. 1: 345–348.

Marconato A, Shapiro DY (1996). Sperm allocation, sperm production and fertilization rates in the bucktooth parrotfish. Animal Behaviour 52: 971–980.

Marconato A, Tessari V, Marin G (1995). The mating system of *Xyrichthys novacula*: sperm economy and fertilization success. Journal of Fish Biology 47: 292–301.

Marconato A, Rasotto MB, Mazzoldi C (1996). On the mechanism of sperm release in three gobiid fishes (Teleostei: Gobiidae). Environmental Biology of Fishes 46: 321–327.

Marconato A, Shapiro DY, Petersen CW, Warner RR, Yoshikawa T (1997). Methodological analysis of fertilization rate in the bluehead wrasse, *Thalassoma bifasciatum*: pair versus group spawns. Marine Ecology Progress Series 161: 61–70.

Markle DF, Waiwood KG (1985). Fertilization failure in gadids: aspects of its measurement. Journal of Northwest Atlantic Fishery Science 6: 87–93.

Mattei X (1988). The flagellar apparatus of spermatozoa in fish. Ultrastructure and evolution. Biology of the Cell 63: 151–158.

Mazzoldi C (2001). Reproductive apparatus and mating system in two tropical goby species (Teleostei, Gobiidae). Journal of Fish Biology 59: 1686–1691.

Mazzoldi C, Rasotto MB (2002). Alternative male mating tactics in *Gobius niger*. Journal of Fish Biology 61: 157–172.

Mazzoldi C, Scaggiante M, Ambrosin E, Rasotto MB (2000). Mating system and alternative male mating tactics in the grass goby, *Zosterisessor ophiocephalus* (Teleostei: Gobiidae). Marine Biology 137: 1041–1048.

Mazzoldi C, Poltronieri C, Rasotto MB (2002). Egg size variability and size-assortative mating in the marbled goby, *Pomatoschistus marmoratus* (Pisces, Gobiidae). Marine Ecology Progress Series 233: 231–239

Mazzoldi C, Petersen CW, Rasotto MB (2005). The influence of mating system on seminal vesicle variability among gobies (Teleostei: Gobiidae). Journal of Zoological Systematics and Evolutionary Research 43: 307–314.

Mazzoldi C, Randieri A, Mollica E, Rasotto MB (2008). Notes on the reproduction of the cardinalfish *Apogon imberbis* from Lachea Island, central Mediterranean, Sicily, Italy. Vie et Milieu 58: 63–66.

McGovern JC, Wyanski DM, Pashuk O, Manooch II CS, Seberry GR (1998). Changes in the sex ratio and size at maturity of gag, *Mycteroperca microlepis*, from the Atlantic coast of the southeastern United States during 1976–1995. Fisheries Bulletin 96: 797–807.

Miller PJ (1984). The tokology of gobioid fishes. In: Potts W, Wootton RJ (eds.), *Fish Reproduction: Strategies and Tactics*. Academic Press, London, pp. 119–153.

Milton DA, Die D, Tenakanai C, Swales S (1998). Selectivity for barramundi (*Lates calcarifer*) in the Fly River, Papua New Guinea: implications for managing gillnet fisheries on protandrous fishes. Marine and Freshwater Research 49: 499–506.

Molloy PP, Goodwin NB, Cote IM, Gage MJG, Reynolds JD (2007). Predicting the effects of exploitation on male-first sex-changing fish. Animal Conservation 10: 30–38.

Morgan MJ, Trippel EA (1996). Skewed sex ratios in spawning shoals of Atlantic cod (*Gadus morhua*). ICES Journal of Marine Science 53: 820–826.

Morris RW (1952). Spawning behavior of the cottid fish *Clinocottus recalvus* (Greeley). Pacific Science 6: 256–258

Morris RW (1956). Clasping mechanism of the cottid fish *Oligocottus synderi* Greeley. Pacific Science 10: 314–317.

Myers RA, Mertz G, Bishop CA (1993). Cod spawning in relation to physical and biological cycles of the northwest Atlantic. Fisheries Oceanography 2: 154–165.

Myers RA, Hutchings JA, Barrowman NJ (1997). Why do fish stocks collapse? The example of cod in Atlantic Canada. Ecological Applications 7: 91–106.

Munehara H (1988). Spawning and subsequent copulating behavior of the elkhorn sculpin *Alcichthys alcicornis* in an aquarium. Japanese Journal of Ichthyology 35: 358–364.

Munehara HK (1996). Sperm transfer during copulation in the marine sculpin *Hemitripterus villosus* (Pisces: Scorpaeniformes) by means of a retractable genital duct and ovarian secretion in females. Copeia 1996: 452–454.

Munehara H, Takano K, Koya Y (1989). Internal gamete association and external fertilization in the elkhorn sculpin, *Alcichthys alcicornis*. Copeia 1989: 673–678.

Munehara H, Takano K, Koya Y (1991). The little dragon sculpin *Blepsias cirrhosus*, another case of internal gamete association and external fertilization. Japanese Journal of Ichthyology 37: 391–394.

Munehara H, Koya Y, Hayakawa Y, Takano K (1997). Extracellular environments for the initiation of external fertilization and micropylar plug formation in a cottid species, *Hemitripterus villosus* (Pallas) (Scorpaeniformes) with internal insemination. Journal of Experimental Marine Biology and Ecology 211: 279–289

Munõz M, Koya Y, Casedevall M (2002). Histochemical analysis of sperm storage in *Helicolenus dactylopterus dactylopterus* (Teleostei: Scorpaenidae). Journal of Experimental Zoology 292: 156–164.

Nakatsuru K, Kramer DL (1982). Is sperm cheap? Limited male fertility and female choice in the lemon tetra (Pisces: Characidae). Science 216: 753–755.

Neat FC (2001). Male parasitic spawning in two species of the triplefin blenny (Tripterigiidae): contrasts in demography, behaviour and gonadal characteristics. Environmental Biology of Fishes 61: 57–64.

Neat FC, Locatello L, Rasotto MB (2003). Reproductive morphology in relation to alternative male reproductive tactics in *Scartella cristata*. Journal of Fish Biology 62: 1381–1391.

Okuda N (2001). The costs of reproduction to males and females of a paternal mouth-brooding cardinalfish *Apogon notatus*. Journal of Fish Biology 58: 776–787

Ota D, Marchesan M, Ferrero EA (1996). Sperm release behaviour and fertilization in the grass goby. Journal of Fish Biology 49: 246–256.

Parker GA (1970). Sperm competition and its evolutionary consequences in the insects. Biological Reviews 45: 525–567.

Parker GA (1990). Sperm competition games: raffles and roles. Proceedings of the Royal Society of London B 242: 120–126

Parker GA (1998). Sperm competition and the evolution of ejaculates: towards a theory base. In: Birkhead TR, Møller AP (eds.), *Sperm Competition and Sexual Selection*. Academic Press, London, pp. 3–54.

Parker GA, Ball MA, Stockley P, Gage MJG (1996). Sperm competition games: individual assessment of sperm competition intensity by group spawners. Proceedings of the Royal Society of London B 263: 1291–1297.

Pennington JT (1985). The ecology of fertilization of echioid eggs: the consequence of sperm dilution, adult aggregation, and synchronous spawning. Biological Bulletin 169: 417–430.

Petersen CW (1991). Variation in fertilization rate in tropical reef fish, *Halichoeres bivattatus*: correlates and implications. Biological Bulletin 181: 232–237.

Petersen CW, Levitan D (2001). The Allee effect: a barrier to recovery by exploited species. In: Reynolds JD, Mace GM, Redford KH, Robinson JG (eds.), *Conservation of Exploited Species*. Cambridge University Press, Cambridge, pp. 281–300.

Petersen CW, Warner RR (1998). Sperm competition in fishes. In: Birkhead TR, Møller AP (eds.), *Sperm Competition and Sexual Selection*. Academic Press, London, pp. 435–463.

Petersen CW, Warner RR (2002). The ecological context of reproductive behavior. In: Sale PF (ed.), *Dynamics and Diversity in a Complex Ecosystem*. Academic Press, San Diego, pp. 103–118.

Petersen CW, Warner RR, Cohen S, Hess H, Sewell A (1992). Variable pelagic fertilization success: implications for mate choice and spatial patterns of mating. Ecology 73: 391–401

Petersen CW, Warner RR, Shapiro DY, Marconato A (2001). Components of fertilization success in the bluehead wrasse, *Thalassoma bifasciatum*. Behavioral Ecology 12: 237–245.

Petersen CW, Mazzoldi C, Zarrella KA, Hale RB (2005). Fertilization mode, sperm characteristics, mate choice and parental care patterns in *Artedius spp.* (Cottidae). Journal of Fish Biology 67: 239–254.

Pilastro A, Bisazza A (1999). Insemination efficiency of two alternative male mating tactics in the guppy (*Poecilia reticulata*). Proceedings of the Royal Society of London B 266: 1887–1891.

Pilastro A, Giacomello E, Bisazza A (1997). Sexual selection for small size in male mosquitofish (*Gambusia holbrooki*). Proceedings of the Royal Society of London B 264: 1125–1129.

Pilastro A, Scaggiante M, Rasotto MB (2002). Individual adjustment of sperm expenditure accords with sperm competition theory. Proceeding of the National Academy of Sciences of the United States of America 99: 9913–9915.

Pilastro A, Simonato M, Bisazza A, Evans JP (2004). Cryptic female preference for colorful males in guppies. Evolution 58: 665–669.

Punt AE, Garratt PA, Govender A (1993). On an approach for applying per-recruit methods to a protogynous hermaphrodite, with an illustration for the slinger *Chrysoblephus puniceus* (Pisces:Sparidae). South African Journal of Marine Science 13: 109–119.

Purchase CF, Hasselman DJ, Weir LK (2007). Relationship between fertilization success and the number of milt donors in rainbow smelt *Osmerus mordax* (Mitchell): implications for population growth rate. Journal of Fish Biology 70: 934–946.

Ragland HC, Fisher EA (1987). Internal fertilization and male parental care in the scalyhead sculpin, *Artedius harringtoni*. Copeia 1987: 1059–1062.

Rakatin A, Ferguson MM, Trippel EA (1999). Sperm competition and fertilization success in Atlantic cod (*Gadus morhua*): effect of sire size and condition factor on gamete quality. Canadian Journal of Fisheries and Aquatic Sciences 56: 2315–2323.

Rakitin A, Ferguson MM, Trippel EA (2001). Male reproductive success and body size in Atlantic cod *Gadus morhua* L. Marine Biology 138: 1077–1085.

Rasotto MB (1995). Male reproductive apparatus of some blennioidei (Pisces: Teleostei). Copeia 1995: 907–914.

Rasotto MB, Mazzoldi C (2002). Male traits associated with alternative reproductive tactics in *Gobius niger*. Journal of Fish Biology 61: 173–184.

Rasotto MB, Sadovy Y (1995). Peculiarities of the male urogenital apparatus of two grunt species (Teleostei, Haemulidae). Journal of Fish Biology 46: 936–948.

Rasotto MB, Marconato A, Shapiro DY (1992). Reproductive apparatus of two jawfish species (Opistognathidae) with description of a juxtatesticular body. Copeia 1992: 1046–1053.

Richtarski U, Patzner RA (2000). Comparative morphology of male reproductive systems in Mediterranean blennies (Blennidae). Journal of Fish Biology 56: 22–36.

Robertson DR (1972). Social control of sex-reversal in a coral-reef fish. Science 177: 1007–1009.

Rowe S, Hutchings JA (2003). Mating systems and the conservation of commercially exploited marine fish. Trends in Ecology and Evolution 18: 567–572.

Rowe S, Hutchings JA, Bekkevold D, Rakitin A (2004). Depensation, probability of fertilization, and the mating system of Atlantic cod (*Gadus morhua* L.). ICES Journal of Marine Science 61: 1144–1150.

Rowe S, Hutchings JA, Skjæraasen JE (2007). Nonrandom mating in a broadcast spawner: mate size influences reproductive success in Atlantic cod (*Gadus morhua*). Canadian Journal of Fisheries and Aquatic Sciences 64: 219–226.

Ruchon F, Laugier T, Quignard JP (1995). Alternative male reproductive strategies in the peacock blenny. Journal of Fish Biology 47: 826–840.

Rurangwa E, Roelants I, Huyskens G, Ebrahimi M, Kime DE, Ollevier F (1998). The minimum effective spermatozoa: egg ratio for artificial insemination and the effects of mercury on sperm motility and fertilization ability in *Clarias gariepinus*. Journal of Fish Biology 53: 402–413.

Scaggiante M, Mazzoldi C, Petersen CW, Rasotto MB (1999). Sperm competition and mode of fertilization in the grass goby *Zosterisessor ophiocephalus* (Teleostei: Gobiidae). Journal of Experimental Zoology 283: 81–90.

Scaggiante M, Rasotto MB, Romvaldi C, Pilastro A (2005). Territorial male gobies respond aggressively to sneakers but do not adjust their sperm expenditure. Behavioral Ecology 16: 1001–1007.

Schärer L, Robertson DR (1999). Sperm and milt characteristics and male v. female gametic investment in the Caribbean reef fish, *Thalassoma bifasciatum*. Journal of Fish Biology 55: 329–343.

Schulte-Hostedde AI, Burness G (2005). Fertilization dynamics of sperm from different male mating tactics in bluegill (*Lepomis macrochirus*). Canadian Journal of Zoology 83: 1638–1642.

Schwalme K, Chouinard GA (1999). Seasonal dynamics in feeding, organ weights, and reproductive maturation of Atlantic cod (*Gadus morhua*) in the southern Gulf of St. Lawrence. ICES Journal of Marine Science 56: 303–319.

Shapiro DY (1980). Serial female changes after simultaneous removal of males from social groups of coral reef fish. Science 209: 1136–1137.

Shapiro DY (1987). Reproduction in groupers. In: Polovina JJ, Ralston S (eds.), *Tropical Snappers and Groupers: Biology and Fisheries Management*. Westview Press, Boulder, Colorado, pp. 295–327.

Shapiro DY, Giraldeau L-A (1996). Mating tactics in external fertilizers when sperm is limited. Behavioral Ecology 7: 19–23.

Shapiro DY, Marconato A, Yoshikawa T (1994). Sperm economy in a coral reef fish. Ecology 75: 1334–1344.

Simmons LW (2001). *Sperm Competition and its Evolutionary Consequences in the Insects*. Princeton University Press, Princeton, New Jersey.

Smith CL (1982). Patterns of reproduction in coral reef fishes. In: Huntsman GR, Nicholson WR, Fox WW Jr (eds.), The biological basis for reef fishery management. U.S. Department of Commerce, NOAA Tech. Memo. NMFS, NOAA-TM-NMFS-SEFC-80, pp. 49–66.

Snook RR (2005). Sperm in competition: not playing by the numbers. Trends in Ecology and Evolution 20: 46–53.

Stockley P, Gage MJG, Parker GA, Møller AP (1997). Sperm competition in fishes: the evolution of testis size and ejaculate characteristics. American Naturalist 149: 933–954.

Suquet M, Billard R, Cosson J, Normant Y, Fauvel C (1995). Artificial insemination in turbot (*Scophthalmus maximus*): determination of the optimal sperm per egg ratio and time of gamete contact. Aquaculture 133: 83–90.

Taborsky M (1998). Sperm competition in fish: "bourgeois" males and parasitic spawning. Trends in Ecology and Evolution 13: 222–227.

Thresher RE (1984). *Reproduction in Reef Fishes*. TFH Publications, Neptune City, New Jersey.

Trippel EA, Morgan MJ (1994). Sperm longevity in pelagic spawning cod (*Gadus morhua*). Copeia 1994: 1025–1029.

Trivers RL (1972). Parental investment and sexual selection. In: Campbell B (ed.), *Sexual Selection and the Descent of Man 1871–1971*. Aldine Press, Chicago, pp. 136–179.

Turner E, Montgomerie R (2002). Ovarian fluid enhances sperm movement in Arctic charr. Journal of Fish Biology 60: 1570–1579.

Uglem I, Galloway TF, Rosenqvist G, Folstad I (2001). Male dimorphism, sperm traits and immunology in the corkwing wrasse (*Symphodus melops* L.). Behavioral Ecology and Sociobiology 50: 511–518.

Urbach D, Folstad I, Rudolfsen G (2005). Effects of ovarian fluid on sperm velocity in Arctic charr (*Salvelinus alpinus*). Behavioral Ecology and Sociobiology 57: 438–444.

Vagelli A (1999). The reproductive biology and early ontogeny of the mouthbrooding Banggai cardinalfish, *Pterapogon kauderni* (Perciformes, Apogonidae). Environmental Biology of Fishes 56: 79–92.

Van Look KJW, Dzyuba B, Cliffe A, Koldewey HJ, Holt WV (2007). Dimorphic sperm and the unlikely route to fertilisation in the yellow seahorse. Journal of Experimental Biology 210: 432–437

Vincent ACJ, Sadovy Y (1998). Reproductive ecology in the conservation and management of fishes. In: Caro T (ed.), *Behavioral Ecology and Conservation Biology*. Oxford University Press, New York, pp. 209–245.

Warner RR, Shapiro DY, Marconato A, Petersen CW (1995). Sexual conflict: males with higher mating success convey the lowest fertilization benefits to females. Proceedings of the Royal Society of London B 262: 135–139.

Watanabe S, Hara M, Watanabe Y (2000). Male internal fertilization and introsperm-like sperm of the seaweed pipefish (*Syngnathus schlegeli*). Zoological Science 6: 759–767

Wilson EO, Bossert WH (1971). *A Primer of Population Biology*. Sinauer, Sunderland, Massachusetts.

Wooninck L, Strassman JE, Queller DC, Fleischer R, Warner RR (1998). Characterization of hypervariable microsatellite markers in the bluehead wrasse, *Thalassoma bifasciatum*. Molecular Ecology 7: 1613–1614.

Wourms JP (1981). Viviparity: the maternal-fetal relationship in fishes. American Zoologist 21: 473–515.

Yanagimachi R, Cherr GN, Pillai MC, Baldwin JD (1992). Factors controlling sperm entry into the micropyles of salmonid and herring eggs. Development Growth & Differentiation 34: 447–461.

Yund PO (2000). How severe is sperm limitation in natural populations of marine free-spawners? Trends in Ecology and Evolution 15: 10–13.

8

Bidirectional Sex Change in Marine Fishes

Philip L. Munday, Tetsuo Kuwamura, and Frederieke J. Kroon

Sex change (sequential hermaphroditism) is well known in fishes, where its occurrence and evolutionary advantage have been the focus of numerous reviews since the early 1960s (e.g., Atz 1964; Ghiselin 1969; Warner 1978, 1988; Kuwamura & Nakashima 1998; Munday et al. 2006a; Sadovy de Mitcheson & Liu 2008). Typically, individuals of sex-changing species either first function as female and then change sex to male (protogynous sex change) or they first function as male and then change to female (protandrous sex change). Bidirectional sex change—where both males and females change sex in the same population—was not thought to occur, either because the conditions favoring sex change by both males and females did not exist within the same species, or because physiological constraints prevented individuals changing sex more than once. The discovery that individuals of some species can change sex more than once has overturned these assumptions and opened up a whole new area of research into the adaptive significance and proximate mechanisms of sex change in fishes (Kuwamura & Nakashima 1998; Munday et al. 2006a).

One of the earliest reports of bidirectional sex change in fishes involved apparent sex change by males of the protogynous grouper *Epinephalus akaara* when kept together in an aquarium (Tanaka et al. 1990). Subsequent experimental manipulations with the hawkfish *Cirrhitichthys aureus* (Kobayashi & Suzuki 1992) and the goby *Trimma okinawae* (Sunobe & Nakazono 1993) confirmed that bidirectional sex change was possible, with both males and females of these species changing sex when kept in same-sex pairs. Bidirectional sex change was then confirmed in natural populations of the coral goby *Paragobiodon echinocephalus* (Kuwamura et al. 1994a) and various species of *Gobiodon*

(Nakashima et al. 1996; Munday et al. 1998) by monitoring known-sex individuals in the field and by experimental manipulations of their social groups. The number of species reported to exhibit bidirectional sex change has continued to grow and although many cases are restricted to animals in captivity, there are also more confirmations that bidirectional sex change occurs in nature (e.g., Manabe et al. 2007a). The detection of bidirectional sex change in natural populations is important, because it demonstrates that this sexual pattern is not peculiar to fish in captivity and because it provides the opportunity to assess the adaptive significance of this mode of sex change by examining the correlates of sex change under natural conditions.

Bidirectional sex changers provide a unique opportunity to test theoretical explanations for the presence of sex change in animals, because the same species exhibits both protogynous and protandrous sex change and the same individual can change sex more than once. This extreme lability in sexual expression means that individuals are likely to be highly sensitive to the environmental conditions favoring either protogynous or protandrous sex change and it should be possible to test predictions of sex-change theory by altering those conditions to induce sex change in either direction (Munday 2002). Examining the social and environmental conditions under which sex change in each direction occurs in the same individual could provide a powerful test of sex allocation theory. Furthermore, some bidirectional sex changers have gonadal sex-cell allocation that is more similar to that of simultaneous hermaphrodites than it is to most sequential hermaphrodites (St. Mary 1993, 1996). Therefore, these species might also be useful for bridging the gap between sex allocation theory related to simultaneous hermaphrodites and theory related to sequential hermaphrodites (St. Mary 1997).

Most examples of bidirectional sex change involve repetitive or serial sex change, where adults change sex more than once, usually female–male–female. These species often appear to be fundamentally protogynous, with the exception that males sometimes revert to functional females. In addition to repetitive sex change by adults, some bidirectional sex changers are capable of maturing into either sex at maturation (Kuwamura et al. 1994a; Hobbs et al. 2004; Liu & Sadovy 2004). Furthermore, Kuwamura et al. (2007) found that primary males of a diandric protogynous wrasse can change sex to female. Therefore, even where alternative reproductive strategies develop early in life, there can remain the potential for sex change in each direction by adults. These examples demonstrate that bidirectional sex change may take a number of different forms and that lability in sexual expression may be present throughout much of an individual's life history.

Here we review the distribution of bidirectional sex change among teleost fishes and examine the patterns of sexual development and sexual expression that these species exhibit. We then use this information in conjunction with behavioral and ecological correlates to consider the adaptive significance of bidirectional sex change. We consider what bidirectional sex change tells us about the evolutionary advantage of sex change in general, and whether it is consistent with existing explanations for the presence of sex change in animals. Finally, we explore the proximate mechanisms controlling sex change in each direction and the physiological changes that occur in the brain, gonads, and endocrine system, which enable such extreme sexual lability to occur.

SEXUAL PATTERNS OF BIDIRECTIONAL SEX CHANGERS

Functional hermaphroditism has been confirmed in 27 families and 94 genera of teleost fishes, with a further 21 families and 31 genera containing unconfirmed examples (Sadovy de Mitcheson & Liu 2008). Many hundreds of species within these families are known to change sex. To date, 6 families, 12 genera, and 25 species are confirmed to be capable of bidirectional sex change (Table 8.1). A similar number of closely related species are likely to exhibit bidirectional sex change, based on the similarity of their ecologies and patterns of gonadal sex allocation with known bidirectional sex changers (Table 8.1). It is almost certain that many more fish species will be found to be bidirectional sex changers once the potential for this sexual pattern is tested further. The families where bidirectional sex change has so far been detected are commonly found on coral reefs and many of the species have a small body size and cryptic lifestyle. Bidirectional sex change seems to be particularly prevalent in the family Gobiidae and these small fishes offer an excellent opportunity to compare and contrast the conditions favoring sex change in each direction.

Bidirectional sex change is commonly detected using laboratory manipulations where pairs or groups of individuals of the same sex are kept together in aquariums. These experiments demonstrate that a species has the capacity for bidirectional sex change and can provide important information on the proximate mechanisms controlling sex change. However, they might not provide an accurate indication of the frequency of sex change in each direction in nature. For example, both male and female *Trimma okinawae* can be induced to change sex by keeping same-sex pairs together in aquariums (Sunobe & Nakazono 1993). However, monitoring of natural populations shows that sex change from male to female occurs less frequently than sex change from female to male (Manabe et al. 2007a). Therefore, even though both males and females have an equal capacity to change sex, this is not reflected in patterns of sex change in nature. Similarly, *Paragobiodon echinocephalus* can be induced to change sex from male to female by placing two males together on a coral head (Nakashima et al. 1995), but sex change by males in the natural population is much less frequent than sex change by females (Kuwamura et al. 1994a).

TABLE 8.1 Species confirmed or likely to exhibit bidirectional sex change

Family	Species	Data type	Habitat	Mating system	Spawning mode	Sexually dichromatic	Primary references
Gobiidae	Trimma okinawae	F & C	CR, cryptic	polygynous	demersal	N	Sunobe and Nakazono 1993; Manabe et al. 2007a
	T. grammistes	C	RR, cryptic	polygynous	demersal	N	Shiobara 2000
	T. kudoi	C	RR, cryptic	polygynous	demersal	N	Manabe et al. 2008
	T. yanagitai	C	RR	unknown	demersal	N	Sakurai et al. 2009
	Paragobiodon echinochephalus	F & C	CR, cryptic	monogamous	demersal	N	Kuwamura et al. 1994a; Nakashima et al. 1995
	Gobiodon histrio	F	CR, cryptic	monogamous	demersal	N	Munday et al. 1998; Munday 2002
	G. erythrospilus[a]	F & C	CR, cryptic	monogamous	demersal	N	Nakashima et al. 1996; Munday 2002
	G. micropus	C	CR, cryptic	monogamous	demersal	N	Nakashima et al. 1996
	G. sp.[b]	C	CR, cryptic	monogamous	demersal	N	Nakashima et al. 1996
	G. quinquestrigatus	C	CR, cryptic	monogamous	demersal	N	Nakashima et al. 1996
	Lythrypnus dalli	F & C	RR	polygynous	demersal	N	St. Mary 1993, 1996; Reavis & Grober 1999; Black et al. 2005a, b; Rodgers et al. 2007
Serranidae	Epinephelus akaara	C	RR	unknown	pelagic	N	Tanaka et al. 1990
	Cephalopholis boenak	C	CR	polygynous	pelagic	N	Liu & Sadovy 2004, 2005
Pseudochromidae	Pseudochromis flavivertex	C	CR, cryptic	pairs	demersal	N	Michael 2004; Wittenrich & Munday 2005
	P. aldabraensis	C	CR, cryptic	pairs	demersal	N	Michael 2004; Wittenrich & Munday 2005
	P. cyanotaenia	C	CR, cryptic	pairs	demersal	Y	Michael 2004; Wittenrich & Munday 2005
Pomacanthidae	Centropyge acanthops	C	CR	polygynous	pelagic	N	Hioki & Suzuki 1996
	C. ferrugata	C	CR	polygynous	pelagic	Y	Sakai et al. 2003
	C. fisheri	C	CR	polygynous	pelagic	N	Hioki & Suzuki 1996
	C. flavissimus	C	CR	polygynous	pelagic	N	Hioki & Suzuki 1996

Family	Species	Data	Habitat	Mating system	Spawning	Bidirectional	Reference
Cirrhitidae	*Cirrhitichthys aureus*	C	CR	unknown	pelagic	N	Kobayashi & Suzuki 1992
	C. falco	F	CR	polygynous	pelagic	N	Kadota (2009)
Labridae	*Labroides dimidiatus*	C	CR	polygynous	pelagic	N	Kuwamura et al. 2002
	Halichoeres trimaculatus	F & C	CR	polygynous	pelagic	Y	Kuwamura et al. 2007
	Pseudolabrus sieboldi	C	RR	polygynous	pelagic	Y	Ohta et al. 2003
Proposed or likely							
Gobiidae	*Paragobiodon xanthosomus*	H	CR, cryptic	monogomous	demersal	N	Lassig 1977; Fishelson 1989
	Gobiodon okinawae	H	CR	monogamous	demersal	N	Cole & Hoese 2001
	G. citrinus	H	CR, cryptic	unknown	demersal	N	Fishelson 1989
	Eviota epiphanes	H	CR, cryptic	unknown	demersal	N	Cole 1990
	Lythrypnus zebra	H & C	RR	unknown	demersal	N	St. Mary 1993, 1996
	L. nesiotes	H	CR, RR	unknown	demersal	N	St. Mary 2000
	Bryaninops 4 species	H	CR, cryptic	monogamous	demersal	N	Fishelson 1989; Munday et al. 2002
	Priolepis 3 species	H	CR, cryptic	monogamous	demersal	N	Cole 1990; Sunobe & Nakazono 1999
	Trimma 3 species	H	CR, cryptic	polygynous	demersal	N	Cole 1990
Pseudochromidae	*Ogilbyina queenslandiae*	F	CR, cryptic	pairs	demersal	Y	Ferrell 1987
Pomacanthidae	*Apolemichthys trimaculatus*	C	CR	unknown	pelagic	N	Hioki & Suzuki 1995
Cirrhitidae	*Neocirrhites armatus*	H	CR, cryptic	polygynous	pelagic	N	Sadovy & Donaldson 1995
	Cirrhitichthys 3 species	H	CR, cryptic	unknown	pelagic	N	Kobayashi & Suzuki 1992; Sadovy & Donaldson 1995
	Cirrhitops hubbardi	H	CR	unknown	pelagic	N	Kobayashi & Suzuki 1992
	Cyprinocirrhites polyactis	H	CR	unknown	pelagic	N	Kobayashi & Suzuki 1992

NOTE: Bidirectional sex change is confirmed or deemed likely based on observational data in the field (F) or in captivity (C), or on gonadal histology (H). Habitat is classified as coral reef (CR) or rocky reef (RR).

[a] *G. rivulatus rivulatus* in Nakashima et al. 1996.

[b] *G. oculolineatus* in Nakashima et al. 1996.

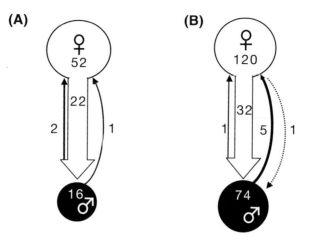

FIGURE 8.1. Frequency of sex change in each direction in natural populations of (A) *Trimma okinawae* and (B) *Paragobiodon echinocephalus*. Numbers of adult males and females in the study population are shown in the large circles. Arrows show the direction of sex change and the number of individuals changing sex. The thin, solid line to the left of the female-to-male sex change arrow indicates females that changed sex to male and then changed back to female during the course of each study. The dotted line in (B) indicates a male that changed sex to female and then back to male. Data from Manabe et al. 2007a and Kuwamura et al. 1994a.

Trimma okinawae and *P. echinocephalus* have been extensively studied and these two species serve as useful examples of different life histories exhibited by bidirectional sex changers. *Trimma okinawae* is a small polygynous goby that inhabits caves, holes and the undersides of corals. The most common social arrangement is a breeding group of one male and one to seven females (Sunobe & Nakazono 1990; Manabe et al. 2007a), although single males are common at the end of the breeding season. The male is usually the largest individual in a social group and protogynous sex change occurs if he disappears. In this circumstance the largest female within the group changes sex to male, or a large female from another social group immigrates and changes sex to male (Manabe et al. 2007a, b). Protogynous sex change also occurs when a female becomes single, either due to the loss of all other females in the group, or her movement from the group. Protandrous sex change only occurs when a solitary male enters a new social group that already contains a dominant male. Therefore, this species is basically protogynous, except that single males revert to female if they reenter a group and are smaller than the dominant male in the new group (Figure 8.1A).

Paragobiodon echinocephalus is a monogamous goby that lives among the branches of the coral *Stylophora pistillata*. The most common social arrangement is a breeding pair (Kuwamura et al. 1993, 1996). Any other individuals in the social group are juveniles. If one member of the breeding pair disappears, a new pair is formed through either (1) the maturation of a juvenile already present on the coral or the immigration and maturation of a juvenile from another coral, or (2) the immigration of an adult from another coral (Kuwamura et al. 1994a). Protogynous sex change occurs in two circumstances: when a single adult female changes sex and forms a pair with a juvenile that has matured as a female, or when adult immigration results in two females cohabiting a coral colony (one of them will change sex to male). In contrast, protandrous sex change only takes place when a male moves and forms a pair with another male - one of the two males will change sex to female. The narrower range of circumstances where males change sex compared to females means that male to female sex change tends to be much less frequent in the population than female to male sex change (Figure 8.1B).

In addition to bidirectional sex change as adults, juvenile coral-dwelling gobies (*Gobiodon* and *Paragobiodon* species) can mature into either sex depending on the sex of their future partner (Kuwamura et al. 1994a; Hobbs et al. 2004). Usually, a juvenile will mature as a female and form a pair with a single adult male or with an adult female that has changed sex to male. New pairs may also form by the maturation of two juveniles on a coral without an existing breeding pair. In this case, one of the juveniles matures as a female and the other as a male.

One of the most unexpected recent discoveries is that primary males of the diandric wrasse *Halichoeres trimaculatus* can change sex to female, and then back to male again (Kuwamura et al. 2007). In diandric species, males are derived either by sex change from female (secondary males) or by direct development before maturation (primary males). It has long been thought that primary males are gonochoristic and do not change sex. However, Kuwamura et al. (2007) observed sex change to female in a primary male in the field and induced sex change in primary males when they were kept together with other males and females in aquaria. This shows that fundamentally protogynous and protandrous life histories can occur within the same species.

In addition to well-described examples of bidirectional sex change in natural populations, such as those discussed above, there is a growing list of species where bidirectional sex change has been demonstrated in captivity (Table 8.1). Many of these species are known to be protogynous sex changers in natural populations, but it is not known if males change sex back to female, even though they clearly have the capacity to do so. The pygmy angelfishes, *Centropyge* species, are a good example. These fishes are haremic, protogynous sex changers (Sakai & Kohda 1997; Michael 2004) and the largest female in a group will change sex to male if the dominant male disappears. Laboratory manipulations of social groups have shown that males have the capacity to change sex back to female (Sakai et al. 2003), but whether they do so in nature is unknown.

Kuwamura et al. (2002) demonstrated the importance of both field and laboratory assessments of bidirectional sex change with the cleaner wrasse *Labroides dimidiatus*. This species is a haremic protogynous hermaphrodite where the dominant female changes sex to male following the loss of the harem male (Robertson 1972; Kuwamura 1984). Experiments in captivity revealed that the smaller male will change sex to female if two males are kept together in an aquarium. However, functional sex change by males could not be confirmed in nature. After removal of females in a low density population, single males formed pairs and exhibited spawning behavior for several days, but the pairs did not persist long enough for functional sex change to occur (Kuwamura et al. 2002). Determining if males of protogynous species known to be capable of sex change to female under laboratory conditions also exhibit reverse sex change in natural populations, and under what circumstances, is an important area for future research.

REPRODUCTIVE CHARACTERISTICS
Size and Age Structure

Signatures of protogynous or protandrous sex change are often present in the age- or size-frequency distribution of males and females (Sadovy & Shapiro 1987). The older age and size classes are expected to be dominated by males in protogynous sex changers. In contrast, older age and size classes are expected to be dominated by females in protandrous sex changers. These patterns are likely to break down in bidirectional sex changers, because larger and older individuals may revert to the original sex. Indeed, age and size distributions of males and females overlap substantially in some bidirectional sex changers (Munday et al. 1998; Manabe et al. 2008)(Figure 8.2A). In other bidirectional sex changers, such as *Trimma okinawae*, there is little overlap in the size-frequency distribution of males and females (Figure 8.2B). This occurs because only the largest individual in a social group becomes male, and males entering new groups change sex to female if they are smaller than the resident male (Manabe et al. 2007a). These different patterns show that size- or age-frequency distributions of the sexes do not always provide clear evidence for the presence or absence of bidirectional sex change.

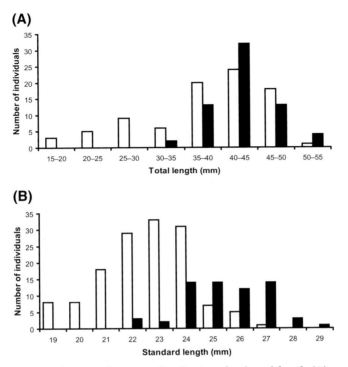

FIGURE 8.2. Size frequency distribution of male and female (A) *Gobiodon histrio* and (B) *Trimma okinawae*. White bars, female; black bars, male. Data from Munday et al. 1998 and Manabe et al. 2007a.

Sexual Dimorphism

Initial observations of bidirectional sex change mostly came from species where males and females had similar coloration and external morphology (except for the shape of the genital papillae, which often differs between males and females). It is easy to imagine that species exhibiting sex change in both directions might have similar male and female phenotypes, because this would reduce the morphological changes necessary to shift from one sexual phenotype to the other. However, sexual dimorphism does not preclude bidirectional sex change and a number of species in which males exhibit much brighter body coloration than females are capable of changing sex in each direction (Table 8.1). Changes in body coloration are closely linked to the timing of sex change in each direction in these species (Sakai et al. 2003; Wittenrich & Munday 2005; Kuwamura et al. 2007).

Gonadal Structure and Sex-Cell Allocation

Gonadal structure and patterns of sex allocation differ greatly among bidirectional sex-changing species. Some bidirectional species (e.g. *Trimma*) have delineated gonads with separate regions of ovarian and testicular tissue that proliferate or regress, depending on the functional sex of the individual (Table 8.2, Figure 8.3). The ovarian and testicular regions are often separated by a thin wall of connective tissue (Cole 1990; Sunobe & Nakazono 1993; Manabe et al. 2008). During sex change from female to male, the ovarian region contracts and ceases producing vitellogenic oocytes, and the testicular region expands and starts producing sperm. The opposite occurs during protandrous sex change (Sunobe & Nakazono 1993; Manabe et al. 2008). *Trimma* can change from one functional sex to the other in less than one week (Table 8.2), and it seems likely that the retention of a regressed gonadal region of either male or female tissue assists with this rapid transition in sexual function.

Other bidirectional sex-changing species do not have distinct separation of ovarian and testicular tissue within the gonad (Table 8.2, Figure 8.4), but tissue of the other sex is sometimes scattered throughout the functional ovary or testis. Testes in particular are likely to contain numerous previtellogenic oocytes. For example, there is no evidence of testicular tissue in mature ovaries of *Gobiodon* and *Paragobiodon*, but up to 20% of the testis can be taken up by previtellogenic oocytes (Cole 1990; Cole & Hoese 2001; Munday 2002)(Figure 8.4). Generally, the proportion of female tissue in the testis is greatest in newly sex-changed males and declines in males that have been in a breeding pair for an extended period (Munday 2002; Kroon et al. 2003). A similar pattern is seen in species of *Pseudochromis*; ovaries have no male tissue, but over 10% of the functional testes may be taken up by previtellogenic oocytes, especially after functional sex change (Wittenrich & Munday 2005). A common feature of all these species is that they do not have mature gametes of both sexes present in the gonad at the same time, nor do they retain regressed regions of ovarian or testicular tissue like that seen in *Trimma* species. During sex change, the reproductive tissue of one sex degenerates, and the reproductive tissue of the other sex develops throughout the gonad. Sex change in each direction often takes more than two weeks in these species (Table 8.2), probably because of the need to restructure the entire gonad.

Lythrypnus comprises an interesting group of hermaphroditic gobies because some species have gonadal sex-cell allocation resembling simultaneous hermaphrodites, even though individuals function as only one sex at time (St. Mary 1993, 1994, 1996, 2000). Gonads of these species can contain both vitellogenic oocytes and spermatozoa at the same time, although allocation to male and female sex cells is usually strongly biased to one sex or the other. More importantly, individuals function solely as male or solely as female over repeated breeding bouts.

TABLE 8.2 Reproductive characteristics of some bidirectional sex-changing fishes

Family/Species	Gonad type	Female to male		Male to female		Primary references
		Relative size	Min. time	Relative size	Min. time	
Gobiidae						
Trimma okinawae	delimited	L	6	S	4	Sunobe & Nakazono 1993
T. kudoi.	delimited	S	16	L	12	Manabe et al. 2008
Paragobiodon echinochephalus	non-delimited	L	27	S	24	Nakashima et al. 1995
Gobiodon histrio	non-delimited	L	<28	S	<28	Munday et al. 1998
G. okinawae	non-delimited	L	<21	S	>21	Cole & Hoese 2001
G. quinquestrigatus	non-delimited	L	30	S	23	Nakashima et al. 1996
Lythrypnus dalli	delimited	L	5–11	S	14	Reavis & Grober 1999; Black et al. 2005b
L. dalli		L	16	S	17	Rodgers et al. 2007
Pseudochromidae						
Pseudochromis flavivertex	non-delimited	L	28	S	52	Wittenrich & Munday 2005
P. aldabraensis	non-delimited	L	18	S	64	Wittenrich & Munday 2005
P. cyanotaenia	non-delimited	L	23	S	67	Wittenrich & Munday 2005
Pomacanthidae						
Centropyge acanthops	non-delimited	L	8	S	91	Hioki & Suzuki 1996
C. ferrugata	non-delimited	L	15	S	47	Sakai et al. 2003
C. fisheri	non-delimited	L	6	S	35	Hioki & Suzuki 1996
Labridae						
Labroides dimidiatus	non-delimited	L	14	S	53	Kuwamura et al. 2002

NOTE: Gonadal form, minimum recorded time to change sex in each direction, and the relative size of the individual(s) that change sex within same-sex groups are reported. Minimum time to sex change is based on the observation of fertilized egg clutches or complete functional sex change assessed by histological examination of gonads.

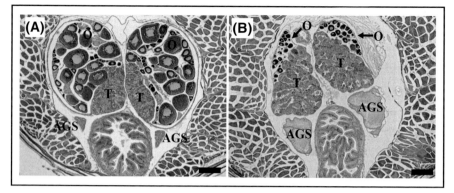

FIGURE 8.3. Gonadal structure of (A) functional female and (B) functional male *Trimma kudoi*. O = ovary. T = testis. AGS = accessory gonadal structure. Gonads of functional females contain vitellogenic oocytes but no spermatozoa in the testicular region and an inactive AGS. Gonads of functional males have spermatozoa and an active AGS, which is characteristic of functional males, but no vitellogenic oocytes. Adapted from Manabe et al. 2008.

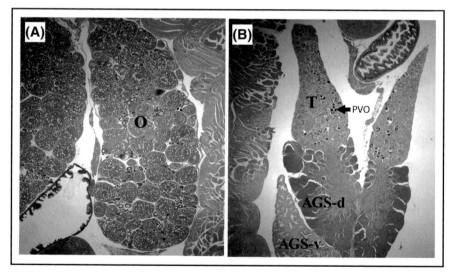

FIGURE 8.4. Gonadal structure of (A) functional female and (B) functional male *Gobiodon erythrospilus*. O = ovary. T = testis. AGS-d = dorsal accessory gonadal structure. AGS-v = ventral accessory gonadal structure. Active, secretory AGS-d is characteristic of a functional male. Females do not have secretory AGS-d. Note previtellogenic oocytes (PVO) scattered throughout the testis.

Therefore, from a functional perspective, these species are sequential hermaphrodites. *L. dalli* changes from one functional sex to the other and can appropriately be considered a bidirectional sex changer (Kuwamura & Nakashima 1998; Rodgers et al. 2007). All species of *Lythrypnus* have a unique gonadal structure with a central area of ovarian tissue between two or three peripheral regions of testicular tissue (St. Mary 1998). During sex change from functional female to functional male, the area of testicular tissue increases, and ovarian tissue declines. The opposite occurs during sex change from functional male to functional female.

Correlates of Gonadal Allocation and Duration of Sex Change

In general, differences in gonadal allocation among genera appear to be broadly related to the probability of unpredictable changes in social organization. The gonads of habitat-specialist species that are highly site attached and rarely move among social groups, such as *Gobiodon* and *Paragobiodon*, are largely dominated by either male or female tissue, and there are never mature gametes of both sexes present. Individuals of these species tend to retain the same partner for considerable periods of time and are therefore less likely to have the need for rapid or repeated sex change. In contrast, species that are less site attached and more likely to move between social groups, such as *Trimma* and *Lythrypnus*, exhibit either a greater allocation to both male and female tissue, or they continue to produce small numbers of mature gametes of the opposite sex, which might aid rapid sex change in each direction. Consequently, it seems that patterns of gonadal allocation among bidirectional sex changers are often associated with the predictability of their social environment.

The time required to change sex in each direction appears to be approximately equal for some species, but much slower from male to female in other species (Table 8.2). Species in which male to female sex change takes much longer than female to male sex change tend to be polygynous species that are known to exhibit protogynous sex change in nature, but where reverse sex change by males has so far not been observed. Strong dominance behaviors by harem or territorial males of these species may increase the time required to initiate sex change in subordinate males. For example, aggressive relationships between male *Centropyge ferrugata* can take several weeks to be settled, even when individuals are in continuous close contact (Sakai et al. 2003). In the field, these strong dominance behaviors may prevent subordinate males entering the territory or harem and thus limit the opportunities for male to female sex change.

There is also considerable variation among individuals in the time required for sex change and some of this variation is related to environmental conditions. For example, sex change in each direction takes approximately the same length of time in *L. dalli* when conditions are similar for groups of males and groups of females (Rodgers et al. 2007), but the time taken to change sex is influenced by temperature, social organization, and season (Black et al. 2005b; Lorenzi et al. 2006). Sex change is more rapid in warmer water (Black et al. 2005b), when dominance hierarchies have been established (Black et al. 2005b; Lorenzi et al. 2006) and earlier in the breeding season (Lorenzi et al. 2006).

Individuals also adjust their allocation to male and female tissue depending on their current reproductive success and the certainty of mating opportunities. Male *L. dalli* that do not receive eggs in their nest retain higher levels of female tissue in their gonads than do males that have received eggs, possibly so that they can rapidly change sex back to female (St. Mary 1994). Similarly, female *L. dalli* delay sex change more near the end of the breeding season, perhaps because the cost of sex change could outweigh the benefits when breeding has nearly ceased (Lorenzi et al. 2006). Both of these observations suggest that individuals assess and adjust their gonadal sex allocation in relation to their expected reproductive success as either a male or a female.

ADAPTIVE SIGNIFICANCE
Bidirectional Sex Change and the Size-Advantage Hypothesis

The size-advantage hypothesis (SAH) predicts that sex change will be favored when the reproductive success of the sexes increases unequally with size, such that an individual reproduces more efficiently as one sex when small and more efficiently as the other sex when large (Ghiselin 1969; Warner 1975). The direction of sex change is often associated with the mating system: protogynous species typically exhibit a polygynous mating system, whereas protandrous species tend to be monogamous or random pair spawners (Munday et al. 2006a). The correlation between the mating system and the direction of sex change is consistent with expectations of the SAH, because male reproductive success tends to increase more steeply with size in polygynous mating systems, thereby favoring sex change from female to male (Figure 8.5A). In contrast, female reproductive success tends to increase more steeply with size in monogamous and random pair-spawning systems, thereby favoring sex change from male to female (Figure 8.5B). However, both polygynous and monogamous mating systems are associated in some instances with bidirectional sex change (Table 8.1), which raises the question of whether the SAH can explain sex change in each direction and, if it can, whether it is capable of providing an explanation for bidirectional sex change in all species.

Bidirectional sex change in the polygynous *T. okinawae* appears to be consistent with the SAH (Nakashima et al. 1995). This species is fundamentally a protogynous sex changer, except that males can change sex back to female. Individuals first breed as female and then change sex to male if they become large enough to control a harem of females (Sunobe & Nakazono 1990). Protogyny is favored

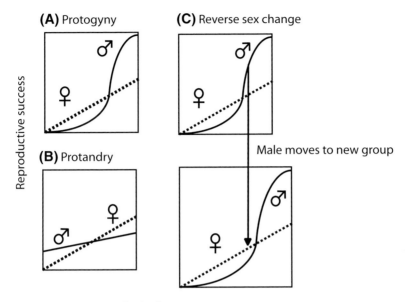

FIGURE 8.5.

A. Protogynous sex change can be favored in polygynous groups because male reproductive success (dark line) tends to increase more steeply with size than does female reproductive success (dotted line).

B. Protandrous sex change can be favored in monogamous groups or random pairing because female reproductive success tends to increases more steeply with size than does male reproductive success.

C. Repetitive sex change is favored if a male that changes sex from female moves to a new group where he is smaller than the resident male. In this case his reproductive success will be improved by changing sex back to female, because the resident male will prevent him breeding as a male. Adapted from Nakashima et al. 1995.

in haremic mating systems, because small males are excluded from breeding by the dominant male and thus have very low reproductive success. In contrast, large males have very high reproductive success because they mate with all the females in the harem. In polygynous societies, individuals can maximize their lifetime reproductive output by first reproducing as female and then changing sex to male when they reach a size large enough to defend a harem. As expected, the largest female in a *T. okinawae* harem changes sex following loss of the dominant male (Sunobe & Nakazono 1990; Manabe et al. 2007a, b). Females even move between social groups to improve their position in the size hierarchy and increase their chance of becoming the dominant female that changes sex following the loss of the male (Manabe et al. 2007b).

Reverse sex change in *T. okinawae* occurs if a single male enters an existing group where he is smaller than the resident male (Manabe et al. 2007a). The larger dominant male is likely to exclude him from reproducing as a male in the new group, but he could reproduce as a female. Therefore, his current reproductive success would be greater if he changed sex and functioned as a female. In effect, the male has slipped back down the size-fecundity slope to a position where his current size-related reproductive success is greater as a female than as a male (Figure 8.5C). As expected, he changes sex back to female. Bidirectional sex change in *T. okinawae* is an excellent test of the SAH, because it shows that females generally change sex when their reproductive success is higher as a male and revert to female if their reproductive success as a male falls below that of a similar sized female in the same social group.

An important question is why a male should enter an existing social group and change sex to female, instead of remaining male. Male *T. okinawae* only enter a new group and change sex to female if they are single (Manabe et al. 2007a). Because of their small size, most gobies are exposed to a very high risk of predation (Munday & Jones 1998) and consequently tend to have short life spans (Hernaman & Munday 2005; Depczynski & Bellwood 2006; Winterbottom & Southcott 2008). High and unpredictable rates of mortality mean that a male might unexpectedly lose all the females in his harem. The high risk of predation also means that a single male is likely to have a limited and uncertain period for reproduction available to him. Consequently, any delay in finding a breeding partner might seriously diminish his reproductive value (expected future reproductive success taking into account effects of growth and the probability of survival)(Munday et al. 2006a). The reproductive value of a single male may often be greater if he changes sex and breeds as a female in an existing group than if he remains male and waits for a new harem to develop.

Bidirectional Sex Change and Size-Fecundity Skew

Adjustment of functional sex in relation to reproductive value also seems to occur in *Trimma kudoi* (Manabe et al. 2008). This species appears to have a polygynous mating system, where males are larger than females and the sex ratio is highly skewed toward females. Therefore, we might expect the largest individual in a social group to always be male. However, Manabe et al. (2008) found that the second largest female usually changed sex to male when small groups of females were kept in aquariums. Muñoz & Warner (2003) showed that the largest female might not benefit from changing sex to male if the combined fecundity of all other females in the group (i.e., her prospective harem after sex change) was less than her existing fecundity. In this case, one of the smaller females may benefit most from changing sex to male. The patterns of sex change observed in *Trimma Kudoi* were broadly consistent with this prediction. Furthermore, when pairs of males or pairs of females were kept together, the larger one usually became/stayed female and the smaller stayed/became male (Manabe et al. 2008). This is exactly the social arrangement expected to maximize reproductive success in a breeding pair. Consequently, it seems that individuals of this species may adjust their sex to suit the precise social circumstances they encounter and it will not always be the largest individual that changes sex to male. Further analysis on the mating system of this species and a comparison of the size of males and females in groups of different sizes are needed to determine if individuals are indeed adjusting their sexual function in relation to fecundity relationships associated with different sized groups, or if there are other benefits to small male size.

Bidirectional Sex Change and the Cost of Movement

In contrast to species of *Trimma*, bidirectional sex change in coral-dwelling gobies (*Gobiodon* and *Paragobiodon*) is not consistent with the SAH, because there is no difference in the rate at which reproductive success of males and females increases with size (Kuwamura et al. 1993; Munday 2002). Coral-dwelling gobies are pair forming and monogamous. The reproductive success of a breeding pair depends on the size of both the male and the female (Kuwamura et al. 1993), and consequently, the male and female in a pair are approximately the same size (Kuwamura et al. 1993; Nakashima et al. 1996; Munday et al. 1998). Why reproductive success depends on the size of both sexes is uncertain, but it is likely that the fecundity of females increases with size and that the ability of males to care for the egg clutch also increases with size. Munday et al. (2006b) demonstrated that if one individual in a pair of *G. histrio* is smaller than the other, the smaller individual accelerates its growth and the larger individual slows its growth until the pair becomes size matched. This adjustment of growth by both individuals, regardless of sex, supports the notion that both male and female size are critical to the reproductive success of the pair.

The advantage of bidirectional sex change in coral-dwelling gobies is that they may need to move among hosts to find a new partner following the loss of a breeding partner, or the death of their host coral colony. Movement is likely to be risky for these small fishes. Furthermore, nearly all suitable coral colonies are occupied by a breeding pair and there are very few single adults in the population (Kuwamura et al. 1994a, b; Hobbs & Munday 2004). Being able to change sex in either direction enables an adult coral goby to mate with any other adult it encounters (Nakashima et al. 1995; Munday et al. 1998; Munday 2002). Nakashima et al. (1995) demonstrated that single *P. echinocephalus* adults prefer to pair with an adult of the same sex on a nearby coral rather than moving farther to find a single partner of the opposite sex. A similar pattern occurs with *G. histrio* and the closely related *G. erythrospilus* on the Great Barrier Reef (Munday unpublished data). Therefore, the benefit of changing sex in each direction appears to be that it reduces the risk of predation associated with searching for a new mate of the appropriate sex in a situation where there is low mate availability.

Juveniles reside either on the same corals as adult pairs (Kuwamura et al. 1994b, Thompson et al. 2007), or in corals that are too small for a breeding pair (Kuwamura et al. 1994a, 1996; Hobbs & Munday 2004). These juveniles will readily form a breeding pair with a single adult (Kuwamura et al. 1994a; Hobbs et al. 2004). Only protogynous sex change would be required if pair formation always occurred by this mechanism—a single female would change sex to male, and the juvenile would mature as a female to form a breeding pair. However, by pairing with a small partner an adult is likely to suffer a significant loss of reproductive success. Instead of forming a pair with a small juvenile, some adults move and pair with another adult (Kuwamura et al. 1994a; Nakashima et al. 1996; Munday et al. 1998; Munday 2002). Protogynous sex change occurs if two females end up cohabiting, and protandrous sex change occurs if two males end up cohabiting.

Coral-dwelling gobies would need to move and might form new partnerships if their host coral dies. Gobies might also move to secure more favorable habitat patches or larger partners. The size of coral-dwelling gobies is often correlated with the size of their host coral and the space between the coral branches. Small, tightly branched coral heads tend to contain small pairs, whereas, large, more openly branched coral heads tend to contain larger pairs (Kuwamura et al. 1994b; Munday 2001; Hobbs & Munday 2004). Because reproductive success of both males and females is size dependent, an individual of either sex might move from an existing partnership if a position becomes available in a superior coral. The benefits of moving to find a larger partner or a more favorable coral colony are, however, offset by the risk of predation involved in searching for a new partner or host coral. This trade-off probably explains why adults rarely move more than a few meters, even when superior breeding opportunities are available elsewhere (Nakashima et al. 1995).

Bidirectional Sex Change and Alternative Mating Strategies

Bidirectional sex change has recently been discovered in primary males of the diandric, protogynous wrasse *Halichoeres trimaculatus* (Kuwamura et al. 2007). Why primary males change sex has not been explicitly tested, but it might be associated with changes in the densities or proportions of different male strategies within the local population, which affect the success of these different strategies. Diandric protogynous hermaphrodites have alternative male mating strategies—individuals either mature as female and then change sex to become a large territorial male later in life, or they mature as a small primary male. In some species, the proportion of primary males in the population varies depending on the reproductive success likely to be associated with that strategy. For example, in the bluehead wrasse *Thalassoma bifasciatum*, there are few primary males on small reefs with small populations because territorial males can control most spawning events under these conditions, and thus primary males have low reproductive success (Warner & Hoffman 1980; Munday et al. 2006c). The number of primary males increases sharply on contiguous reefs with large populations, because primary males have high reproductive success under these conditions (Warner & Hoffman 1980). It is possible that sex change by primary males is associated with changes in local densities or sex ratios that alter the favorability of a primary male strategy. Primary males may change sex to female if social conditions change such that their reproductive success becomes higher as a female than as a primary male (Kuwamura et al. 2007).

Alternative male strategies have also been described in the bidirectional sex-changing goby *L. dalli* (Drilling & Grober 2005). This species could provide an excellent opportunity to test if: (1) the proportion of small sneaker males varies with male density or other components of the social environment that influence the mating success of this male strategy; and (2) if these sneaker males change sex to female when their reproductive success becomes seriously constrained by competition with larger males.

Toward a General Explanation for Bidirectional Sex Change

The unifying principal favoring bidirectional sex change in the previous examples is that it can be beneficial when social conditions experienced by males might change dramatically and unpredictably and when a delay in finding a new partner sex could result in a substantial loss of reproductive output. Bidirectional sex change in other small species, such as *Lythrypnus*, also appears to be related to unpredictable changes in social organization and the potential loss of reproductive output (St. Mary 1994). Bidirectional sex change is most prevalent among small coral reef fishes and is particularly well developed in gobies, many of which are known to experience high rates of mortality (Hernaman & Munday 2005). Bidirectional sex change is also common among habitat-specialist fishes where there may be few options to form new partnerships following mate loss. The advantage of bidirectional sex change in larger, habitat generalist species, such as serranids, remains uncertain.

In some instances, bidirectional sex change is clearly consistent with the SAH; in other cases, it is not. In general, however, sex change appears to occur when it is likely to increase an individual's reproductive value. Bidirectional sex change enables individuals to tailor their sexual function to suit unpredictable circumstances that have significant impacts on reproductive value. The fact that some individuals change sex more than once and that they do so when it is likely to increase their reproductive value suggests that individuals can assess their immediate environment to make decisions about the costs and benefits of changing sex, or not.

PROXIMATE MECHANISMS AND
PHYSIOLOGICAL CONTROL

Social Factors

The social environment has a strong influence on the timing of sex change in many hermaphroditic fishes (Shapiro 1984; Warner 1988; Ross 1990). A common observation is that the presence of a dominant male prevents females from changing sex in protogynous species, whereas the presence of a dominant female prevents males changing sex in protandrous species. Social conditions also influence the timing and direction of sex change in bidirectional sex changers, although not in a way consistent with a simple suppression model of sex change (Kuwamura & Nakashima 1998). In bidirectional sex changers, the presence of a large, dominant individual can either suppress or induce sex change in a smaller subordinate of the same sex. In most species studied to date, the larger individual changes sex to male when two or more females are kept together in a group. In contrast, one or more of the smaller individuals change sex to female when two or more males are kept together in a group (e.g., Sunobe & Nakazono 1993; Nakashima et al. 1995; Reavis & Grober 1999; Kuwamura et al. 2002; Munday 2002; Sakai et al. 2003; Wittenrich & Munday 2005; but see Manabe et al. 2008). This indicates that individuals adjust their sex according to their position in a social hierarchy.

In polygynous species the choice of male phenotype by the largest individual in a group and female phenotype by smaller individuals in the group is easily understood by the relative reproductive success of males and females of different sizes in this mating system—the largest individual has higher reproductive success as a male and smaller individuals have higher reproductive success as female. Rodgers et al. (2007) demonstrated that individuals of *L. dalli* determined their sexual phenotype based on a simple operational principle: be female if subordinate and male if dominant. This simple rule of thumb is likely to maximize individual reproductive success within a polygynous mating system, because large males have the highest reproductive success (St. Mary 1994). Dominance relationships are usually size based but may be related to other factors among similar-sized individuals. Consequently, a slightly smaller individual may change sex to male, or retain the dominant male position, if it has already established dominance within the group. The timing of sex change was also influenced by the strength of dominance relationships within groups. Sex change occurred rapidly when one individual was larger than the others and was therefore clearly dominant. Sex change took longer when individuals were of similar size and dominance relationships needed to be established independently of body size (Black et al. 2005b; Rodgers et al. 2007).

As observed in other species, the larger individual becomes/stays male and the smaller individual becomes/stays female when two coral-dwelling gobies of the same sex cohabit a coral colony (Kuwamura et al. 1994a; Nakashima et al. 1996; Munday 2002). In this case, however, males and females have similar size-related reproductive success and, therefore, larger individuals do not appear to choose the male phenotype simply because it increases their fertility. Single females always change sex to male (Kroon et al. 2005), which suggests that there might be some other benefit to adopting the male phenotype when dominant. For example, males might have lower mortality rates than females (Munday unpublished data). Pairs might also benefit from the larger individual being male and the smaller female because females tend to grow faster than males (Kuwamura et al. 1994a; Munday 2002). Both individuals in the pair would benefit by having the smaller partner adopt the faster-growing female phenotype because the pair will become size-matched more quickly (Kuwamura et al. 1994a; Munday et al. 2006b).

Interestingly, sex change does not occur in heterosexual sex pairs of coral-dwelling gobies, even if the male is smaller than the female (Kuwamura et al. 1994a; Munday 2002). This suggests that individuals refrain from changing sex when it is not necessary and the costs outweigh the benefits. In another exception to the common pattern of the larger individual changing sex to male, Manabe et al. (2008) found that it was usually the smaller individual that changed sex to male when two female *Trimma Kudoi* were kept together. Furthermore, an intermediate-sized individual usually changes sex to male in small groups of females (Manabe et al. 2008). These examples demonstrate that a simple suppression model of sex change based on dominant male behavior alone cannot explain the proximate control of sex change in all bidirectional sex changers. Individuals may often adjust their sexual function to suit the precise social conditions they experience and the proximate mechanisms that control sex change must therefore be sensitive to these conditions.

Linking Social Factors to Endocrine Control Mechanisms

Sexual development in teleost fishes is largely determined by reciprocal interactions in the brain-pituitary-gonadal (BPG) axis (Francis 1992; Perry & Grober 2003; Frisch 2004). In species with social control of sex change, such as bidirectional sex changers, the brain is hypothesized to determine the gonadal sex because social events can only affect the gonads through the brain (Kobayashi et al. 2009). The hormonal component of the BPG axis in the brain is gonadotropin releasing hormone (GnRH), which is synthesized primarily in the hypothalamus, preoptic area, and nervus terminalis. This hormone stimulates the release of pituitary gonadotropin (GtH) into the blood. GtH stimulates gametogenesis and steroidogenesis in the gonad. The gonadal steroid hormones, of which testosterone (T), 11-ketotestosterone (11-KT), and estradiol (E_2) are the most studied, are carried in the blood and play an important role in gametogenesis and in regulating reproductive behavior and development of secondary sexual characteristics (Fostier et al. 1983; Liley & Stacey 1983).

Kroon et al. (2003, 2005) demonstrated that a single enzymatic pathway can regulate both female and male sexual differentiation in the coral-dwelling goby *G. erythrospilus*. E_2 concentration in females was twice that in males, while concentrations of T did not differ between the sexes (Kroon et al. 2003). Manipulating E_2 levels via the aromatase (P450arom) pathway induced adult sex change in each direction under natural social conditions (Kroon et al. 2005). The presence and activity of P450arom controls the androgen/estrogen ratio by catalyzing the irreversible conversion of T into E_2. Specifically, an increase in E_2 resulted in protandrous sex change and a decrease in E_2 resulted in protogynous sex change (Kroon et al. 2005).

Gene isoforms for the cytochrome P450arom enzyme have been identified in ovarian (*CYP19A1*) and brain (*CYP19A2*) tissue of *G. histrio* (Gardner et al. 2003, 2005), further supporting a role for E_2 and the aromatase pathway in mediating sex change in each direction. Variations in levels of T and 11-KT do not appear to play a role in regulating sex change in *Gobiodon* species (Kroon et al. 2003, 2009). In fact, a lack of sex-specific differences in 11-KT concentrations may permit serial adult sex change in bidirectional sex-changing species, such as *Gobiodon* (Kroon et al. 2009).

At least in *Gobiodon* species, it seems likely that behavioral interactions between individuals could mediate sex change by the regulation of E_2 synthesis through the aromatase pathway (Kroon et al. 2005). This mechanism of socially controlled sex change could operate through behavioral modulation of cortisol concentrations and subsequent regulation of the glucocorticoid response element (GRE) on the *CYP19A1* isoform (Gardner et al. 2005). The patterns of sex change in this genus could thus be explained if exposure to male behavior activates cortisol synthesis, which in turn facilitates E_2 production via the GRE transcriptional factor on *CYP19A1*, suppressing protogynous sex change in females and inducing protandrous sex change in subordinate males. Conversely, the absence of male behavior would deactivate cortisol synthesis and reduce E_2 production, inducing protogynous sex change in dominant or single females. Further research is needed to link behavioral patterns in *Gobiodon* species to changes in steroid concentrations that could mediate sex change in each direction.

Gonadal steroidogenic pathways involved in bidirectional sex change have also been examined in *T. okinawae*. During sex change in either direction, three steroidogenic enzymes required for converting cholesterol into gonadal steroids (P450 cholesterol side-chain cleavage, P450scc; 3β-hydroxysteroid dehydrogenase, 3β-HSD; and P450arom) were detected in gonadal tissue (Sunobe et al. 2005a, 2005b). Sunobe et al. (2005a) hypothesized that the activity of the P450arom enzyme in the ovary, as well as of other enzymes involved in steroidogenesis, may mediate sex change in either direction. A transcription factor involved in the regulation of the three steroidogenic enzymes, Ad4-binding protein (Ad4BP)/

steroidogenic factor-1(SF-1), was found in ovarian, testicular, brain, and kidney tissue (Kobayashi et al. 2005). During adult serial sex change, expression of Ad4BP/SF-1 was observed in ovarian tissue but not in testicular tissue and was higher in females than in males, suggesting a direct relationship between Ad4BP/SF-1 expression and the female phase in *T. okinawae* (Kobayashi et al. 2005).

Gonadal steroids such as 11-KT and E_2 have been documented in *L. dalli* (Carlisle et al. 2000; Black et al. 2005b; Rodgers et al. 2006). In females undergoing protogynous sex change, 11-KT concentrations in urine samples were correlated with the percentage of testicular tissue, size of the accessory gonadal structure (AGS), and male behavior (Black et al. 2005b). Furthermore, female *L. dalli* implanted with 11-KT developed enlarged testicular tissue, regressed ovarian tissue, and active AGS (Carlisle et al. 2000). 11-KT concentrations and behavioral interactions were correlated, but it is unclear whether 11-KT concentrations influences behavior or vice versa (Black et al. 2005b). Relationships between E_2 concentrations and behavior and reproductive morphology were not presented.

The potential importance of the aromatase pathway in mediating protogynous sex change in *L. dalli* has been explored by Black et al. (2005a). Aromatase activity in both brain and gonadal tissues was significantly higher in females than in males. However, changes in aromatase activity in brain and gonadal tissues during the process of protogynous sex change were dissimilar, suggesting differential regulation of aromatase activity in these two tissues (see, for example, Gardner et al. 2005). Black et al. (2005a) hypothesize that changes in brain aromatase activity affect steroid concentrations, resulting in both behavioral and morphological sex change in *L. dalli*.

Overall, these studies suggest that changes in E_2 concentrations play a key role in mediating sexual function in bidirectional sex changers. Examination of gonadal steroids and steroidogenic pathways demonstrate that a single enzymatic pathway can regulate both female and male sexual differentiation (Kroon et al. 2005). Specifically, activation and deactivation of the P450arom pathway appears to initiate serial sex change in bidirectional sex-changing fish. It is likely that the activity of the aromatase pathway regulates sex change by increasing or decreasing the availability of E_2 precursors, such that these become available (or not) for androgen pathways such as 17β-hydroxysteroid dehydrogenase (Ohta et al. 2003). Further research is needed to understand how changes in social organization interact with the BPG axis in different species to influence the endocrine pathways mediating sex change in each direction.

CONCLUSIONS

Research on the reproductive strategies of hermaphroditic fishes over the past two decades has overturned the assumption that individuals change sex only once in their life and revealed a remarkable complexity and diversity of sexual strategies that fishes use to maximize their reproductive success (Kuwamura & Nakashima 1998; Munday et al. 2006a). It is now clear that a considerable number of tropical marine fishes are capable of bidirectional sex change and at least some of them exhibit this trait in nature. Bidirectional sex change is most prevalent among small coral-reef fishes, especially from the family Gobiidae (gobies). The unpredictability of their social circumstances, combined with a high risk of predation experienced due to their small body size, appears to favor bidirectional sex change in these species. The benefits of bidirectional sex change in larger species, such as some groupers, is uncertain.

Sex change in each direction appears to be consistent with the SAH for some polygynous species, but not for pair-forming species. In polygynous species, large females usually change sex to male when they have the opportunity to defend a harem. Males will change sex back to female if they move to a new group where they are smaller than the resident male. It seems that males will opt to join a new group and change sex to female in circumstances where they have low reproductive value as a male (e.g., if they are single). Interestingly, in at least one polygynous species, it is usually one of the smaller females that changes sex in pairs or small groups of females (Manabe et al. 2008). This suggests that the even among polygynous species the conditions favoring sex change in each direction may differ.

Bidirectional sex change among pair-forming gobies is not consistent with the SAH. In these species, it seems that the low density of available partners and the high risk of searching for a new partner favor the ability to change sex in either direction to facilitate the formation of new breeding pairs. A general explanation for all bidirectional sex-changing species is that individuals change sex when it is likely to increases their reproductive value (expected future reproductive success accounting for size-based fecundity, growth and mortality). At least some species appear to have evolved the capacity to change sex in each direction because their local environmental conditions can change dramatically and unpredictably in a way that favors repeated sex change.

The message for sex allocation theory in general is that very complex sexual patterns can evolve when they are favored by variable and unpredictable environmental conditions. Existing explanations for the evolution of sex change are generally sufficient to account for the presence of bidirectional sex change, but no single hypothesis (e.g., SAH) can account for all cases. The overarching principle is that individuals usually change sex when it increases their reproductive value and individuals of many species clearly have reliable mechanisms for assessing their reproductive value as either a male or a female, and adjusting their sexual function accordingly.

FUTURE RESEARCH AND DIRECTIONS

The capacity for bidirectional sex change has most often been demonstrated in laboratory experiments where pairs or small groups of males or females are kept together in aquariums. How often sex change in each direction occurs in nature is unknown for most species. Studying bidirectional sex change under natural conditions in the field must become a priority in order to properly document the occurrence of this sexual pattern and understand its adaptive significance.

Correlating ecological and environmental conditions with patterns of sex change within and among species will help assess the advantages of bidirectional sex change in fishes. More experimental manipulations are also needed to explicitly test specific hypotheses about the benefits of sex change in each direction. Further comparative analyses and experiments using species with highly labile sexual function, such as species of *Lythrypnus*, should be particularly promising because these species are likely to exhibit the greatest flexibility in sexual function in relation to environmental conditions. The recent discovery of sex change by primary males of a protogynous sex-changing wrasse opens up another new avenue of research into bidirectional sex change and indicates that sexual patterns in many fishes are likely to be even more flexible than currently appreciated.

The endocrine mechanisms responsible for translating environmental stimuli into functional sex change are still not fully understood in fishes. Bidirectional sex changers provide a unique opportunity to assess the biochemical pathways involved in sex change and recent research has pointed to a key role of the aromatase pathway as a primary means of regulating sex change in each direction. Although we are gaining a better understanding of how steroid hormones act to regulate gonadal sex change, we still lack a clear perspective on how social behavior and brain function interact with steroidal hormones to induce or prevent sex change in fishes. Linking behavior and brain function with steroidal pathways is a key challenge for understanding sex change in fishes, especially in light of the different social environments of bidirectional sex-changing species.

Hermaphrodites can be classified according to two different characters—anatomy or function. This duality in the way that hermaphrodites are diagnosed has sometime caused confusion about whether a species should be called a sequential or simultaneous hermaphrodite. It is clear that a gradation exists in patterns of gonadal sex allocation (anatomy), ranging from species whose gonads have only male or only female sex cells present at one time to species that have equal allocations of male and female sex-cells in the gonads at the same time. In some of these species with mixed gonadal allocation, there are only ever mature gametes of one sex present. Gonads of some other species, however, may contain mature gametes of both sexes, even though individuals only function as one sex at a time. This gradation in sex-cell allocation, which is not always closely linked to an individual's functional sex, means that gonadal allocation (anatomy) alone is not sufficient to describe a species as a sequential hermaphrodite or a simultaneous hermaphrodite. We encourage the use of direct observations of breeding behavior and sexual function in conjunction with assessments of gonadal sex-cell allocation to aid the description of sexual strategies in hermaphroditic fishes (see also Sadovy de Mitcheson & Liu 2008).

The capacity for individuals to change sex more than once and also in association with changes in the social environment demonstrates the remarkable flexibility in reproductive function that many fishes possess and that this flexibility is an adaptive response to variation in local environmental conditions. Determining the conditions under which sex change in each direction occurs and the mechanisms controlling sexual function in these species will continue to be an exciting area of research, with important implications for understanding the functional role and evolutionary advantage of sex change in general.

ACKNOWLEDGMENTS

We are grateful to Hisaya Manabe and Tomoki Sunobe for providing data and photomicrographs of *Trimma* species, Sue Riley for assistance with histology and photomicrographs of *Gobiodon* species, and Mathew Grober and Tatsuru Kadota for unpublished data on *Lythrypnus dalli* and *Cirrhitichthys falco*. Tamoki Sunobe and Marian Wong provided helpful comments on the manuscript.

REFERENCES

Atz JW (1964). Intersexuality in fishes. In: Armstrong CN, Marshall AJ, (eds.), *Intersexuality in Vertebrates Including Man*. Academic Press, London, pp. 145–232.

Black MP, Balthazart J, Baillien M, Grober MS (2005a). Socially induced and rapid increases in aggression are inversely related to brain aromatase activity in a sex-changing fish, *Lythrypnus dalli*. Proceedings of the Royal Society B 272: 2435–2440.

Black MP, Moore B, Canario AVM, Ford D, Reavis RH, Grober MS (2005b). Reproduction in context: field testing a laboratory model of socially controlled sex change in *Lythrypnus dalli* (Gilbert). Journal of Experimental Marine Biology and Ecology 318: 127–143.

Carlisle SL, Marxer-Miller SK, Canario AVM, Oliveira RF, Canario L, Grober MS (2000). Effects of 11-Ketotestosterone on genital papilla morphology in the sex changing fish *Lythrypnus dalli*. Journal of Fish Biology 57: 445–456.

Cole KS (1990). Patterns of gonad structure in hermaphroditic gobies (Teleostei: Gobiidae). Environmental Biology of Fishes 28: 125–142.

Cole KS, Hoese DF (2001). Gonad morphology, colony demography and evidence for hermahroditism in *Gobiodon okinawae* (Teleostei: Gobiidae). Environmental Biology of Fishes 61: 161–173.

Depczynski M, Bellwood DR (2005). Shortest recorded vertebrate lifespan found in a coral reef fish. Current Biology 15: R288–R289.

Drilling CC, Grober MS (2005). An initial description of alternative male reproductive phenotypes in the bluebanded goby, *Lythrypnus dalli* (Teleostei, Gobiidae). Environmental Biology of Fishes 72: 361–372.

Ferrell DG (1987). Population structure of *Ogilbyina queenslandiae* on One Tree Island. Master's thesis, Department of Marine Biology, Sydney University, Queensland, Australia.

Fishelson L (1989). Bisexuality and pedogenesis in gobies (Gobiidae: Teleostei) and other fish, or, why so many little fish in tropical seas? Senckenbergiana Maritima 20: 147–169.

Fostier EP, Jalabert B, Billard R, Breton B, Zohar Y (1983). The gonadal steroids. In: Hoar WS, Randall DJ, Donaldson EM (eds.), *Fish Physiology*, vol. 9A. Academic Press, London, pp. 277–372.

Francis RC (1992). Sexual lability in teleosts: developmental factors. The Quarterly Review of Biology 67: 1–18.

Frisch A (2004). Sex-change and gonadal steroids in sequentially hermaphroditic teleost fish. Reviews in Fish Biology and Fisheries 14: 481–499.

Gardner L, Anderson TA, Place AR, Elizur A (2003). Sex change strategy and the aromatase genes. Fish Physiology and Biochemistry 28: 147–148.

Gardner L, Anderson T, Place AR, Dixon B, Elizur A (2005). Sex change strategy and the aromatase genes. Journal of Steroid Biochemistry and Molecular Biology 94: 395–404.

Ghiselin MT (1969). The evolution of hermaphroditism among animals. The Quarterly Review of Biology 44: 189–208.

Hernaman V, Munday PL (2005). Life history characteristics of coral reef gobies I. Growth and lifespan. Marine Ecology Progress Series 290: 207–221.

Hioki S, Suzuki K (1995). Spawning behavior, eggs, larvae, and hermaphroditism of the angelfish, *Apolemichthys trimaculatus*, in captivity. Bulletin of the Institute of Oceanic Research & Development, Tokai University 16: 13–22 (in Japanese).

Hioki S, Suzuki K (1996). Sex changing from male to female on the way of protogynous process in three *Centropyge* angelfishes (Pomacanthidae: Teleostei). Bulletin of the Institute of Oceanic Research & Development, Tokai University 17: 27–34 (in Japanese).

Hobbs J-PA, Munday PL (2004). Intraspecific competition controls spatial distribution and social organisation of the coral dwelling goby, *Gobiodon histrio*. Marine Ecology Progress Series 278: 253–259.

Hobbs J-PA, Munday PL, Jones GP (2004). Social induction of maturation and sex determination in a coral reef fish. Proceedings of the Royal Society B 271: 2109–2114.

Kadota T (2009). Ecological study on the mating system and sexual pattern of hawkfishes (Pisces: Cirrhitidae) on reefs of Kuchierabu-jima Island, Southern Japan. Doctoral thesis, Hiroshima University, Hiroshima, Japan.

Kobayashi K, Suzuki K (1992). Hermaphroditism and sexual function in *Cirrhitichthys aureus* and other Japanese Hawkfishes (Cirrhitidae: Teleostei). Japanese Journal of Ichthyology 38: 397–410.

Kobayashi Y, Sunobe T, Kobayashi T, Nagahama Y, Nakamura M (2005). Promotor analysis of two aromatase genes in the serial-sex changing gobiid fish, *Trimma okinawae*. Fish Physiology and Biochemistry 31: 123–127.

Kobayashi Y, Nakamura M, Sunobe T, Usami T, Kobayashi T, Manabe H, Paul-Prasanth B, Suzuki N, Nagahama Y (2009). Sex-change in the gobiid fish is mediated through rapid switching of gonadotropin receptors from ovarian to testicular protion or vice-versa. Endocrinology 150: 871–878.

Kroon FJ, Munday PL, Pankhurst NW (2003). Steroid hormone levels and bi-directional sex change in Gobiodon histrio. Journal of Fish Biology 62: 153–167.

Kroon FJ, Munday PL, Westcott DA, Hobbs, J-PA, Liley NR (2005). Aromatase pathway mediates sex change in each direction. Proceedings of the Royal Society B 272: 1399–1405.

Kroon FJ, Munday PL, Westcott DA (2009). Equivalent whole-body concentrations of 11-ketotestosterone in female and male coral goby (Gobiodon erythrospilus), a bi-directional sex-changing fish. Journal of Fish Biology 75: 685–692.

Kuwamura T (1984). Social structure of the protogynous fish Labroides dimidiatus. Publications of the Seto Marine Biological Laboratory 29: 117–177.

Kuwamura T, Nakashima Y (1998). New aspects of sex change among reef fishes: recent studies in Japan. Environmental Biology of Fishes 52: 125–135.

Kuwamura T, Yogo Y, Nakashima Y (1993). Size-assortative monogamy and paternal egg care in a coral goby Paragobiodon echinocephalus. Ethology 95: 65–75.

Kuwamura T, Nakashima Y, Yogo Y (1994a). Sex change in either direction by growth-rate advantage in the monogamous coral goby, Paragobiodon echinocephalus. Behavioral Ecology 5: 434–438.

Kuwamura T, Yogo Y, Nakashima Y (1994b). Population dynamics of goby Paragobiodon echinocephalus and host coral Stylophora pistillata. Marine Ecology Progress Series 103: 17–23.

Kuwamura T, Nakashima Y, Yogo Y (1996). Plasticity in size and age at maturity in a monogamous fish: effect of host coral size and frequency dependence. Behavioral Ecology and Sociobiology 38: 365–370.

Kuwamura T, Tanaka N, Nakashima Y, Karino K, Sakai Y (2002). Reversed sex-change in the protogynous reef fish, Labroides dimidiatus. Ethology 108: 443–450.

Kuwamura T, Suzuki S, Tanaka N, Ouchi E, Karino K, Nakashima Y (2007). Sex change of primary males in a diandric labrid Halichoeres trimaculatus: coexistence of protandry and protogyny within a species. Journal of Fish Biology 70: 1898–1906.

Lassig BR (1977). Socioecological strategies adopted by obligate coral-dwelling fishes. In: Taylor DL (ed.), Proceedings of the Third International Coral Reef Symposium. Vol. 1, Biology. Rosenstiel School of Marine and Atmospheric Science, Miami, Florida, pp. 566–570.

Liley N, Stacey NE (1983). Hormones, pheromones, and reproductive behaviour in fish. In: Hoar WS, Randall DJ (eds.), Fish Physiology, vol. 9A. Academic Press, New York, pp. 1–63.

Liu M, Sadovy Y (2004). The influence of social factors on adult sex change and juvenile sexual differentiation in a diandric, protogynous epinepheline, Cephalopholis boenak (Pisces, Serranidae). Journal of Zoology London 264: 239–248.

Liu M, Sadovy Y (2005). Habitat association and social structure of the chocolate hind, Cephalopholis boenak (Pisces: Serranidae: Epinephelinae), at Ping Chau Island, northeastern Hong Kong waters. Environmental Biology of Fishes 74: 9–18.

Lorenzi V, Earley RL, Grober MS (2006). Preventing behavioural interactions with a male facilitates sex change in female blueband gobies, Lythrypnus dalli. Behavioral Ecology and Sociobiology 59: 715–722.

Manabe H, Ishimura M, Shinomiya A, Sunobe T (2007a). Field evidence for bi-directional sex change in the polygynous gobiid fish *Trimma okinawae*. Journal of Fish Biology 70: 600–609.

Manabe H, Ishimura M, Shinomiya A, Sunobe T (2007b). Inter-group movement of females of the polygynous gobiid fish *Trimma okinawae* in relation to timing of protogynous sex change. Journal of Ethology 25: 133–137.

Manabe H, Matsuoko M, Goto K, Dewa S, Shinomiya A, Sakurai M, Sunobe T (2008). Bi-directional sex change in the gobiid fish *Trimma* sp.: does size-advantage exist? Behaviour 145: 99–113.

Michael SW (2004). *Basslets, Dottybacks and Hawkfishes*. TFH Publications, Neptune City, New Jersey.

Munday PL (2001). Fitness consequences of habitat selection and competition among coral-dwelling fish. Oecologia 128: 585–593.

Munday PL (2002). Bi-directional sex change: testing the growth-rate advantage model. Behavioral Ecology and Sociobiology 52: 247–254.

Munday PL, Jones GP (1998). The ecological implications of small body size among coral-reef fishes. Oceanography and Marine Biology: An Annual Review 36: 373–411.

Munday PL, Buston PM, Warner RR (2006a). Diversity and flexibility of sex-change strategies in animals. Trends in Ecology and Evolution 21: 89–95.

Munday PL, Cardoni AM, Syms C (2006b). Cooperative growth regulation in coral dwelling fishes. Biology Letters 2: 355–358.

Munday PL, White JW, Warner RR (2006c). A social basis for the development of primary males in a sex-changing fish. Proceedings of the Royal Society B 273: 2845–2851.

Muñoz RC, Warner RR (2003). A new version of the size-advantage hypothesis for sex change: incorporating sperm competition and size-fecundity skew. American Naturalist 161: 749–761.

Nakashima Y, Kuwamura T, Yogo Y (1995). Why be a both-ways sex changer. Ethology 101: 301–307.

Nakashima Y, Kuwamura T, Yogo Y (1996). Both-ways sex change in monogamous coral gobies, *Gobiodon* spp. Environmental Biology of Fishes 46: 281–288.

Ohta K, Sundaray JK, Okida T, Sakai M, Kitano T, Yamaguchi A, Takeda T, Matsuyama M (2003). Bi-directional sex change and its steroidogenesis in the wrasse, *Pseudolabrus sieboldi*. Fish Physiology and Biochemistry 28: 173–174.

Perry AN, Grober MS (2003). A model for social control of sex change: interactions of behaviour, neuropeptides, glucocorticoids, and sex steroids. Hormones and Behaviour 43: 32–38.

Reavis RH, Grober MS (1999). An integrative approach to sex change: social, behavioural and neurochemical changes in *Lythrypnus dalli* (Pisces). Acta Ethologica 2: 51–60.

Robertson DR (1972). Social control of sex reversal in a coral-reef fish. Science 177: 1007–1009.

Rodgers EW, Earley RL, Grober MS (2006). Elevated 11-ketotestosterone during paternal behavior in the Bluebanded goby (*Lythrypnus dalli*). Hormones and Behavior 49: 610–614.

Rodgers EW, Earley RL, Grober MS (2007). Social status determines sexual phenotype in the bi-directional sex changing blueband goby *Lythrypnus dalli*. Journal of Fish Biology 70: 1660–1668.

Ross RM (1990). The evolution of sex-change mechanisms in fishes. Environmental Biology of Fishes 29: 81–93.

Sadovy Y, Donaldson TJ (1995). Sexual pattern of *Neocirrhites armatus* (Cirrhitidae) with notes on other hawkfish species. Environmental Biology of Fishes 42: 143–150.

Sadovy Y, Shapiro DY (1987). Criteria for the diagnosis of hermaphroditism in fishes. Copeia 1987: 136–156.

Sadovy de Mitcheson Y, Liu M (2008). Functional hermaphroditism in teleosts. Fish and Fisheries 9: 1–43.

Sakai Y, Kohda M (1997). Harem structure of the protogynous angelfish, *Centropyge ferrugatus* (Pomacanthidae). Environmental Biology of Fishes 49: 333–339.

Sakai Y, Karino K, Kuwamura T, Nakashima Y, Maruo Y (2003). Sexually dichromatic protogynous angelfish *Centropyge ferrugata* (Pomacanthidae) males can change back to females. Zoological Science 20: 627–633.

Sakurai M, Nakakoji S, Manabe H, Dewa S, Shinomiya A, Sunobe T (2009). Bi-directional sex change and gonad structure in the gobiid fish *Trimma yanagitai*. Ichthyological Research 56: 82–86.

Shapiro DY (1984). Sex reversal and sociodemographic processes in coral reef fishes. In: Potts GW, Wootton RJ (eds.), *Fish Reproduction*. Academic Press, London, pp. 103–117.

Shiobara Y (2000). Reproduction and hermaphroditism of the gobiid fish, *Trimma grammistes*, from Suruga Bay, central Japan. Science Reports of the Museum, Tokai University 2: 19–30.

St. Mary CM (1993). Novel sexual patterns in two simultaneously hermaphroditic gobies, *Lythrypnus dalli* and *Lythrypnus zebra*. Copeia 1993: 1062–1072.

St. Mary CM (1994). Sex allocation in a simultaneous hermaphrodite, the blueband goby (*Lythrypnus dalli*): the effects of body size and behavioral gender and the consequences for reproduction. Behavioral Ecology 5: 304–313.

St. Mary CM (1996). Sex allocation in a simultaneous hermaphrodite, the zebra goby *Lythrypnus zebra*: insights gained through a comparison with its sympatric congener, *Lythrypnus dalli*. Environmental Biology of Fishes 45: 177–190.

St. Mary CM (1997). Sequential patterns of sex allocation in simultaneous hermaphrodites: do we need models that specifically incorporate this complexity? The American Naturalist 150: 73–97.

St. Mary CM (1998). Characteristic gonad structure in the gobiid genus *Lythrypnus* with comparisons to other hermaphroditic gobies. Copeia 1998: 720–724.

St. Mary CM (2000). Sex allocation in *Lythrypnus* (Gobiidae): variations on a hermaphroditic theme. Environmental Biology of Fishes 58: 321–333.

Sunobe T, Nakazono A (1990). Polygynous mating system of *Trimma okinawae* (Pisces: Gobiidae) at Kagoshima, Japan with a note on sex change. Ethology 84: 133–143.

Sunobe T, Nakazono A (1993). Sex change in both directions by alteration of social dominance in *Trimma okinawae* (Pisces: Gobiidae). Ethology 94: 339–345.

Sunobe T, Nakazono A (1999). Mating system and hermaphroditism in the gobiid fish, *Priolepis cincta*, at Kagoshima, Japan. Ichthyological Research 46: 103–105.

Sunobe T, Nakamura M, Kobayashi Y, Kobayashi T, Nagahama Y (2005a). Aromatase immunoreactivity and the role of enzymes in steroid pathways for inducing sex change in the hermaphrodite gobiid fish *Trimma okinawae*. Comparative Biochemistry and Physiology a-Molecular and Integrative Physiology 141: 54–59.

Sunobe T, Nakamura M, Kobayashi Y, Kobayashi T, Nagahama Y (2005b). Gonadal structure and P450scc and 3ß-Hsd immunoreactivity in the gobiid fish *Trimma okinawae* during bidirectional sex change. Ichthyological Research 52: 27–32.

Tanaka H, Hirose K, Nogami K, Hattori K, Ishibashi N (1990). Sexual maturation and sex reversal in red spotted grouper, *Epinephelus akaara*. Bulletin of the National Research Institute of Aquaculture 17: 1–15 (in Japanese).

Thompson VJ, Munday PL, Jones GP (2007). Habitat patch size and mating system as determinants of social group size in coral-dwelling fishes. Coral Reefs 26: 165–174.

Warner RR (1975). The adaptive significance of sequential hermaphroditism in animals. The American Naturalist 109: 61–82.

Warner RR (1978). The evolution of hermaphroditism and unisexuality in aquatic and terrestrial vertebrates. In: Reese ES, Lighter FJ (eds.), *Contrasts in Behaviour*. John Wiley & Sons, New York, pp. 78–101.

Warner RR (1988). Sex change and the size-advantage model. Trends in Ecology and Evolution 3: 133–136.

Warner RR, Hoffman SG (1980). Population density and the economics of territorial defense in a coral reef fish. Ecology 61: 772–780.

Winterbottom R, Southcott L (2008). Short lifespan and high mortality in the western Pacific coral reef goby *Trimma nasa*. Marine Ecology Progress Series 366: 203–208.

Wittenrich ML, Munday PL (2005). Labile sex allocation in three species of *Pseudochromis* (Pseudochromidae): an experimental evaluation. Zoological Science 22: 797–803.

9

Neuroendocrine Regulation of Sex Change and Alternate Sexual Phenotypes in Sex-Changing Reef Fishes

John Godwin

Our understanding of the diversity of sexual strategies occurring in nature has increased dramatically in the last 25 years. This is particularly true in the area of determination of an individual's sex by environmental signals. Progress in understanding environmental sex determination (ESD) has been greatest in the areas of temperature dependent sex determination in reptiles (Crews 2003) and social determination of sex and sexual expression in fishes, the primary topic of this chapter. Related to the question of how the same genotype may direct the development of different sexes in response to environmental signals is that of what physiological mechanisms generate discrete sexual phenotypes within a sex. Alternate male phenotypes are found in many marine fishes, and these males are typically "mosaics" of morphological and physiological features found in females and territorial male phenotypes. I will discuss physiological and neural mechanisms underlying the expression of alternate male phenotypes in sex-changing fishes here as well, both because these are closely related to those underlying sex change and because individual life histories in a number of species involve transitions both between sexes and between sexual phenotypes within a sex. Can study of physiological and neurobiological mechanisms inform our understanding of evolutionary patterns of sexual expression? Put another way, will "How" approaches inform "Why" questions? Although a definitive answer to this question is not yet possible, the strong conservation of basic mechanisms underlying sexual development argues that the answer is likely to be yes, and extremely useful from a practical standpoint for study. This conservation is also important because adaptations uncovered in "extreme" examples of environmental effects on reproductive physiology, such as socially mediated sex change, are likely to also yield insights into social influences on neuroendocrine systems subserving sexual expression more generally. This has already occurred to some extent thanks to advances in measurement and manipulation of both steroid hormone and neural signaling systems beginning in the late 1980s (e.g., Goodson & Bass 2000; Black et al. 2005). As genomic information and technologies become available for species that are not traditional biomedical models, the pace of such contributions is likely to accelerate.

This chapter is organized in a series of subsections intended to review what is known and what are promising areas for inquiry. First, I briefly review the vertebrate and teleost neuroendocrine axis as well as key steroid and neural signaling systems known to affect reproductive function, reproductive behavior, and aggression in fishes. Then, I review the known involvement of these systems in regulating sex change and sexual behavior for the best studied groups of tropical marine fishes in this respect: the gobies (Gobiidae); the basslets and groupers (Serranidae); and the wrasses, parrotfishes, and damselfishes (Labridae, Scaridae, and Pomacentridae). Finally, I will briefly examine the state of research in this area as well as prospects for future research.

INTEGRATING ENVIRONMENTAL INFORMATION: NEUROENDOCRINE AND NEURAL SIGNALING SYSTEMS IN FISHES

Both the overall organization and the functioning of the neuroendocrine system subserving reproduction are remarkably conserved across vertebrate animals (Figure 9.1). The nervous system is also the logical site of integration for social information, although direct environmental influences on the gonads are likely important for forms of ESD, such as temperature dependent sex determination (e.g., Crews 1993; Luckenbach 2005).

Across vertebrate animals, the endocrine control of reproduction centers on what is termed the hypothalamo-pituitary-gonadal (HPG) axis. Hypothalamic gonadotropin releasing hormone (GnRH) neurons integrate a variety of internal and environmental influences and control secretion of gonadotropins from the pituitary gland. Teleost fishes have three distinct types of GnRH, with the form expressed in the hypothalamus and controlling pituitary gonadotropin secretion being most similar to one originally described in chickens (chicken II GnRH). Other GnRH forms are expressed in the terminal nerve and tegmentum respectively, where they play neuromodulatory roles in neural processing (Soga et al. 2005; Maruska & Tricas 2007). Unlike tetrapods, the teleost pituitary is said to be hard-wired, with direct innervation of anterior pituitary target cells rather than release of hypothalamic releasing hormones into a portal system. As in tetrapods, teleosts have two gonadotropins (GtH I and GtH II) with functions very compa-

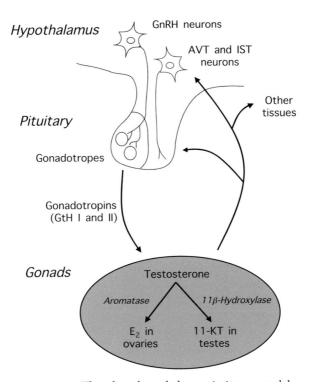

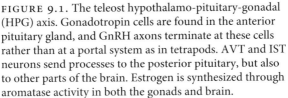

FIGURE 9.1. The teleost hypothalamo-pituitary-gonadal
(HPG) axis. Gonadotropin cells are found in the anterior
pituitary gland, and GnRH axons terminate at these cells
rather than at a portal system as in tetrapods. AVT and IST
neurons send processes to the posterior pituitary, but also
to other parts of the brain. Estrogen is synthesized through
aromatase activity in both the gonads and brain.

rable to follicle-stimulating hormone (FSH) and luteinizing hormone (LH) in
tetrapods, respectively. These gonadotropins regulate development and activity
of the gonads generally and steroid production and conversion, specifically.

The steroid hormones in fishes are generally similar to those of tetrapods but
also show important differences. Testosterone (T) is a major product of both the
ovaries and testes and functions both as an important androgenic signal and,
more importantly, as a prohormone in the biosynthesis of the predominant circulating estrogen in females, estradiol 17β (E_2), and of a key androgen, 11-
ketotestosterone (11-KT), in males (Borg 1994). Both E_2 and 11-KT have been
implicated in the control of gonadal sex differentiation and sexual behavior, and
control of the synthesis of these hormones appears likely to play a key role in the

process. The rate-limiting enzyme in estrogen synthesis is a member of the cytochrome p450 family of proteins and is most commonly referred to as aromatase. Aromatase enzymes catalyze the conversion of androgens to estrogens (Callard et al. 1990, 2001). As with a number of other important gene products relative to patterns in tetrapods, fishes produce two forms of aromatase that are distinct gene products, a gonadal form, termed gonadal aromatase, p450AromA, or cyp19A1; and a brain form, variously termed brain aromatase, p450AromB, and cyp19A2. The key synthetic enzyme regulating levels of 11-KT is 11β-hydroxylase. Unlike testosterone, 11-KT cannot be aromatized and this may contribute to androgenic effects as, at least in brain tissue, teleosts exhibit very high aromatase levels (Callard et al. 2001; Forlano et al. 2006).

In addition to studies of the HPG axis in sex-changing fishes, there is increasing interest in other neural and endocrine signaling systems that may influence both sexual expression and especially behavior. The endocrine stress system, mainly centered on the hypothalamo-pituitary-interrenal gland (HPI) axis, may play an important role in linking social behavior and interactions with gonadal function including sex change (Perry & Grober 2003). As with the HPG axis, this system is fundamentally similar to that found in tetrapods, while exhibiting some differences. Fishes do not have adrenal glands and the primary steroid hormone associated with stress responses in fishes, cortisol (chemically identical to that in humans), is produced by the interrenal gland. Cortisol secretion is stimulated by adrenocorticotropin (ACTH) from the anterior pituitary gland, as in tetrapods. ACTH secretion is in turn stimulated by corticotropin releasing hormone from the hypothalamus.

REEF FISH MODELS
OF SOCIALLY CONTROLLED SEX CHANGE

Hermaphroditism is exhibited in at least seven orders and 27 families of teleost fishes (Mank et al. 2006; Sadovy de Mitcheson & Liu 2008). A complete review of physiological correlates of sex change across teleosts is beyond the scope of this chapter, but excellent reviews of much of this literature have been previously published (Baroiller et al. 1999; Devlin & Nagahama 2002; Frisch 2004). Of the groups that exhibit sex change, however, relatively few are well studied at either the behavioral or the physiological level, and only four families of fishes have received significant attention at both levels. These are the primarily or commonly reef-associated gobies, basslets and groupers, wrasses, and parrotfishes (Labridae and Scaridae). Fortunately, for the goal of discerning general principles underlying sexual phenotype expression, these families exhibit some substantial and useful contrasts in body size, space use patterns, and mating systems. I review studies from these four families below. Gobies are covered first, followed by serranids, and then wrasses and parrotfishes. This sort of discussion is typically organized by physiological system rather than by taxonomic group. This chapter will present findings on physiological correlates and experimental manipulations of sex change within groups, with the goal of helping readers appreciate the integration of systems across physiological levels of organization within the same or similar species.

The Gobies

The gobies (family Gobiidae) are very common on both tropical and temperate reefs and have many features that make them attractive as models for both field and laboratory studies of sex change. These features include space use patterns often characterized by strong site fidelity and small territorial or home range areas, small body sizes, and high population densities that are advantageous in making statistically robust field experiments feasible. Gobies often show excellent adaptability to captivity, facilitating the study of their behavior and physiology under tightly controlled laboratory conditions. Finally, some goby species also display bidirectional sex change, allowing testing of hypothesized mechanisms of both protogynous and protandrous sex change in the same species.

The first goby species to be studied in detail with respect to sex change was the blackeye goby, *Coryphopterus nicholsi*. Cole described spawning behavior and reproductive success in *C. nicholsi* as well as identifying this species as a sex changer (1983). This initial work was followed with both ecological and physiological studies focused on steroid hormone metabolism. Kroon & Liley (2000) measured levels of E_2, testosterone, and 11-KT in whole-body extracts of mature males and females captured in the fall outside the summer breeding season. Testosterone and 11-ketotestosterone levels were significantly higher in males while estradiol levels were significantly higher in females. In order to test hypothesized roles for these steroids in mediating sexual phenotype and sex change, these authors treated females with slow-release elastomer implants containing 17α-methyltestosterone, 11-ketoadrenosterone, 11-ketotestosterone, or no hormone as a blank control treatment. Neither the control nor the 17α-methyltestosterone treatments induced gonadal changes. However, treatment with either of the 11-oxygenated androgens induced complete or nearly complete sex change over six weeks in all treated females. A second experiment exposed females to an aromatase inhibitor, fadrozole, and found a dose-response relationship where increasing concentrations of the drug produced increasing development of testicular morphology with the highest dose inducing testicular development in 87% of treated individuals.

Other goby species show similar variations in steroid metabolism and effects of estrogen synthesis on sexual development. Estradiol levels are higher in females than males in *Gobiodon histrio*, a species that exhibits bidirectional sex change, while testosterone levels were not different (Kroon et al. 2003). Levels of 11-KT were too low to permit measurement in most individuals of this small-bodied species and these authors suggest low 11-KT levels may be important in its ability to change sex in either direction. Experimental support for the role of estrogen synthesis in regulating sex change in *Gobiodon* was provided in a follow up study with *Gobiodon erythrospilus* (Kroon et al. 2005). In heterosexual experimental pairs, female *G. erythrospilus* treated with implants containing the aromatase inhibitor fadrozole changed sex to become male, while males treated with estradiol (the product of the aromatase enzyme) changed sex to become females. Conversely, sex change was not observed in this paired treatment for females treated with estrogen or males treated with the aromatase inhibitor. Unpaired individuals in this species become male, but estradiol treatment prevented this change in unpaired females. Unpaired males treated with estradiol changed sex to become female, while control males did not (Kroon et al. 2005). Unpaired females who were handled but not implanted did show sex change. These observations and those in *C. nicholsi* described above strongly implicate estradiol synthesis in the regulation of gonadal sex in gobies.

A series of studies in another gobiid implicates steroid hormone metabolism in controlling various aspects of sexual phenotype ranging from gonadal function to behavior. The blue-banded goby, *Lythrypnus dalli,* is native to warm, temperate regions of the eastern Pacific where it is found in relatively shallow, rocky areas associated with crevices or sometimes sea urchins for shelter. This species is a bidirectional sex changer that very interestingly displays shifts in allocation associated with social interactions. These shifts are associated with increases in dominance related behaviors, morphological changes including increases in body size, length of the dorsal fin, and changes in the genital papilla, in addition to the change in gonadal allocation between ovarian and testicular tissue (Carlisle et al. 2000). Individuals assume only one behavioral sex at a time with socially dominant individuals becoming males (St. Mary 1994; Rodgers et al. 2005; Lorenzi et al. 2006). At least some of these shifts are associated with changes in steroid hormone metabolism. Consistent with results for the bidirectionally changing *G. histrio* described above, comparisons of steroid levels excreted into holding water indicated a difference between male and female *L. dalli* in estradiol, although no differences were found between the sexes for testosterone, 11-ketotestosterone, or cortisol in this study (Lorenzi et al. 2008). As in some other sex changing species (see below), levels of 11-ketotesterone vary in male *L. dalli* with reproductive activity, being elevated in experienced brooding males (Rodgers et al. 2006). Treatment with this potent androgen causes the genital papilla to become more male-like in form (Carlisle et al. 2000). There also appears to be within-sex variation in reproductive tactics in male *L. dalli* (Drilling & Grober 2005), but endocrine correlates of this variation have not yet been explored.

All vertebrate animals produce or convert various types of steroid hormones in the central nervous system (Callard et al. 1990). In rats, testosterone masculinizes the brain around the time of birth, but only after conversion to estradiol by the enzyme aromatase (Clemens & Gladue 1978). Steroid hormone conversion by the brain also appears to play a critical role in controlling sex change and sexual phenotype development in fishes, but masculinization appears be blocked by estrogen synthesis rather than stimulated by it as in rats. Male removal from social groups in *L. dalli* induced the rapid development of a dominance-related behavior known as *displacement* in the largest female (Black et al. 2005). This increase in male-typical aggressive behavior was closely correlated with decreases in brain aromatase activity, suggesting an important functional link. Interestingly, while female *L. dalli* have higher aromatase activity in both the brain and gonads than males, alterations in these levels with assumption of social dominance were more rapid in the brain (Figure 9.2). This is consistent with social cues being the critical regulator of this process. What is as yet unknown for this system and other teleost systems are the mechanisms by which changes or differences in aromatase activity influence behavioral phenotype development. Two possibilities have been suggested (Schlinger et al. 1999; Black et al. 2005). First, aromatase in the brain may prevent masculinization by converting testosterone to estradiol and thereby preventing this hormone from acting as an androgen or, potentially, from being converted to the potent androgen 11-ketotestosterone. Alternatively, estradiol—produced either by the ovaries or in the brain through aromatization of testosterone—may inhibit male development and behavioral display. I return to this issue in the discussion of wrasses.

The biochemical regulation of steroid metabolism is beginning to be addressed in the bidirectionally sex changing goby *Trimma okinawae*. As in other teleosts, this species expresses two forms of the aromatase enzyme, cytochrome p450AromA predominantly in the gonads and p450AromB predominantly in the brain (Kobayashi et al. 2004). Immunoreactivity for the ovarian form has been found in the gonad, as predicted (Sunobe et al. 2005). The upstream promoter regulatory region has been examined for the gonadal form of aromatase (p450AromA) in *T. okinawae* and found to contain response elements for several key mediators, including several estrogen response elements, or EREs (Kobayashi et al. 2005). This finding is consistent with regulation of aromatase by estrogen and with the potential stimulation of aromatase expression and induction of male-to-female sex change in gobies as described for *G. erythrospilus* above, although this effect of exogenous estrogen has not been specifically tested at this writing. The promoter regions of both the ovarian and brain forms of aromatase were examined in *G. histrio* by Gardner et al. (2005). These authors, by contrast, identified putative EREs for the brain form and not the ovarian form. Interestingly, a response element for glucocorticoids (GRE), key mediators of the

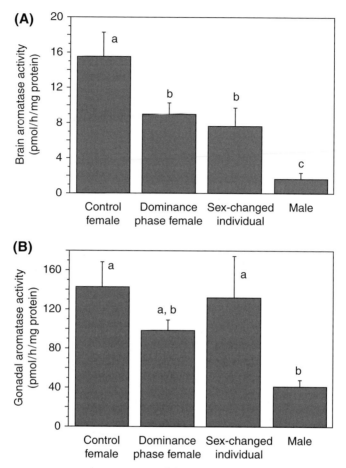

FIGURE 9.2. Changes in gonadal aromatase activity in brain and gonad between sexual phenotypes and during sex change in the blue-banded goby, *Lythrypnus dalli*. Groups designated with different letters over the bars are significantly different. Redrawn from Black et al. 2005.

endocrine stress response, was suggested for the promoter region of ovarian aromatase in *G. histrio*. These were not described in the other two species they examined (*Lates calcarifer* and *Cromileptes altivelis*), or for *T. okinawae*. This is potentially important, as glucocorticoid stress hormones have been proposed as important mediators of socially controlled sex change (Perry & Grober 2003).

What might be the consequences of these differences in steroid hormone influences on the neural substrates of reproductive function and behavior in gobies? Neuropeptide hormone systems have been the subject of intense research in the area of social neuroscience in recent years, and these signaling systems are implicated in key social behaviors in a number of mammalian, avian, and amphibian models (for recent reviews, see Goodson & Bass 2001; Rose & Moore 2002; De Vries & Panzica 2006). The neuropeptides arginine vasotocin (AVT) and isotocin (IST) have been the subject of substantial research in fishes, as have the homologous hormones in other vertebrate groups. This area is just beginning to be explored in gobies, but the extraordinary diversity found in this family suggests that it is likely to be a fruitful area of research.

Isotocin is the teleost homolog of oxytocin in mammals, which has a wide array of described functions, including a common association with female-typical sociosexual behaviors. Consistent with this overall pattern from mammals, female *L. dalli* show larger numbers of isotocin neurons than do males. By contrast, another bidirectionally sex-changing goby, *Trimma okinawae*, shows larger vasotocin neurons in male-phase than female-phase individuals (Grober & Sunobe 1996). Although not a sex-changing species, comparisons of AVT neurons in another gobiid species, the halfspotted goby (*Asterropteryx semipunctata*), indicated sex differences that are dependent on when sampling was performed (Maruska et al. 2007).

Gonadotropin-releasing hormone (GnRH) has been a focus of research in teleost reproduction, but it has not yet been well studied in sex-changing gobies. Studies in non-sex changing gobies suggest differences across sexual phenotypes, but with substantial variation in these patterns across species. In the study cited above on half spotted gobies, sex differences in GnRH neurons were evident and, as with AVT, these varied across sampling periods (Maruska et al. 2007). Scaggiante et al. (2006) studied GnRH neural phenotype in two European goby species, the grass goby (*Zosterisessor ophiocephalus*) and the black goby (*Gobius niger*), comparing sexual phenotypes and seasonal changes. As with the halfspotted goby, these authors found a complex pattern of differences, with grass gobies displaying dimorphism in forebrain GnRH expression both intrasexually across male reproductive tactics and intersexually. Black gobies, by contrast, displayed only intersexual differences. Interpretation of such divergence of among species is complicated by other potentially confounding differences (as noted by Scaggiante et al. 2006) and the lack of a comprehensive phylogenetic framework. Fortunately, the speciose nature of the Gobiidae and variation both within and among groups is favorable for this type of work going forward.

Basslets and Groupers

The groupers and basslets (family Serranidae) are the second of the groups of sex-changing species that has been well studied at both the behavioral and physiological levels. In addition to providing intriguing contrasts in body size and sexual pattern, including especially the presence of simultaneous hermaphroditism in the family, the serranids are ecologically and economically important because groupers support important fisheries in the tropics and subtropics. A better understanding of sexual function and sex change in groupers therefore has the strong potential of contributing to captive breeding and aquaculture efforts. Serranids are somewhat unusual in the context of this chapter because most of the behavioral work has focused on one part of the family, the basslets and hamlets, while most of the endocrine work to date has focused on the typically larger groupers. There are some exceptions and these will be highlighted, but it is also likely that there are strong similarities in mechanisms regulating both gonadal function and behavior across the family.

Much of the initial ethological study of hermaphroditism in the serranids focused on the basslet *Pseudanthias squamipinnis* (formerly *Anthias*, Shapiro, 1980, 1981). While this behavioral work has not been followed by physiological studies in *Pseudanthias*, it did help stimulate work in other small serannids, including studies of reproductive endocrinology. The belted sandbass, *Serranus subligarius*, is a simultaneous hermaphrodite native to the Gulf of Mexico. Despite having both functional ovarian and testicular tissue, individuals vary in their likelihood of spawning in the female or male role, with larger fish spawning more often as males (Oliver 1997). A follow-up study by the same researcher showed that this shift from predominantly female-role spawning to male-role spawning with increasing body size is mirrored by increasing plasma concentrations of both 11-KT and 20βS, a progestin-class steroid hormone that induces final oocyte maturation in some fishes (Cheek et al. 2000). Studies in other teleosts show that 20βS binds a G-protein coupled membrane steroid receptor (Zhu et al. 2003a, b), suggesting the possibility for rapid, non-genomic effects of this steroid on behavior.

Endocrine studies of gonadal function and gonadal sex change in serranids have focused primarily on groupers, especially members of the genus *Epinephelus*. This genus has a circumtropical distribution and extends into many warm temperate regions. *Epinephelus* species are the focus of a number of economically important fisheries and, increasingly, are of interest as potential aquaculture cultivars. These relatively large predators are often also very important ecologically and there is considerable interest in the potential of stock enhancement to replenish populations that have declined due to fishing pressure (Sadovy & Domeier 2005). These factors have provided considerable impetus to understanding reproductive endocrinology in groupers and substantial progress is being made.

The steroid hormone correlates of sexual phenotype have now been characterized for a number of protogynous groupers. The pattern of sexual development in groupers appears to be largely monandric, with males developing only from mature females, although there is at least one example of diandry (in which males may develop directly from juveniles or by sex change from females) from the

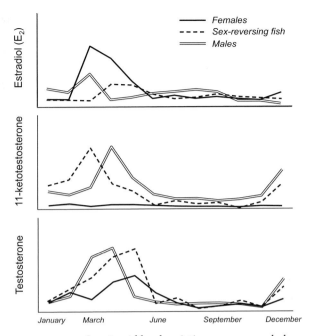

FIGURE 9.3. Sex steroid level variation across sexual phenotype and season in the grouper *Epinephelus akaara*.
Redrawn from Li et al. 2007.

genus *Epinephelus* (Fennessy & Sadovy 2002). Females typically have higher
plasma levels of E_2 and lower levels of 11-KT (Alam et al. 2006).

Plasma steroid profiles in male and female groupers have been investigated in
five species, three in the genus *Epinephelus*, and two others. The consistent pattern in the breeding season is higher E_2 in females and higher 11-KT in males,
with the predicted shifts in these levels in fish undergoing sex change (Figure 9.3).
Plasma T (testosterone) levels do not show these consistent sex differences, possibly because T is the biochemical precursor for both E_2 and 11-KT. Higher estrogen levels for female groupers in the breeding season are reflected in, and likely a
result of, higher aromatase expression as described in the red-spotted grouper,
Epinephelus akaara, associated with both naturally occurring and hormonally induced sex change (Li et al. 2006a, b, 2007). Aromatase activity and mRNA levels were also reduced in the orange-spotted grouper with androgen-induced sex
change (Zhang et al. 2007).

Male grouper show higher 11-KT synthetic capacity than females. Female
honeycomb grouper (*Epinephelus merra*) have low 11-KT levels and some ovarian expression of the key enzyme in 11-KT synthesis, 11β-hydroxylase, which

may be important in stimulating sex change as the process proceeds (Alam et al. 2006). Male honeycomb grouper have higher 11β-hydroxylase-like immunoreactivity in the gonads and significantly higher production of 11-KT from gonadal fragments in vitro.

As in the gobies discussed above and the wrasses and parrotfishes discussed in the next section, both steroid hormone manipulations—hormone implants and inhibitors of steroidogenesis—are effective in inducing sex change in groupers. Two groups have now described successful induction of protogynous sex change through inhibition of aromatase, specifically using the drug fadrozole, as in the goby studies referenced earlier (Bhandari et al. 2004, 2006; Li et al. 2006a, b). Several groups have also described successful induction of sex change with androgen implants in groupers and another large serranid (Yeh et al. 2003a, b; Benton & Berlinsky 2006; Li et al. 2006a; Sarter et al. 2006; Zhang et al. 2007). The initiating events of sex change have been difficult to discern because decreases in aromatase expression and E_2 production at the onset of the process are temporally confounded with increases in 11-KT. However, Bhandari et al. (2005) showed that coadministering E2 could prevent sex change that would otherwise be induced by inhibiting aromatase. These authors suggest that a decline in estrogen biosynthesis is the key event in the initiation of sex change and this hypothesis is consistent with findings and suggestions put forward for other fishes (Black et al. 2005; Kroon et al. 2005; Marsh 2007).

Little information is so far available regarding endocrine aspects of sexual function in other parts of the HPG axis, but this should change rapidly thanks to the development of a number of molecular tools for various grouper species. These include the molecular cloning of the different subunits of the gonadotropins (Shein et al. 2003; Li et al. 2005), a GnRH receptor from the orange-spotted grouper (He et al. 2006), and the brain form of aromatase (p450AromB)(Zhang et al. 2004).

The Labroidei (Wrasses, Parrotfishes, and Damselfishes)

Wrasses, parrotfishes, and damselfishes are conspicuous and ecologically important members of tropical and many temperate zone reef communities. These families are closely related (Kaufman & Liem 1982; see also Streelman & Karl 1997), and they are related to the cichlids of tropical fresh waters, which are the subject of considerable physiological research and a current genome sequencing effort. Another valuable feature of many wrasse and parrotfish species for exploring the physiological bases of sexual phenotype development is the presence of both functional protogynous sex change and discrete alternate male phenotypes. These alternate male phenotypes include typically large and colorful terminal phase (TP) males and smaller initial phase (IP) males that are often nearly indistinguishable externally from females (Warner 1984; Cardwell 1991a). This display of "three sexes" allows comparisons both across sexes and within a sex where behavioral phenotype differs strongly, but gonadal sex does not. The damselfishes (Pomacentridae) are useful models because both protogyny and protandry are exhibited in the family. This group is reviewed briefly at the end of the chapter, but it has received much less attention from a physiological standpoint than the wrasses and parrotfishes.

STEROID HORMONE CORRELATES
OF SEXUAL PHENOTYPE AND SEX CHANGE

Among the three main groups of sex changing reef fishes that are the focus of this chapter, steroid hormone correlates of sexual phenotype and sex change have been most extensively explored in the wrasses and parrotfishes. Reinboth & Becker (1984) focused on steroid hormone synthesis in studying testosterone conversion by the gonads of the Mediterranean wrasse *Coris julis*. A key finding of this study was that production of 11β-hydroxylated androgens by the gonads was largely restricted to secondary males. This finding is supported by a variety of other studies in wrasses and parrotfishes. Nakamura et al. (1989) explored differences in gonad structure and ultrastructure as well as circulating gonadal steroid levels in the Hawaiian saddleback wrasse, *Thalassoma duperrey*. Consistent with the findings of Reinboth & Becker (1984), terminal phase males had significantly greater circulating levels of 11-ketotestosterone than females (Figure 9.4). Conversely, females had higher circulating levels of E_2 and testosterone (likely related to testosterone's role as a substrate in E_2 synthesis). A follow-up study in *T. duperrey* found that 11-KT synthesis was greater *in vitro* by the testes of TP males than those of IP males and that circulating 11-KT levels were also higher in TP males. The saddleback wrasse has also been a very useful model in behavioral studies of sex change and in understanding monoamine neurotransmitter contributions to the process. We return to these findings below.

The bamboo wrasse, *Pseudolabrus seboldi,* exhibits sexual phenotype differences similar to those found in the *Thalassoma* species with both diandry and protogynous sex change. As in the saddleback wrasse, terminal phase male bamboo wrasses have higher plasma 11-KT levels than females overall, and levels within individuals are correlated with the degree of development of sexually dimorphic fin coloration (Ohta et al. 2008). Surprisingly, no differences were found in plasma E_2 levels in this study, but the authors suggest this may be attributable to sampling outside the spawning season.

The Caribbean stoplight parrotfish, *Sparisoma viride*, is another common and conspicuous species that exhibits both protogynous sex change and discrete alternate male phenotypes. Cardwell & Liley (1991a) compared circulating steroid levels between sexual phenotypes in *S. viride* using samples from field-caught

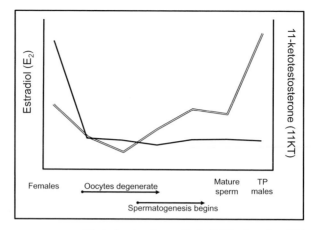

FIGURE 9.4. Variation in plasma E_2 (solid line) and 11-KT (double line) between females, TP males, and over the course of natural sex change in the saddleback wrasse, *Thalassoma duperrey*. Redrawn from Nakamura et al. 2005.

females, IP males, and TP males as well as some sex-changing individuals. As with saddleback wrasses, E_2 levels were higher in females, and 11-KT levels were elevated in TP males relative to both females and IP males. A follow-up study examined steroid hormone correlates of sexual phenotype and social status among males, finding that territorial TP males had elevated 11-KT levels relative to TP males that did not hold territories (Cardwell & Liley 1991a).

Little is known about potential differences in the expression of receptors for the key androgens and E_2. Kim et al. (2002) cloned portions of a nuclear androgen receptor (AR) and the α subtype of the estrogen receptor (ERα) in threespot wrasses (*H. trimaculatus*) and compared expression of these genes across sexual phenotypes using semi-quantitative RT-PCR (reverse transcription-polymerase chain reaction). No differences were found between females and terminal-phase males in ERα expression in either the gonads or brain, or in AR mRNA expression in the gonads. TP males did show higher AR mRNA expression in the brain suggesting potentially greater androgen sensitivity in this morph type. No information is available on other estrogen receptor subtypes known to be expressed in fishes (Hawkins et al. 2000, 2005). It is also worth noting that PCR-based techniques give little anatomical resolution, so sexual phenotype differences in more restricted regions of the gonads or brain remain to be critically examined.

As described above for gobies and groupers, estrogen synthesis through the activity of aromatase appears to be a key regulator of sex change in wrasses and parrotfish. Following up on the studies detailing differences in plasma concentrations of key gonadal steroids, immunocytochemistry (which localizes the enzyme proteins in tissues) was used to examine the expression of aromatase and the key enzyme in 11-KT synthesis, 11β-hydroxylase, across sexual phenotypes in the saddleback wrasse, *T. duperrey* (reviewed in Nakamura & Kobayashi 2005)(Figure 9.4). Aromatase expression was high in the ovary of *T. duperrey* and declined with sex change, while 11β-hydroxylase was expressed in the ovary, declined at the onset of sex change, and then steadily increased with testicular development. These changes in gonadal steroidogenic enzyme expression correlated well with concurrent changes in E_2 and 11-KT, respectively.

STEROID HORMONE CONTROL OF SEXUAL PHENOTYPE DIFFERENCES: MANIPULATIVE STUDIES

A number of studies in both wrasses and parrotfishes have taken a manipulative approach to addressing the role of gonadal steroid hormones in controlling sex change. A key emerging issue has been experimentally addressing the relative importance of estrogen inhibition and androgen stimulation in protogynous sex change.

The bluehead wrasse, *Thalassoma bifasciatum*, was the subject of a series of studies addressing androgen effects on sex change. Several authors reported that injections of androgens induced bluing of the head, a feature characteristic of terminal phase males, and some degree of gonadal sex change (Stoll 1955; Roede 1972; Reinboth 1975). In contrast, neither testosterone nor dihydrotestosterone implants administered to female bluehead wrasses induced testicular development in a later study, although the fish did exhibit discernible color changes within five days that progressed and "approached that which exists in the wild" (Kramer et al. 1988). It is noteworthy that this study was conducted with captive fish; the investigators were unable to assess the initial sex of treated individuals at the beginning of the experiments (female or initial phase male) and therefore inferred effects from sex ratios following sacrifice at the end of the study; and that T is a precursor for both E_2 and 11-KT. The last point complicates interpretation, because implanted T could act in at least three distinct ways: as an androgen through the nuclear androgen receptor as T or following conversion to 11-KT, or as an estrogen through one of the nuclear estrogen receptors following conversion by aromatase (biochemically distinct membrane receptors for steroids in teleosts add additional possibilities)(Thomas et al. 2006).

The non-aromatizable 11-KT does reliably induce sex change in the labroids in which this has been tested. Grober & Bass (1991) induced sex and color change in captive bluehead wrasses with 11-KT implants, and studies in the author's laboratory have replicated these results in free-living bluehead wrasses (Semsar & Godwin 2004; Austin et al. unpublished data)(Figure 9.5).

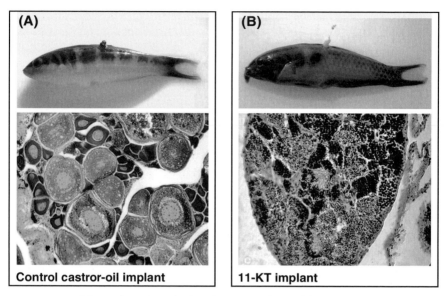

FIGURE 9.5. Effects of 11-KT on body coloration and sexual development in the blue-head wrasse, *Thalassoma bifasciatum*. Females were implanted for two weeks in their natural habitat in the presence of dominant TP males, receiving either (A) a control castor oil implant or (B) an implant containing 11-KT dissolved in castor oil.

Behavioral sex change in the absence of gonads is not accompanied by permanent color changes, presumably because of a lack of endogenous 11-KT (Godwin et al. 1996; discussed below). Following their demonstration of sex differences in plasma 11-KT in stoplight parrotfish, Cardwell & Liley (1991a) were also able to induce sex change in females held in a reef enclosure with 11-KT injections. Control injections did not induce sex change. Feeding female threespot wrasses 11-KT induced complete sex change, but the aromatase inhibitor fadrozole used in goby and grouper studies referenced above was equally effective (Higa et al. 2003). An important and informative addition to these studies on threespot wrasses were treatments where E_2 was coadministered with either 11-KT or fadrozole. This coadministration of E_2 blocked the sex change that both 11-KT and fadrozole induced when administered alone. These results are consistent with those discussed previously for gobies and groupers, but the effects of estrogen supplementation suggest that estrogen inhibition may be more important than androgen stimulation in the initiation of sex change. It is also possible that a key action of 11-KT in inducing sex change is through inhibition of aromatase function, as has been demonstrated in another perciform teleost (Braun & Thomas 2003).

NEURAL CORRELATES OF SEXUAL PHENOTYPE
IN *THALASSOMA* WRASSES

Sex and role change involve widespread changes in sexual phenotype. This process is largely socially regulated in wrasses and parrotfishes, indicating that transduction of social cues must occur through the central nervous system. This is a complex problem that will be challenging to address, but progress is being made in understanding the neural correlates of sexual phenotype development.

The best-studied group of labroids in terms of linking brain and behavior are wrasses in the genus *Thalassoma*. This genus of approximately 30 species has a worldwide distribution in tropical and many warm temperate seas and members are often conspicuous components of the reef fish community where they occur. Members of this genus, especially the saddleback wrasse *T. duperrey* and bluehead wrasse *T. bifasciatum,* have also been well studied in terms of behavior and ecology (see Warner et al. 1975; Warner 1984; Ross 1986, 1987). This provides a rich context in which to interpret neuroendocrine and neurobiological patterns. The saddleback wrasse has proven amenable to studies of socially mediated sex change in captivity (Ross et al. 1983, 1990). The bluehead wrasse is well suited to manipulative studies of sex change under field conditions due to ease of capture, site attachment on small coral patch reefs that facilitates observations and recapture, and because the process is very rapid under field conditions with behavioral changes occurring almost immediately and full gonadal sex change being completed in 8 to 10 days (Warner & Swearer 1991; Godwin et al. 1996). Finally, the histological, ultrastructural, and endocrine correlates of gonadal function have been well studied in the saddleback wrasse as discussed above, and to a lesser extent in the bluehead wrasse (Nakamura et al. 1989; Hourigan et al. 1991; Shapiro & Rasotto 1993, 1998). This work provides critical background for understanding how neuropeptide and neurotransmitter systems contribute to sex and role change.

Following their study investigating the effects of T implants on sex change in bluehead wrasses, Kramer and colleagues tested the effects of manipulating the HPG axis at the pituitary and hypothalamic levels. The first of these experiments involved injecting human chorionic gonadotropin (hCG) at one of two dosages (20 or 40 international units/g body weight) either once or three times weekly or a saline control (Koulish & Kramer 1989). Fish were sacrificed at one, two, four, and six weeks after the initiation of treatments. Substantially higher proportions of individuals were found with both degenerating ovarian tissue and cysts of proliferating spermatogenic tissue in the low-dose hCG treatments than in the control treatment and few females remained, while there was a large number of females and no individuals undergoing sex change in the saline control group. Interestingly, the authors indicated that only about 40% of the individuals diagnosed as undergoing sex reversal showed evidence of permanent color change, but these individuals may have been very early on in the process. Pituitary gonadotropin secretion is primarily under the control of gonadotropin releasing hormone (GnRH), and the next study in this series addressed the effect of exogenous synthetic GnRH coadministered with the dopamine receptor antagonist domperidone (coadministered because dopamine antagonizes gonadotropin release in some teleosts and the authors wished to avoid this potential confound)(Kramer et al. 1993). After six weeks of treatment, 93% of treated females exhibited early signs of sex reversal, defined as the presence of spermatogenic-like cysts of cells, but more advanced sex change was not observed. The last in this series of studies addressed a potential role for the neuropeptide hormone NPY because of potential interactions with the GnRH system (Kramer & Imbriano 1997). As with the hCG and synthetic GnRH studies, NPY injections three times a week over eight weeks induced signs of sex change in approximately 82% of treated females. Also as in the previous studies, most of these sex-reversing individuals were in the early stages of the process as measured by spermatogenic tissue development.

This is a useful series of studies, but several points are noteworthy. First, the females were held individually in aquaria, so it is not known whether the treatments administered could overcome the social inhibition normally experienced by female bluehead wrasses in the wild. Second, if estrogenic inhibition is important as suggested by several studies, experiments under captive conditions may lower this inhibition if ovaries are regressed. Finally, while these experiments provided unambiguous evidence of spermatogenic tissue development in females, the time course was very slow relative to the eight to ten days required to complete sex change in field experiments (Warner & Swearer 1991). Nevertheless, the conclusions are consistent with a model where changes in gonadotropin production and release are critical for regulating sex change.

Do different sexual phenotypes in the bluehead wrasse exhibit differences in GnRH neural phenotype? This question was addressed in two studies where it was found that (i) TP male bluehead wrasses exhibited higher numbers of GnRH neurons in the preoptic area of the hypothalamus than either females or IP males, and (ii) a TP male-like GnRH neuronal phenotype could be induced in both females and IP males with 11-KT implants (Grober & Bass 1991; Grober et al. 1991). A similar pattern was described in the Ballan wrasse (*Labrus berggylta*), a monandric protogynous species in which males have larger numbers of GnRH neurons than do females (Elofsson et al. 1999). A recent study in saddleback wrasses adds to these results with demonstrations of extensive GnRH innervation well outside the hypothalamus, reflecting the variety of functions subserved by this hormone (Maruska & Tricas 2007). As with the effects of the manipulative studies described above, these results also point to a role for GnRH neurons in regulating sex and role change. More generally, such a role would also be consistent with the differences observed across sexual phenotypes in gonadal function and sex steroid profiles. Still lacking, unfortunately, is a detailed model of precisely how changes in GnRH neuron function could drive the sex change process. Detailed studies of production and release of GnRH as well as, ideally, GnRH receptor expression throughout the sex change process could be very valuable in this regard.

What is the relationship between gonadal and behavioral changes during sex change? One of the earliest descriptions of socially controlled sex change, in the cleaner wrasse *Labroides dimidiatus*, noted that behavioral changes in sex-changing females could occur in a matter of minutes. Warner & Swearer (1991) demonstrated socially controlled sex change in the bluehead wrasse and described a similar very rapid change in behavior following removal of dominant TP males. These observations call into question a role for gonadal hormones, and especially increases in plasma androgens, in driving at least the initial behavioral shifts. A gonadal role in behavioral changes is impossible to rule out in females with gonads. Therefore, working on patch reefs in the U.S. Virgin Islands, we surgically removed the ovaries in bluehead wrasse sex change candidates prior to TP male removal to test their necessity in the process (Godwin et al. 1996). We found that ovariectomized female bluehead wrasses can indeed exhibit complete behavioral sex change upon becoming behaviorally dominant and maintain that dominant status and associated territorial and courtship behaviors for at least several weeks. Behavioral profiles for these newly dominant ovariectomized females were not discernibly different from those of dominant sham controls, although permanent blue color development did not occur due presumably to the lack of testes and 11-KT.

What does mediate behavioral sex change? Based on the wealth of data implicating arginine vasotocin (AVT) in male-typical behaviors across vertebrates (see references above), we examined the expression of this neuropeptide across sexual phenotypes and sex change in bluehead wrasses. We found that AVT mRNA expression in the preoptic area of the hypothalamus was relatively low in females, high in TP males and intermediate in IP males (Godwin et al. 2000). Importantly, AVT mRNA abundances increased rapidly with the assumption of male behavior during sex change, consistent with a hypothesized role in the behavioral changes the fish were undergoing. Since expression of AVT and its mammalian homologue AVP are strongly androgen dependent in a number of species, it was unclear from these findings whether increases in AVT expression were related to the changing behavioral profile of sex changers or perhaps simply in response to the changing gonadal hormone environment over sex change. To address this, we compared AVT mRNA and AVT neuron size in females who were either ovariectomized or left intact (sham control) and either made socially dominant through TP males removal or remained in the presence of an aggressively dominant TP male (Semsar &

Godwin 2003). We found that AVT mRNA abundances were dependent on social status and that gonadal status had no effect (i.e., the presence of either an ovary in subordinate animals or an intersexual gonad in sham control dominants). Consistent with this lack of gonadal influence, we also found that neither castration of TP males or 11-KT implants in ovariectomized females affected hypothalamic AVT mRNA abundances.

The correlations between TP male-typical behavioral profiles and AVT expression are consistent with a key role for this neuropeptide in determining behavioral phenotype in bluehead wrasses. We next tested whether AVT was either necessary or sufficient for inducing the display of these behaviors in bluehead wrasses. On larger patch reefs in the U.S. Virgin Islands, approximately one-third to one-half of the TP males present are not territory holders (NT-TP). These individuals are typically larger than resident IP males and smaller than territory-holding TP males. When a territorial TP male (T-TP) disappears, an NT-TP male typically quickly takes over that territory. We found that NT-TP males showed increases in territorial aggression and courtship on days when they received an intraperitoneal injection of AVT in saline, but not on days when they received only the saline control (Semsar et al. 2001). Administering an AVT blocker to T-TP males reduced both their displays of territorial aggression and courtship, and it resulted in approximately half of these males abandoning their territories for that day's spawning period (Manning compound, an AVP V1 receptor antagonist, was used here). In order to test whether AVT was necessary for the assumption rather than simply maintenance of territorial status, we captured and injected NT-TP males with either the antagonist or a saline as a control and then captured T-TP males immediately before the daily spawning period, opening their territories for occupancy by the treated NT-TP males (Semsar & Godwin 2004). NT-TP males treated with saline all successfully occupied territories while only 25% of the NT-TPs treated with the receptor antagonist did. Finally, in order to test the role of AVT in mediating behavioral transitions during behavioral sex change, we treated the largest females in social groups on small reefs with either saline or the AVT receptor antagonist and then created a social environment permissive to sex change by removing TP males. As with the NT-TP males, saline-treated females responded to this social "opportunity" by exhibiting behavioral sex change, while females treated with the AVT receptor antagonist did not.

Do steroid hormones play a role in determining behavioral phenotypes in bluehead wrasses? Several lines of evidence now suggest they do. One intriguing aspect of the AVT manipulation experiments described above was that while we could induce territorial behavior in NT-TP males with AVT injections and reduce or block it in T-TP males, NT-TP males, and female sex change candidates with an AVT receptor antagonist, we could not induce TP male-typical behaviors in a socially inhibitory environment in either females or IP males with exogenous

AVT. This finding is consistent with those from some other species (e.g., Goodson & Bass 2000) but raised the question of how sensitivity to AVT is mediated. Hypothesizing that 11-KT might increase sensitivity to AVT in TP males, we implanted ovariectomized females with 11-KT and compared their behavioral responses to exogenous AVT with those of ovariectomized females receiving blank implants (Semsar & Godwin 2004). The 11-KT implants induced full color change, but did not induce contesting or defense of territories or increase the behavioral responsiveness to AVT. Interestingly however, 11-KT implants did increase the display of *opportunistic* courtship behavior, as seen when implanted fish would encounter gravid females away from established spawning sites and exhibit TP male-typical courtship displays directed towards these females. Blank-implanted females did not display any courtship behavior. This finding suggests that 11-KT may mediate development of some components of a TP male behavioral phenotype (such as courtship), but is not responsible for the development of aggressive territoriality.

As with gonadal change, it appears that estrogenic inhibition is also critical in regulating behavioral aspects of sex change. The brain form of aromatase is strongly expressed in the preoptic area of the hypothalamus in bluehead wrasses, with aromatase-immunoreactive glial cells in very close proximity to both AVT neurons and tyrosine hydroxylase-immunoreactive neurons (putatively dopaminergic, Marsh et al. 2006). Cloning of a portion of the brain form of aromatase from bluehead wrasses allowed localization and quantification of aromatase mRNA through in situ hybridization. This work localized aromatase mRNA to the preoptic area of the hypothalamus and showed that expression was higher in females than in TP males, with IP males displaying intermediate values (Marsh 2007; Marsh et al. under review). Finally, we found that estradiol implants both increased aromatase mRNA abundances in the preoptic area, as assessed by *in situ* hybridization, and effectively blocked behavioral sex change in female bluehead wrasses under socially permissive conditions, while blank-implanted females did exhibit complete behavioral sex change.

Evidence for estrogen and AVT influences on TP male behavior and changes in these systems during sex change lead logically to questions about how perception of the social environment is transduced into changes in neuroendocrine phenotype. This area is still poorly explored, but some progress has been made. The monoamine neurotransmitters play important roles in sociosexual behavior and neuroendocrine function in fishes as in other vertebrates (e.g., Summers & Winberg 2006; Nelson & Trainor 2007). Working with saddleback wrasses undergoing sex change in experimental pens, Larson et al. (2003a) documented changes in serotonergic, noradrenergic, and dopaminergic metabolism over a number of brain areas. These changes were most pronounced during the first week of sex reversal, and a decline in serotonergic activity was mirrored by an increase in noradrenergic activity in the preoptic area of the hypothalamus. These reciprocal changes in this key brain region for the regulation of reproductive function and sexual behavior were suggested to be particularly important for initiating gonadal sex reversal. A second study employed a variety of pharmacological agents to either interfere with or augment the actions of the serotonergic, dopaminergic, and noradrenergic monoamine neurotransmitter systems (Larson et al. 2003b). These agents were administered under experimental environments that were either socially permissive or socially inhibitory for sex change. Briefly summarized, the conclusions from this study suggested that both dopamine and serotonin (5-hydroxytryptamine, or 5-HT) are important influences inhibiting females from entering into the sex reversal process while serotonin also inhibits the completion of sex reversal. By contrast, the results suggest that noradrenergic influences stimulate both phases of sex change. These saddleback wrasse experiments were performed in floating 1-m^3 pens, a design previously used very successfully to study sex change in this species (e.g., Ross et al. 1983). In a preliminary study with bluehead wrasses, our laboratory implanted the same dopaminergic inhibitor used by Larson and colleagues, haloperidol—a nonselective dopamine receptor antagonist—into free-living females on coral patch reefs and found that this treatment could stimulate sex change in an inhibitory social environment in the wild as well.

Serotonergic metabolism also affects at least male-typical aggression in bluehead wrasses. Fluoxetine administered either acutely in the field or chronically in the laboratory decreases aggression against territorial intruders by TP males (Perreault et al. 2003). Fluoxetine (trade name: Prozac) is a selective serotonin reuptake inhibitor that should increase serotonergic signaling and so these behavioral results are consistent with a general pattern of serotonergic inhibition of aggression in fishes (Winberg & Nilsson 1993). However, perhaps somewhat surprisingly, the mechanisms of action of this drug are not completely understood even in well-studied mammalian models. The inhibition of aggression in TP male bluehead wrasses was accompanied by a decrease in AVT mRNA abundances in the preoptic area of the hypothalamus, which could also account for the inhibition of aggression in this experiment (Semsar et al. 2004). It is also possible that this type of fluoxetine effect is due to changes in neural synthesis of steroid hormones and indirect effects on signaling through the gamma-aminobutyric acid (GABA) neurotransmitter system rather than changes in serotonin signaling (discussed further in Semsar et al. 2004; also see Pinna et al. 2003).

Experimental results with saddleback wrasses suggest that further characterization of monoamine neurotransmitter effects is likely to be a fruitful line of investigation. Distributions of the rate-limiting enzyme in dopamine synthesis, tyrosine hydroxylase, were described for the brain of the bluehead wrasse (Marsh et al. 2006). However, little is otherwise known about innervation patterns of other key monoaminergic systems (e.g., see Khan & Thomas 1993), and non-aminergic neurotransmitter systems likely to be important influences on reproductive function and behavior (especially the GABAergic and glutamatergic systems) or distributions of receptors for each.

THE POMACENTRIDAE
(*DASCYLLUS* AND *AMPHIPRION*)

The gobies have been excellent models in part because of the bidirectionality of sex change in a number of species, enabling the study of sex change in both directions. While this degree of plasticity in a single species is not known or well characterized in other groups currently, the damselfishes (Pomacentridae) present opportunities here because protogyny is observed in the genus *Dascyllus* (reviewed in Godwin 1995) and protandry in the genus *Amphiprion* (Miura et al. 2008). Species in these genera are useful models because they are common on many west Pacific reefs, typically very site-attached and easily captured (facilitating field studies), relatively small bodied and adapt well to captivity, and because this family is the sister taxon to the well-studied cichlids that are focus of a current genome sequencing effort. Sexual patterns in the genus *Dascyllus* have been very carefully explored in a series of studies by K. Asoh (e.g., Asoh 2003, 2005), setting the stage very well for physiological work.

Amphiprion are developmentally protogynous, but functionally protandrous, as juvenile animals display an immature ovarian morphology, males possess an ovotestis with immature ovarian tissue and mature spermatogenic tissue, and females possess only ovarian tissue (Reinboth 1980; Godwin 1994b; Miura et al. 2008). Godwin & Thomas (1993) described increases in E_2 and decreases in 11-KT with protandrous sex change in *Amphiprion melanopus*. Levels of T and androstenedione followed a similar pattern to that of E_2 and, significantly, decreased during gonadal sex change before rising again in females. Overall, androgen levels were not strongly correlated with the display of aggressive behavior, declining over the course of sex change while the fish became much more aggressive (Godwin 1994a). Miura and coworkers found greater numbers of cells immunoreactive for the steroidogenic p450 cholesterol side-change cleavage enzyme in *Amphiprion clarkii* at the ovarian stage relative to individuals with spermatogenic tissue, which is potentially consistent with the higher E_2 levels and trend toward higher T and androstenedione found by Godwin & Thomas (1993). The only study to examine neural differences in *Amphiprion* found greater numbers of GnRH neurons in males than in females (Elofsson et al. 1997), a result consistent with that found in wrasses, despite the opposite direction of sex change exhibited. The damselfishes are labroids like the wrasses and parrotfishes, but relatively little work has been done with them.

CONCLUSIONS AND DIRECTIONS

In a 1980 review of the environmental control of sex in fishes, Reinboth remarked, "The term 'social control' is nothing better than a 'black box'" (Reinboth 1980, p. 54). The box remains pretty dark, but substantial progress has been made in the last two decades. The findings described above for three divergent groups of coral reef fishes exhibiting socially mediated sex change show some remarkable consistencies. Estrogens in general and E_2 in particular appear likely to play a controlling role in the onset of sex change with E_2 levels being dependent on the expression and activity of the aromatase enzymes in the ovaries and brain. Research in several species also suggests regulation of these changes in estrogen metabolism occurs through gonadotropin signaling, with gonadotropes being under the control of GnRH neurons in the preoptic area of the hypothalamus. Despite these advances, a number of key questions remain to be addressed.

Prominent among these questions is the precise mechanism by which a change in GnRH and gonadotropin signaling could lead to sex change. The rapid degeneration of more advanced oocytes at the initiation of protogynous sex change suggests a loss of gonadotropin support for the ovary. The observation that sex change is more common in the nonbreeding season in at least some species also suggests that the differentiation of testicular tissue in the ovary may be more likely when gonadotropin and estradiol levels are low. Beyond these observations, however, detailed models for how gonadotropic mediation of sex change might work are still lacking.

The mechanisms by which social information is transduced into neural and the neuroendocrine events that underlie changes in reproductive physiology and behavior also remain very poorly understood in both sex-changing and gonochoristic species. This is a challenging problem, but also a very important one, as many of the mechanisms are likely conserved across both fishes and vertebrates more generally. Sex-changing species provide favorable systems for addressing it because of the dramatic and often very manipulatable nature of the transformation, because the process is usually relatively rapid, and because it can be experimentally studied in an adult animal. The relationship between monoamine metabolism in the brain and social dominance as well as the documented effects of monoamine manipulations on sex change in saddleback wrasses (Larson et al. 2003b) suggest these neurotransmitter systems are likely to be an important part of this puzzle. The support for dopaminergic innervation of AVT neurons in bluehead wrasses (Marsh et al. 2006) and serotonergic innervation of GnRH neurons in a gonochoristic teleost, the Atlantic croaker, *Micropogonias undulatus* (Khan & Thomas 1993), are consistent with such a role but much more work needs to be done in this area. The neuroanatomy of these systems in the teleost brain presents some logistical advantages because neuropeptidergic neurons are distributed periventricularly in the preoptic area of the hypothalamus. This feature has been exploited in elegant studies of AVT and isotocin neurons in trout (Saito & Urano 2001; Saito et al. 2004), but so far, it has not been examined in any sex-changing species.

Another area that should be fruitful for further inquiry is the relationship between social stress, stress-related hormones, and gonadotropin signaling. Perry & Grober (2003) developed a detailed model of how glucocorticoid hormones might control sexual phenotype in the bluehead wrasse with applicability to sex-changing fishes more generally. Briefly, they proposed that chronic social stress produces elevated glucocorticoid levels in subordinate individuals and that these elevated glucocorticoids may inhibit processes necessary for sex and/or role change at several levels, including the sorts of direct interactions with the HPG axis that have been described in other species, inhibition of AVT expression, and reductions in 11-KT levels through competitive inhibition of enzymatic reactions critical to 11-KT synthesis. Few data are available to assess the model to date. Godwin & Thomas (1993) found elevated cortisol levels during protandrous sex change in *A. melanopus*, but this result is not inconsistent with Perry and Grober's model since *A. melanopus* is protandrous rather than the protogynous case the model addresses. Contrary to the predictions of this model, cortisol implants in female sandperch (*Parapercis cylindrica*) did not inhibit sex change (Frisch et al. 2007). This model and other potential links between glucocorticoid metabolism and sex change nevertheless deserve further study both because of the potential complexity of the links between cortisol and social interactions (see Abbott et al. 2003 for an example from primates) and a large literature describing effects of stress on reproduction.

The next five to ten years should see a substantial increase in our understanding of socially controlled sex change. A number of gobies, wrasses, and groupers have now been established as excellent laboratory models and an understanding of their basic reproductive biology is in place. Advances in molecular endocrinology, neurobiology, and genomics will greatly facilitate investigations and studies in these different systems should continue to aid in discerning general principles. Progress in our understanding of sex change will be important for insights into a basic problem in reproductive biology. These advances will also be of practical value in aquaculture, and potentially in stock enhancement efforts, because many ecologically and economically important reef species are hermaphrodites.

REFERENCES

Abbott DH, Keverne EB, Bercovitch FB, Shively CA, Medoza SP, Saltzman W, Snowdon CT, Ziegler TE, Banjevic M, Garland T Jr, Sapolsky RM (2003). Are subordinates always stressed? A comparative analysis of rank differences in cortisol levels among primates. Hormones and Behavior 43: 67–82.

Alam MA, Bhandari RK, Kobayashi Y, Nakamura S, Soyano K, Nakamura M (2006). Changes in androgen-producing cell size and circulating 11-ketotestosterone level during female-male sex change of honeycomb grouper *Epinephelus merra*. Molecular Reproduction and Development 73: 206–214.

Asoh K (2003). Reproductive parameters of female Hawaiian damselfish *Dascyllus albisella* with comparison to other tropical and subtropical damselfishes. Marine Biology 143(4): 803–810.

Asoh K (2005). Gonadal development and diandric protogyny in two populations of *Dascyllus reticulatus* from Madang, Papua New Guinea. Journal of Fish Biology 66(4): 1127–1148.

Baroiller JF, Guigen Y, Fostier A (1999). Endocrine and environmental aspects of sex differentiation in fish. Cellular and Molecular Life Sciences 55(6–7): 910–931.

Benton CB, Berlinsky DL (2006). Induced sex change in black sea bass. Journal of Fish Biology 69(5): 1491–1503.

Bhandari RK, Higa M, Nakamura S, Nakamura M (2004). Aromatase inhibitor induces complete sex change in the protogynous honeycomb grouper (*Epinephelus merra*). Molecular Reproduction and Development 67(3): 303–307.

Bhandari RK, Alam MA, Higa M, Soyano K, Nakamura M (2005). Evidence that estrogen regulates the sex change of honeycomb grouper (*Epinephelus merra*), a protogynous hermaphrodite fish. Journal of Experimental Zoology A 303A(6): 497–503.

Bhandari RK, Alam MA, Soyano K, Nakamura M (2006). Induction of female-to-male sex change in the honeycomb grouper (*Epinephelus merra*) by 11-ketotestosterone treatments. Zoological Science 23(1): 65–69.

Black MP, Reavis RH, Grober MS (2004). Socially induced sex change regulates forebrain isotocin in *Lythrypnus dalli*. Neuroreport 15: 185–189.

Black MP, Balthazart J, Baillien M, Grober MS (2005). Socially induced and rapid increases in aggression are inversely related to brain aromatase activity in a sex-changing fish, *Lythrypnus dalli*. Proceedings of the Royal Society B 272: 2435–2440.

Borg B (1994). Androgens in teleost fishes. Comparative Biochemistry and Physiology C 109(3): 219–245.

Braun AM, Thomas P (2003). Androgens inhibit estradiol-17beta synthesis in Atlantic croaker (*Micropogonias undulatus*) ovaries by a nongenomic mechanism initiated at the cell surface. Biology of Reproduction 69(5): 1642–1650.

Callard G, Schlinger B, Pasmanik M, Corina K (1990). Aromatization and estrogen action in brain. Progress in Clinical and Biological Research 342: 105–111.

Callard GV, Tchoudakova AV, Kishida M, Wood E (2001). Differential tissue distribution, developmental programming, estrogen regulation and promoter characteristics of cyp19 genes in teleost fish. Journal of Steroid Biochemistry and Molecular Biology 79(1–5): 305–314.

Cardwell JR, Liley NR (1991a). Androgen control of social-status in males of a wild population of stoplight parrotfish, *Sparisoma viride* (Scaridae). Hormones and Behavior 25(1): 1–18.

Cardwell JR, Liley NR (1991b). Hormonal-control of sex and color-change in the stoplight parrotfish, *Sparisoma viride*. General and Comparative Endocrinology 81(1): 7–20.

Carlisle SL, Marxer-Miller SK, Canario AVM, Oliveira RF, Carneiro L, Grober MS (2000). Effects of 11-ketotestosterone on genital papilla morphology in the sex changing fish *Lythrypnus dalli*. Journal of Fish Biology 57: 445–456.

Cheek AO, Thomas P, Sullivan CV (2000). Sex steroids relative to alternative mating behaviors in the simultaneous hermaphrodite *Serranus subligarius* (Perciformes: Serranidae). Hormones and Behavior 37(3): 198–211.

Clemens LG, Gladue BA (1978). Feminine sexual behavior in rats enhanced by prenatal inhibition of androgen aromatization. Hormones and Behavior 11(2): 190–201.

Cole KS (1983). Protogynous hermaphroditism in a temperate zone territorial marine goby, *Coryphopterus nicholsi*. Copeia 1983(3): 809–812.

Crews D (1993). The organizational concept and vertebrates without sex chromosomes. Brain, Behavior and Evolution 42(4–5): 202–214.

Crews D (2003). Sex determination: where environment and genetics meet. Evolution and Development 5(1): 50–55.

Devlin RH, Nagahama Y (2002). Sex determination and sex differentiation in fish: an overview of genetic, physiological, and environmental influences. Aquaculture 208: 191–364.

De Vries GJ, Panzica GC (2006). Sexual differentiation of central vasopressin and vasotocin systems in vertebrates: different mechanisms, similar endpoints. Neuroscience 138(3): 947–955.

Drilling CC, Grober MS (2005). An initial description of alternative male reproductive phenotypes in the bluebanded goby, *Lythrypnus dalli* (Teleostei: Gobiidae). Environmental Biology of Fishes 72: 361–372.

Elofsson U, Winberg S, Francis RC (1997). Number of preoptic GnRH-immunoreactive cells correlates with sexual phase in a protandrously hermaphroditic fish, the dusky anemonefish (*Amphiprion melanopus*). Journal of Comparative Physiology A 181(5): 484–492.

Elofsson UO, Winberg S, Nilsson GE (1999). Relationships between sex and the size and number of forebrain gonadotropin-releasing hormone-immunoreactive neurones in the Ballan wrasse (*Labrus berggylta*), a protogynous hermaphrodite. The Journal of Comparative Neurology 410(1): 158–170.

Fennessy ST, Sadovy Y (2002). Reproductive biology of a diandric protogynous hermaphrodite, the serranid *Epinephelus andersoni*. Marine and Freshwater Research 53(2): 147–158.

Forlano PM, Schlinger BA, Bass AH (2006). Brain aromatase: new lessons from non-mammalian model systems. Frontiers in Neuroendocrinology 27(3): 247–274.

Frisch A (2004). Sex-change and gonadal steroids in sequentially-hermaphroditic teleost fish. Reviews in Fish Biology and Fisheries 14(4): 481–499.

Frisch AJ, McCormick MI, Pankhurst NW (2007a). Reproductive periodicity and steroid hormone profiles in the sex-changing coral-reef fish, *Plectropomus leopardus*. Coral Reefs 26(1): 189–197.

Frisch AJ, Walker SPW, McCormick ML, Solomon-Lane TK (2007b). Regulation of protogynous sex change by competition between corticosteroids and androgens: an experimental test using sandperch, *Parapercis cylindrica*. Hormones and Behavior 52(4): 540–545.

Gardner L, Anderson T, Place AR, Dixon B, Elizur A (2005). Sex change strategy and the aromatase genes. Journal of Steroid Biochemistry and Molecular Biology 94(5): 395–404.

Godwin J (1994a). Behavioural aspects of protandrous sex change in the anemonefish, *Amphiprion melanopus*, and endocrine correlates. Animal Behaviour 48: 551–567.

Godwin J (1994b). Histological aspects of protandrous sex-change in the anemonefish *Amphiprion melanopus* (Pomacentridae, Teleostei). Journal of Zoology 232: 199–213.

Godwin J (1995). Phylogenetic and habitat influences on mating system structure in the humbug damselfishes (*Dascyllus*, Pomacentridae). Bulletin of Marine Science 57(3): 637–652.

Godwin J, Crews D, Warner RR (1996). Behavioural sex change in the absence of gonads in a coral reef fish. Proceedings of the Royal Society of London Series B 263(1377): 1683–1688.

Godwin J, Sawby R, Warner RR, Crews D, Grober MS (2000). Hypothalamic arginine vasotocin mRNA abundance variation across sexes and with sex change in a coral reef fish. Brain Behavior and Evolution 55(2): 77–84.

Godwin JR, Thomas P (1993). Sex change and steroid profiles in the protandrous anemonefish, *Amphiprion melanopus* (Pomacentridae, Teleostei). Endocrinology 91: 144–157.

Goodson JL, Bass AH (2000). Forebrain peptides modulate sexually polymorphic vocal circuitry. Nature 403(6771): 769–772.

Goodson JL, Bass AH (2001). Social behavior functions and related anatomical characteristics of vasotocin/vasopressin systems in vertebrates. Brain Research Reviews 35(3): 246–265.

Grober MS, Bass AH (1991). Neuronal correlates of sex/role change in labrid fishes: LHRH-like immunoreactivity. Brain, Behavior and Evolution 38(6): 302–312.

Grober MS, Sunobe T (1996). Serial adult sex change involves rapid and reversible changes in forebrain neurochemistry. Neuroreport 7(18): 2945–2949.

Grober MS, Jackson IMD, Bass AH (1991). Gonadal steroids affect LHRH preoptic cell number in sex/role changing fish. Journal of Neurobiology 22(7): 734–741.

Hawkins MB, Thornton JW, Crews D, Skipper JK, Dotte A, Thomas P (2000). Identification of a third distinct estrogen receptor and reclassification of estrogen receptors in teleosts. Proceedings of the National Academy of Sciences of the United States of America 97(20): 10751–10107.

Hawkins MB, Godwin J, Crews D, Thomas, P (2005). The distributions of the duplicate oestrogen receptors ER-beta a and ER-beta b in the forebrain of the Atlantic croaker (*Micropogonias undulatus*): evidence for subfunctionalization after gene duplication. Proceedings of the Royal Society B 272(1563): 633–641.

He Q, Li W, Lin H (2006). Molecular cloning and functional characterization of the gonadotropin-releasing hormone receptor in orange-spotted grouper, *Epinephelus coioides*. Journal of Experimental Zoology A 305A(2): 132.

Higa M, Ogasawara K, Sakaguchi A, Nagahama Y, Nakamura M (2003). Role of steroid hormones in sex change of protogynous wrasse. Fish Physiology and Biochemistry 28(1–4): 149–150.

Hourigan TF, Nakamura M, Nagahama Y, Yamauchi K, Grau EG (1991). Histology, ultrastructure, and *in vitro* steroidogenesis of the testes of two male phenotypes of the protogynous fish, *Thalassoma duperrey* (Labridae). General and Comparative Endocrinology 83: 193–217.

Kaufman LS, Liem KF (1982). Fishes of the suborder Labroidei (Pisces: Perciformes): phylogeny, ecology and evolutionary significance. Brevortia 472: 1–19.

Khan IA, Thomas P (1993). Immunocytochemical localization of serotonin and gonadotropin-releasing-hormone in the brain and pituitary gland of the Atlantic croaker *Micropogonias undulatus*. General and Comparative Endocrinology 91(2): 167–180.

Kim SJ, Ogasawara K, Park JG, Takemura A, Nakamura M (2002). Sequence and expression of androgen receptor and estrogen receptor gene in the sex types of protogynous wrasse, *Halichoeres trimaculatus*. General and Comparative Endocrinology 127(2): 165–173.

Kobayashi Y, Kobayashi T, Nakamura M, Sunobe T, Morrey CE, Suzuki N, Nagahama Y (2004). Characterization of two types of cytochrome P450 aromatase in the serial-sex changing gobiid fish, *Trimma okinawae*. Zoological Science 21(4): 417–425.

Kobayashi Y, Sunobe T, Kobayashi T, Nagahama Y, Nakamura M (2005). Promoter analysis of two aromatase genes in the serial-sex changing gobiid fish, *Trimma okinawae*. Fish Physiology and Biochemistry 31(2–3): 123–127.

Koulish S, Kramer CR (1989). Human chorionic-gonadotropin (hcg) induces gonad reversal in a protogynous fish, the bluehead wrasse, *Thalassoma bifasciatum* (Teleostei: Labridae). Journal of Experimental Zoology 252(2): 156–168.

Kramer CR, Imbriano MA (1997). Neuropeptide Y (NPY) induces gonad reversal in the protogynous bluehead wrasse, *Thalassoma bifasciatum* (Teleostei: Labridae). Journal of Experimental Zoology A 279: 133–144.

Kramer CR, Koulish S, Bertacchi PL (1988). The effects of testosterone implants on ovarian morphology in the bluehead wrasse, *Thalassoma bifasciatum* (Bloch) (Teleostei: Labridae). Journal of Fish Biology 32: 397–407.

Kramer CR, Caddell, MT, Bubenheimerlivolsi, L (1993). SGnRH-A [(D-Arg6,Pro9,net-) LHRH] in combination with domperidone induces gonad reversal in a protogynous fish, the bluehead wrasse, *Thalassoma bifasciatum*. Journal of Fish Biology 42(2): 185–195.

Kroon FJ, Liley NR (2000). The role of steroid hormones in protogynous sex change in the blackeye goby, *Coryphopterus nicholsii* (Teleostei: Gobiidae). General and Comparative Endocrinology 118(2): 273–283.

Kroon FJ, Munday PL, Pankhurst NW (2003). Steroid hormone levels and bi-directional sex change in *Gobiodon histrio*. Journal of Fish Biology 62(1): 153–167.

Kroon FJ, Munday PL, Westcott DA, Hobbs JPA, Liley NR (2005). Aromatase pathway mediates sex change in each direction. Proceedings of the Royal Society B 272(1570): 1399–1405.

Larson ET, Norris DO, Grau EG, Summers CH (2003a). Monoamines stimulate sex reversal in the saddleback wrasse. General and Comparative Endocrinology 130(3): 289–298.

Larson ET, Norris DO, Summers CH (2003b). Monoaminergic changes associated with socially induced sex reversal in the saddleback wrasse. Neuroscience 119(1): 251–263.

Li CJ, Zhou L, Wang Y, Hong YH, Gui JF (2005). Molecular and expression characterization of three gonadotropin subunits common alpha, FSH beta and LH beta in groupers. Molecular and Cellular Endocrinology 233(1–2): 33–46.

Li GL, Liu XC, Lin HR (2006a). Effects of aromatizable and nonaromatizable androgens on the sex inversion of red-spotted grouper (*Epinephelus akaara*). Fish Physiology and Biochemistry 32(1): 25–33.

Li GL, Liu XC, Zhang Y, Lin HR (2006b). Gonadal development, aromatase activity and P450 aromatase gene expression during sex inversion of protogynous red-spotted grouper *Epinephelus akaara* (Temminck and Schlegel) after implantation of the aromatase inhibitor, fadrozole. Aquaculture Research 37(5): 484–491.

Li GL, Liu XC, Lin HR (2007). Seasonal changes of serum sex steroids concentration and aromatase activity of gonad and brain in red-spotted grouper (*Epinephelus akaara*). Animal Reproduction Science 99(1–2): 156–166.

Lorenzi V, Earley RL, Grober MS (2006). Preventing behavioural interactions with a male facilitates sex change in female bluebanded gobies, *Lythrypnus dalli*. Behavioral Ecology and Sociobiology 59(6): 715–722.

Lorenzi V, Earley RL, Rodgers EW, Pepper DR, Grober MS (2008). Diurnal patterns and sex differences in cortisol, 11-ketotestosterone, testosterone, and 17 beta-estradiol in the bluebanded goby (*Lythrypnus dalli*). General and Comparative Endocrinology 155(2): 438–446.

Luckenbach JA, Godwin J, Daniels HV, Borski RJ (2003). Gonadal differentiation and effects of temperature on sex determination in southern flounder (*Paralichthys lethostigma*). Aquaculture 216(1–4): 315–327.

Luckenbach JA, Early LW, Rowe AH, Borski RJ, Daniels HV, Godwin J (2005). Aromatase Cytochrome P450: cloning, intron variation, and ontogeny of gene expression in southern flounder (*Paralichthys lethostigma*). Journal of Experimental Zoology Part A: Comparative Experimental Biology 303: 643–656.

Mank JE, Promislow DEL, Avise JC (2006). Evolution of alternative sex-determining mechanisms in teleost fishes. Biological Journal of the Linnean Society 87(1): 83–93.

Marsh KE (2007). Neuroendocrine transduction of social cues in the bluehead wrasse, *Thalassoma bifasciatum*. Ph.D. dissertation, North Carolina State University, Raleigh.

Marsh KE, Creutz LM, Hawkins MB, Godwin J (2006). Aromatase immunoreactivity in the bluehead wrasse brain, *Thalassoma bifasciatum*: immunolocalization and co-regionalization with arginine vasotocin and tyrosine hydroxylase. Brain Research 1126(1): 91–101.

Maruska KP, Tricas TC (2007). Gonadotropin-releasing hormone and receptor distributions in the visual processing regions of four coral reef fishes. Brain, Behavior and Evolution 70(1): 40–56.

Maruska KP, Mizobe MH, Tricas TC (2007). Sex and seasonal co-variation of arginine vasotocin (AVT) and gonadotropin-releasing hormone (GnRH) neurons in the brain of the halfspotted goby. Comparative Biochemistry and Physiology A 147(1): 129–144.

Miura S, Nakamura S, Kobayashi Y, Piferrer F, Nakamura M (2008). Differentiation of ambisexual gonads and immunohistochemical localization of P450 cholesterol side-chain cleavage enzyme during gonadal sex differentiation in the protandrous anemonefish, *Amphiprion clarkii*. Comparative Biochemistry and Physiology B 149(1): 29–37.

Nakamura M, Kobayashi Y (2005). Sex change in coral reef fish. Fish Physiology and Biochemistry 31(2–3): 117–122.

Nakamura M, Hourigan TF, Yamauchi K, Nagahama Y, Grau EG (1989). Histological and ultrastructural evidence for the role of gonadal steroid hormones in sex change in the protogynous wrasse *Thalassoma duperrey*. Environmental Biology of Fishes, 24(2): 117–136.

Nelson RJ, Trainor BC (2007). Neural mechanisms of aggression. Nature Reviews Neuroscience 8(7): 536–546.

Ohta K, Hirano M, Mine T, Mizutani H, Yamaguchi A, Matsuyama M (2008). Body color change and serum steroid hormone levels throughout the process of sex change in the adult wrasse, *Pseudolabrus sieboldi*. Marine Biology 153(5): 843–852.

Oliver AS (1997). Size and density dependent mating tactics in the simultaneously hermaphroditic seabass *Serranus subligarius* (Cope, 1870). Behaviour 134: 563–594.

Perreault HA, Semsar K, Godwin J (2003). Fluoxetine treatment decreases territorial aggression in a coral reef fish. Physiology and Behavior 79(4–5): 719–724.

Perry AN, Grober MS (2003). A model for social control of sex change: interactions of behavior, neuropeptides, glucocorticoids, and sex steroids. Hormones and Behavior 43(1): 31–38.

Pinna G, Dong E, Matsumoto K, Costa E, Guidotti A (2003). In socially isolated mice, the reversal of brain allopregnanolone down-regulation mediates the anti-aggressive action of fluoxetine. Proceedings of the National Academy of Sciences of the United States of America 100(4): 2035–2040.

Rasotto MB, Shapiro DY (1998). Morphology of gonoducts and male genital papilla, in the bluehead wrasse: implications and correlates on the control of gamete release. Journal of Fish Biology 52(4): 716–725.

Reinboth R (1975). Spontaneous and hormone-induced sex-inversion in wrasses (Labridae). Publications Stazione Zoologie Napoli 39: 550–573.

Reinboth R (1980). Can sex inversion be environmentally induced? Biology of Reproduction 22(1): 49–59.

Reinboth R, Becker B (1984). In vitro studies on steroid metabolism by gonadal tissues from ambisexual teleosts. I. conversion of [14C]testosterone by males and females of the protogynous wrasse *Coris julis* L. General and Comparative Endocrinology 55(2): 245–250.

Rodgers EW, Drane S, Grober MS (2005). Sex reversal in pairs of *Lythrypnus dalli*: behavioral and morphological changes. Biological Bulletin 208(2): 120–126.

Rodgers EW, Earley RL, Grober MS (2006). Elevated 11-ketotestosterone during paternal behavior in the bluebanded goby (*Lythrypnus dalli*). Hormones and Behavior 49(5): 610–614.

Roede MJ (1972). Color as related to size, sex, and behaviour in seven Caribbean labrid fish species (genera *Thalassoma, Halichoeres, Hemipteronotus*). Studies of the Fauna of Curacao and other Caribbean Islands XLII: 1–264.

Rose JD, Moore FL (2002). Behavioral neuroendocrinology of vasotocin and vasopressin and the sensorimotor processing hypothesis. Frontiers in Neuroendocrinology 23(4): 317–341.

Ross RM (1986). Social organization and mating system of the Hawaiian reef fish *Thalassoma duperrey* (Labridae). In: Uyeno T, Arai R, Taniuchi T, Matsuura K (eds.), Indo-Pacific Fish

Biology. Proceedings of the International Conference on Indo-Pacific Fishes. Ichthyological Society of Japan, Tokyo, pp. 794–802.

Ross RM (1987). Sex-change linked growth acceleration in a coral-reef fish, *Thalassoma duperrey*. Journal of Experimental Zoology 244(3): 455–461.

Ross RM, Losey GS, Diamond M (1983). Sex change in a coral-reef fish: dependence of stimulation and inhibition on relative size. Science 221(4610): 574–575.

Ross RM, Hourigan TF, Lutnesky MMF, Singh I (1990). Multiple simultaneous sex-changes in social groups of a coral reef fish. Copeia 1990 (2): 427–433.

Sadovy Y, Domeier M (2005). Are aggregation-fisheries sustainable? Reef fish fisheries as a case study. Coral Reefs 24(2): 254–262.

Sadovy de Mitcheson Y, Liu M (2008). Functional hermaphroditism in teleosts. Fish and Fisheries 9(1): 1–43.

Saito D, Urano A (2001). Synchronized periodic Ca^{2+} pulses define neurosecretory activities in magnocellular vasotocin and isotocin neurons. Journal of Neuroscience 21(RC178): 1–6.

Saito D, Komatsuda M, Urano A (2004). Functional organization of preoptic vasotocin and isotocin neurons in the brain of rainbow trout: central and neurohypophysial projections of single neurons. Neuroscience 124(4): 973–984.

Sarter K, Papadaki M, Zanuy S, Mylonas CC (2006). Permanent sex inversion in 1-year-old juveniles of the protogynous dusky grouper (*Epinephelus marginatus*) using controlled-release 17 alpha-methyltestosterone implants. Aquaculture 256(1–4): 443–456.

Scaggiante M, Grober MS, Lorenzi V, Rasotto MB (2006). Variability of GnRH secretion in two goby species with socially controlled alternative male mating tactics. Hormones and Behavior 50(1): 107–117.

Schlinger BA, Creco C, Bass AH (1999). Aromatase activity in the hindbrain vocal control region of a teleost fish: divergence among males with alternative reproductive tactics. Proceedings of the Royal Society of London Series B 266(1415): 131–136.

Semsar K, Godwin J (2003). Social influences on the arginine vasotocin system are independent of gonads in a sex-changing fish. Journal of Neuroscience 23(10): 4386–4393.

Semsar K, Godwin J (2004). Multiple mechanisms of phenotype development in the bluehead wrasse. Hormones and Behavior 45(5): 345–353.

Semsar K, Kandel FLM, Godwin J (2001). Manipulations of the AVT system shift social status and related courtship and aggressive behavior in the bluehead wrasse. Hormones and Behavior 40(1): 21–31.

Semsar K, Perreault HA, Godwin J (2004). Fluoxetine-treated male wrasses exhibit low AVT expression. Brain Research 1029(2): 141–147.

Shapiro DY (1980). Serial female sex changes after simultaneous removal of males from social groups of a coral reef fish. Science 209: 1136–1137.

Shapiro DY (1981). Intragroup behavioural changes and the initiation of sex reversal in a coral reef fish in the laboratory. Animal Behaviour 29: 1199–1212.

Shapiro DY, Rasotto MB (1993). Sex-differentiation and gonadal development in the diandric, protogynous wrasse, *Thalassoma bifasciatum* (Pisces, Labridae). Journal of Zoology 230: 231–245.

Shein NL, Takushima M, Nagae M, Chuda H, Soyano K (2003). Molecular cloning of go-nadotropin cDNA in sevenband grouper, *Epinephelus septemfasciatus*. Fish Physiology and Biochemistry 28(1–4): 107–108.

Soga T, Ogawa S, Millar RP, Sakuma Y, Parhar IS (2005). Localization of the three GnRH types and GnRH receptors in the brain of a cichlid fish: insights into their neuroendocrine and neuromodulator functions. Journal of Comparative Neurology 487(1): 28–41.

St. Mary CM (1994). Sex allocation in a simultaneous hermaphrodite, the blue banded goby (*Lythrypnus dalli*): the effect of body size and behavioral gender and the consequences for reproduction. Behavioral Ecology 5: 304–313.

Stoll LM (1955). Hormonal control of the sexually dimorphic pigmentation of *Thalassoma bifasciatum*. Zoologica 40: 125–131.

Summers CH, Winberg S (2006). Interactions between the neural regulation of stress and aggression. The Journal of Experimental Biology 209: 4581–4589.

Streelman JT, and Karl SA (1997). Reconstructing labroid evolution with single-copy nuclear DNA. Proceedings of the Royal Society of London B 264: 1011–1020.

Sunobe T, Nakamura M, Kobayashi Y, Kobayashi T, Nagahama Y (2005). Aromatase immunoreactivity and the role of enzymes in steroid pathways for inducing sex change in the hermaphrodite gobiid fish *Trimma okinawae*. Comparative Biochemistry and Physiology A 141: 54–59.

Thomas P, Dressing G, Pang YF, Berg H, Tubbs C, Benninghoff A, Doughty K (2006). Progestin, estrogen and androgen G-protein coupled receptors in fish gonads. Steroids 71(4): 310–316.

Warner RR (1984). Mating behavior and hermaphroditism in coral-reef fishes. American Scientist 72(2): 128–136.

Warner RR, Swearer SE (1991). Social-control of sex-change in the bluehead wrasse, *Thalassoma bifasciatum* (Pisces, Labridae). Biological Bulletin 181(2): 199–204.

Warner RR, Robertson DR, Leigh EG (1975). Sex change and sexual selection. Science 190(4215): 633–638.

Winberg S, Nilsson GE (1993). Roles of brain monoamine neurotransmitters in agonistic behavior and stress reactions, with particular reference to fish. Comparative Biochemistry and Physiology C 106(3): 597–614.

Yeh SL, Kuo CM, Ting YY, Chang CF (2003a). Androgens stimulate sex change in protogynous grouper, *Epinephelus coioides*: spawning performance in sex-changed males. Comparative Biochemistry and Physiology C 135: 375–382.

Yeh SL, Kuo CM, Ting YY, Chang CF (2003b). The effects of exogenous androgens on ovarian development and sex change in female orange-spotted protogynous grouper, *Epinephelus coioides*. Aquaculture 218: 729–739.

Zhang WM, Zhang Y, Zhang LH, Zhao HH, Li X, Huang H (2007). The mRNA expression of P450 aromatase, gonadotropin beta-subunits and FTZ-F1 in the orange-spotted grouper (*Epinephelus coioides*) during 17 alpha-methyltestosterone-induced precocious sex change. Molecular Reproduction and Development 74: 665–673.

Zhang Y, Zhang WM, Zhang LH, Zhu TY, Tian J, Li X (2004). Two distinct cytochrome P450 aromatases in the orange-spotted grouper (*Epinephelus coioides*): cDNA cloning and differential mRNA expression. Journal of Steroid Biochemistry and Molecular Biology 92(1–2): 39–50.

Zhu Y, Bond J, Thomas P (2003a). Identification, classification, and partial characterization of genes in humans and other vertebrates homologous to a fish membrane progestin receptor. Proceedings of the National Academy of Sciences of the United States of America 100: 2237–2242.

Zhu Y, Rice CD, Pang YF, Pace M, Thomas P (2003b). Cloning, expression, and characterization of a membrane progestin receptor and evidence it is an intermediary in meiotic maturation of fish oocytes. Proceedings of the National Academy of Sciences of the United States of America 100: 2231–2236.

10

Acoustical Behavior
of Coral Reef Fishes

Phillip S. Lobel, Ingrid M. Kaatz, and Aaron N. Rice

The soundscapes of the ocean have been underappreciated for far too long. The notion that the sea was a silent world was conveyed early in the history of scuba diving (Cousteau & Dumas 1953), and the idea has persisted. This perspective was reinforced by the fact that human hearing is poor underwater and that the sounds of many fishes are not easily heard. Scuba divers are especially at a disadvantage for hearing underwater because of the near constant stream of noisy bubbles running over their ears. Exhaled bubbles and boat noise can often mask our being able to hear sound-producing fishes (Lobel 2001b, 2005; Radford et al. 2005).

Many reef fishes are acoustic, but in most cases, this is not yet a particularly well studied aspect of their behavior. The obstacle to studying underwater bioacoustics has been a combination of technology limitations for underwater recording and the fact that many reef fishes appear to be very discrete about where and when they produce sounds. For these reasons, until recently, underwater acoustic ecology has been largely overlooked. Advances in scuba technologies such as closed-circuit re-breathers newly expose divers to experience with increased awareness the natural structure and tempo of the ambient acoustic world underwater (Lobel 2001b).

There is now an emerging awareness that many fishes produce specific courtship and spawning sounds (e.g., Lobel 1992, 2001a, 2002) and that a coral reef is also a "choral" reef. Research on underwater sound and its ecological role has accelerated in recent years as the result of several scientific and technical developments. First, new technology in the form of camcorders, hydrophones, and computers has made the task of recording and analyzing underwater animal sounds and behavior much easier (Lobel 2001b, 2005). Second, discoveries that fishes produce species-specific courtship and spawning sounds opened the feasibility for the development of passive acoustic monitoring for documenting reproductive patterns (Lobel 1992, 2001b, 2005; Mann & Lobel 1995; Rountree et al. 2006; Luczkovich et al. 2008a). Third, loud noises from ships, sonar, seismic surveys, and global climate experiments such as ATOC (Acoustic Thermometry of Ocean Climate, http://atoc.ucsd .edu) have raised real concern about potential adverse impacts of loud underwater sounds on marine animals (McCauley et al. 2003; Popper 2003; Popper et al. 2003b; Vasconcelos et al. 2007; Popper & Hastings 2009). Last, new research is showing that larval reef fishes may be using the sounds emanating from coral reefs to navigate during their migration from the open ocean to benthic habitat (Simpson et al. 2004; Leis & Lockett 2005; Mann et al. 2007; Radford et al. 2008b). All of this demonstrates that underwater acoustic ecology is important to know in order to better manage ocean resources and for better understanding of how acoustics influences marine animal behavior, ecology, and evolution.

Sound production by fishes is most often associated with two behavioral contexts: reproduction and aggression. Sound production is typically more intense during the breeding season (Bass & McKibben 2003). Even so, the courtship and mating behaviors of many fishes have been elusive, and consequently, their acoustic behavior has been poorly known. Acoustic behavior is often aimed discretely at nearby prospective mates. Some of the quietest sounds are produced by carapids (pearlfishes, which are internal symbionts of sea cucumbers), syngnathids (seahorses), and gobiids (gobies). Only a few fishes make sounds loud enough to be easily heard by a diver, and it is these loudest of the reef fishes that have been the focus of most studies. These notably include the pomacentrids (damselfish), holocentrids (squirrelfish), sciaenids (drums), and batrachoidids (toadfish). Other fishes, such as scarids, can produce audible adventitious sounds while feeding (parrotfish; Sartori & Bright 1973; Takemura et al. 1988). A few fish have been found to also produce sounds specifically associated with the mating act (i.e., gamete release, Lobel 2002). The emerging pattern is that many fishes are sound producers, although the exact role of this behavior for mate selection and in predator-prey interactions is still being explored.

Hearing in fishes is an early evolutionary development in vertebrates and most likely preceded active acoustic behavior (Fay & Popper 2000; Ladich 2000; Bass et al. 2008). Investigations of the fish auditory system and their hearing abilities have been conducted for well over a century, including early detailed studies of the inner ear anatomy and physiology (e.g., Parker 1903; Marage 1906; Lafite-Dupont 1907; Bernoulli 1910; Bierbaum 1914; Warner 1932), clearly demonstrating behavioral and neurophysiological responses of fishes to acoustic stimuli. These studies were a prelude to the famous work on fish hearing by Karl von Frisch (1936, 1938), which led to general acceptance of the belief that fish can hear. However, the level at which fish communicate using acoustic signals is still relatively unknown and is still actively debated.

The inner ear morphological structure of three semicircular ear canals and basic sensitivity to a range of low to mid frequency sounds is an ancestral evolutionary innovation that is a shared character in gnathostomes (Lauder & Liem 1983). The emerging notion is that hearing evolved primarily as a mechanism for monitoring the ambient acoustic soundscape with particular regard to detecting predators or potential prey (Ladich 2000). Some species with advanced hearing specializations are particularly vulnerable prey species, such as freshwater ostariophysians (Ladich 2000) and marine clupeids (Mann et al. 1997). Other advanced taxa have also independently expanded their hearing range beyond low- to midrange frequencies, evolving high-frequency hearing sensitivity (Braun & Grande 2008). While all living fishes, so far as known, maintain their hearing abilities, there are many examples of fishes that have lost their sight, suggesting the overall adaptive significance of the auditory sense.

Sound producing mechanisms in fishes are highly varied and derived from diverse morphological adaptations. However, not all fish families with acoustic species share sound producing abilities. Some fish apparently remain mute (Hawkins & Myrberg 1983; Chen & Mok 1988; Ladich 2000). Fish will also become momentarily silent if threatened. Sound production may attract predators, and some species have been found to go immediately silent when potential predators are detected (Luczkovich et al. 2000; Remage-Healey et al. 2006). The pattern of sound production by fishes in general is not at all clear. Surveys to date have found that some groups of fishes have widespread occurrence of sound production throughout the family (e.g., batrachoidids, pomacentrids, and sciaenids), while other sympatric families make sounds very sparingly. Other fish families include species that appear to have secondarily lost acoustic ability entirely while related fishes still make sounds. These patterns point to an unknown underlying complexity at the core of the evolution of acoustic communication in fishes in general, including coral reef fishes.

This review was developed to provide a summary guide to the literature of shallow, tropical marine (i.e., coral reef) fish bioacoustics and to outline the fish families known for sound production. The hope is that this status report will encourage further research in fish acoustic communication and underwater acoustic ecology. As the reader will note in this chapter, the findings so far indicate that many coral reef fishes are most acoustically active during reproduction. It seems that some species produce sounds only when courting and mating and not at all during other times.

BRIEF HISTORY OF ACOUSTICAL OUTPUT IN FISH

Reports that fishes produce sounds date back to Aristotle 350 BCE, who observed that fishes "emit certain inarticulate sounds and squeaks." Aristotle also wrote that "the apparent voice in all these fishes is a sound caused in some cases by a rubbing motion of their gills, which by the way are prickly, or in other cases by internal parts about their bellies; for they all have air or wind inside them, by rubbing and moving which they produce the sounds" (Aristotle 1910, Book 4, Part 9).

Darwin was also keenly aware of fish sounds, known to him primarily from the work of Dufossé (1862, 1874a, b; cited in Pauly 2004). Dufossé reported that some fish sounds could be voluntarily produced (by pharyngeal bones and swim bladder vibration)(Dufossé 1874a, b). Darwin noted that in fish, as in insects, these sounds could play a role in sexual selection (Pauly 2004).

One of the earliest scientific reports of sound from a tropical reef fish appears to be Burkenroad (1930) describing the sounds of grunts (Haemulidae) in the Caribbean. Advanced research on the sounds of wild fishes was initiated after WWII from the need of the U.S. Navy to understand ambient ocean sounds. The Navy's task was to detect the sounds from enemy ships when sonar became an operational technology (Marshall 1962). The early studies by Fish (1948; Fish et al. 1952, 1954), Griffin (1950), Moulton (1958), and Tavolga (1960) grew out of this effort and established the foundations for modern fish bioacoustics. The specific study of reef fishes in their natural environment was pioneered in the 1960s at the Lerner Marine Laboratory and later, in the underwater habitat, Tektite (e.g., Bright 1972; Bright & Sartori 1972; Collette & Earle 1972; Sartori & Bright 1973).

In 1963, a camera and hydrophone system was deployed on the reef at about 20 m depth off the Lerner Lab on Bimini Island, Bahamas. The system was linked by cable to the shore lab where a room full of electronics was required at the time to process the audio-video signals (see photos in Kronengold et al. 1964). This setup did record a variety of sounds, but it was frustrating because the fixed camera usually did not catch the sound producer (Kumpf 1964). It did, however, document that there were temporal patterns to distinctive marine animal sounds (Cummings et al. 1964). A few years later, experiments were conducted using playback of sounds to test attraction to bony fishes and sharks (Richard 1968; Myrberg et al. 1969). Myrberg initiated his acoustic study of damselfishes at the Lerner Lab (Myrberg 1980, 1996). He and his colleagues continued the study of pomacentrid acoustic behavior and established the importance of pulse-repetition rate as a basis for interspecies recognition (Myrberg & Riggio 1985; Myrberg 1997a). Myrberg also led the development of the "model bottle" technique to elicit sounds and other agonistic behavioral responses from territorial fishes (Myrberg & Thresher 1974), and this simple technique is still a powerful method for experimentally inducing fishes to vocalize (Santangelo & Bass 2006; Tricas et al. 2006).

In 1968, Charles Breder wrote that "very little work on the sonic ecology and its relation to the life history and behavior of any species has been reported." Most astutely, Breder commented that study of fish behavior "has usually been treated as though fishes were both deaf and mute" (Breder 1968, p. 329). Breder's message

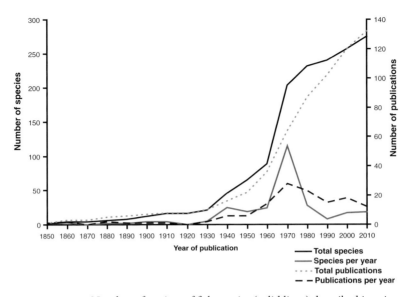

FIGURE 10.1. Number of sonic reef fish species (solid lines) described in scientific publications (dashed lines) in recent history.

is worth repeating today, as little has changed after 40 years. Many papers are still published describing fish behaviors that ignore the potential occurrence or role of sounds. This is not entirely due to neglect; recording and analyzing underwater sounds has been nontrivial. However, there has been a steady increase in fish acoustic studies over the years. To date, fewer than 300 coral reef species have been reported on in about 130 publications (Figure 10.1).

The Tektite program in 1969 was the first to use divers to record fishes. The Tektite scientific divers used Navy diving rebreathers so that scuba bubbles were not an acoustic interference. They customized an underwater housing for an 8-mm movie camera in tandem with a tape recorder and hydrophone. The camera and tape recorder were operated independently but simultaneously. The benefits of using a rebreather to record the sounds of fishes were clearly noted by Bright (1972). The Tektite report (Collette & Earle 1972) was also notable, because it contained a small, flexible 33-RPM record of the marine animal recordings.

In spite of a history of scientific observations and interest, fish acoustic behavior has been a difficult topic to accurately evaluate. This is mainly due to the fact that scuba divers can only hear the loudest of sound producing fishes such as damselfish (Pomacentridae), toadfish (Batrachoididae), groupers (Serranidae), and drums (Sciaenidae). Other fishes that produce sound are relatively quiet and not easily audible without the aid of a hydrophone (Lobel 1992). The challenge has been to clearly record fish sounds, to correlate these sounds to specific individuals and to discriminate responsive behaviors. New technologies for silent diving and underwater video-acoustic synchronous recording has helped overcome past difficulties (Lobel 2001b). Another significant historical problem has been the technical analysis of the sounds. This has evolved rapidly in the last few years. At present, there are a number of software programs that make it easy to do the acoustical analyses on ever more powerful and inexpensive computers. One of the most accessible software programs made especially for scientific analysis of animal sounds is Raven, produced by the Cornell Laboratory of Ornithology (Charif et al. 2008).

Today it is mostly a matter of a scientific diver's skill (and some luck) when making underwater observations, than it is the recording technology, to be successful at obtaining good acoustic recordings from wild fishes (Lobel 2003a, b, 2005). Of course, the alternative to the study of fishes in the wild is in an aquarium. Small aquariums do create acoustic problems, but if the aquarium is large enough and appropriately insulated from ambient noises, quality recordings are possible (Parvulescu 1967; Akamatsu et al. 2002; Okumura et al. 2002). The problem associated with captive fish studies concerns mostly the issue of understanding the full interaction of acoustic fishes in their natural acoustic landscape, including the presence of predators.

RESEARCH OBJECTIVES

The following basic research questions are the main issues in the study of the acoustical ecology of fishes:

- What is the pattern of sound production and associated behavior (especially in a reproductive context)? Are specific sounds repeatedly associated with specific behavior (e.g., courtship and the mating act)? Can we elicit reactions from fishes in response to specific call playbacks (such as attracting mates to a spawning site)?
- How similar are fish sounds among sympatric species, specifically the male courtship-associated sounds? The test is to determine if species will respond differently to playback of conspecific verses sibling species sounds. Differences between species or individuals would indicate that sound display could play an important role in assortative mating, female mate choice, and reproductive isolation of populations based on species-specific sound signals.
- What is the functional morphology of sound production, and how does it affect the acoustic signal?

Environmental background sounds can be both biologically and ecologically meaningful (Simpson et al. 2005; Mann et al. 2007; Radford et al. 2008b). Background noises also create the ambient interference that can limit a fish's ability to discriminate sounds. Hydrodynamic sounds of fishes swimming can be detected by conspecifics and by potential predators (Moulton 1960). The mechanical sounds of marine animals disturbing substrate could contribute to the acoustic features characterizing specific coastal habitats and are hypothesized as cues to larval fishes for finding suitable settlement locations (Radford et al. 2008a). Adventitious sounds, such as scarids grinding food, can signal competitors that there is a food source available (Sartori & Bright 1973) and in pufferfishes these sounds attract conspecifics (Breder & Clark 1947). The ecology of sound has ramifications in how fishes behaviorally use it. The key issue here is the balance between the need to communicate and the risk of being overheard by a predator or a competitor (Myrberg et al. 1969; Myrberg 1981; Luczkovich et al. 2000; Gannon et al. 2005; Remage-Healey et al. 2006).

One important practical application for using fish sound is for tracking where and when fish spawn (Mann & Lobel 1995; Luczkovich et al. 1999, 2008b). The scientific and technical challenge has been to develop methods that allow measurement of fish reproduction synchronously with time-series measurement of temperature, salinity, and other physical oceanographic variables that are easily recorded using modern devices (Mann & Grothues 2009). If we can associate specific sound patterns that are exclusively correlated with specific fish species and behaviors (such as the courtship call of a damselfish or the mating sound of the hamletfish), then we can develop listening devices to document their occurrence in time and space (Lobel 2001b, 2002; Lammers et al. 2008). A passive acoustic detection device that is programmed to monitor fish courtship and mating sounds becomes a "spawn-o-meter" that records data in a way that is comparable to that from other meters (e.g., a current meter, conductivity-temperature-depth meter, and temperature logger) at the same site (Mann & Lobel 1995; Lobel 2001b, 2005). The acoustical information thus generated can be used to evaluate how climate change (sea temperature, salinity, flow, etc.) may offset the timing and location(s) of fish reproduction. Overall, it is clear that being able to monitor fish mating activities is an essential tool for the successful management of fisheries and for associated conservation efforts (Rountree et al. 2006). The application of a spawn-o-meter is that such a device could be used to: (1) define important breeding habitats relevant to establishing protected areas; (2) establish existing relationships between physical oceanography and the timing of fish reproduction; and (3) define a critical endpoint measurement in pollution studies where the courtship vigor of a fish can be monitored and related to its health and fitness (Lobel 2001b).

REVIEW OF REVIEWS: AN ANNOTATED GUIDE

There have been a series of excellent reviews of fish behavior and sound production over the past several decades, and none are obsolete. The most recent update with an overview discussion of sound production mechanisms and sound structure is by Kasumyan (2008). It is detailed and complete and is thus a great introduction. Moreover, predecessor reviews all offer important insights as well as collated information cataloging the diversity of sounds and proposed sound producing mechanisms. Rather than repeat this information, we outline the most significant of these reviews, especially if they refer directly to coral reef fishes in particular. Our chapter is slanted particularly toward reproductive aspects and therefore fills a gap in current acoustic review papers. These review books and papers serve as the foundations for underwater bioacoustics as a science and thus are required reading for everyone interested in this field. Collectively, the reviews listed below summarize the many descriptive studies of individual fish sounds and associated behaviors. A chronological reading of this literature provides a perspective of how the field has developed, how fish produce sounds mechanistically, the diversity of hearing sensitivities and the relationship between sounds and behavior. The fundamental question concerning the role of sound in a fish's behavior is a core theme throughout these publications. It is important to note the date of publications as technology of the day limited what was achievable scientifically at the time.

The first series of papers that defined fish bioacoustics as a research field and raised the scientific awareness concerning the importance of underwater acoustic ecology appeared between 1960 and 1964. Tavolga's 1960 seminal paper reviewed sound production and underwater communication in fishes (Tavolga 1960). This paper was followed by reviews by Marshall (1962) and Moulton (1963, 1964a, b). Collectively, these papers provided the first detailed inventory and assessment of widespread acoustic behavior in fishes, with an emphasis on marine species. They raised the questions of how important sound is in fish behavior and the role of acoustics in animal ecology. The first broad synthesis of marine bioacoustics was published from the proceedings of a 1963 symposium held at the Lerner Marine Laboratory, Bimini, Bahamas (Tavolga 1964). These early studies are remarkable as the scientists were not only working in a new environment—underwater— but also adapting large and cumbersome electronics of the day to record a new source of sounds.

The encyclopedic compendium of fish sounds by Fish & Mowbray (1970) was a landmark publication. They reviewed the acoustic abilities of 220 fishes in 59 families from along the Atlantic coast and Caribbean. It was a dramatic display of the broad range of fishes capable of sound production and many were the first examples for the species. They illustrated the sounds using 160 spectrographs and 329 oscillographs for 153 species in 36 families of fishes. In terms of primary research and recordings, no subsequent work has been able to repeat the sheer diversity and number of fishes examined, and despite being almost 40 years old, *Sounds of the Fishes of the Western North Atlantic* is still an indispensable reference.

The review by Demski et al. (1973) delved into the mechanisms of sound production in fishes. They defined the three categories of sound producing mechanisms as: (1) hydrodynamic sounds from swimming, (2) stridulatory mechanisms of hard structures rubbing, and (3) swimbladder sounds produced by muscle contraction. The field of fish bioacoustics was still in its infancy at the time of their paper. They reported that although sound production has been reported in a number of diverse fishes, the biological significance was known only in a few examples. They concluded at the time that it was possible that many sounds produced by fishes "may have no biological significance but may be incidental to other aspects of the fish's behavior" (p. 1142). Their paper clearly set forth the criteria for research to not only identify the sounds being made but also how such sounds may or may not be directly a part of the fish's behavior.

In terms of sheer comprehensiveness, the seminal review by Fine et al. (1977) is still the most valuable starting point for a student of fish bioacoustics. Simply titled "Communication in fishes," it was published as a chapter in the massive tome edited by Sebeok (1977) titled *How Animals Communicate*. They emphasized all the modalities that could be involved in fish communication, including chemical, visual, and behavioral. Their table on message and modulation of fish vocalization and that from Fish et al. (1952) were the models for our updated table describing all reported occurrences of sounds in coral reef fishes (see Table 10.1 at end of chapter). The Fine et al. (1977) review was comprehensive and accurately established the bioacoustics research agenda that is still appropriate today. The synthesis chapter by Fine et al. (1977) was complimented by the publication of two volumes that collated a series of the most influential primary research papers on sound production and reception in fishes up to that time (Tavolga 1976, 1977).

The review by Myrberg et al. (1978) clearly focused on the issue of communication and proposed that potential information content in fish sounds was determined by signal timing. Based upon extensive field studies of pomacentrids, the authors proposed that the pulsed patterns of fish sounds contain temporal information (i.e., pulse interval and pulse number) that is likely used by these fishes for species identification.

Tavolga et al. (1981) brought together the next major synthesis volume on hearing and communication in fishes. In his review in this book, Myrberg (1981) expanded on his view of how to test the concepts of communication in fishes. He defined how communication in fishes could be observed and he provided a framework for analyzing interactions. He also introduced the concept of acoustic interception, where an unintended recipient receives the information. Interception can be used by potential competitors and by predators. This chapter was followed by Hawkins & Myrberg (1983) and Hawkins (1986), both of which review the physical aspects of underwater sound and how these special conditions influence the way fishes can acoustically communicate. Schwarz (1985) included in her review an emphasis on the underwater environmental noises as they may affect fish behavior. She also concluded that for purposes of communication, fishes should use pulsed signal patterns rather than continuous sounds or varying frequencies. Bass & Clark (2003) present the most recent updated review covering the physical acoustic properties of underwater sound as basis for better understanding how sound communication by fishes can function in water.

In 1997, the journal *Marine and Freshwater Behaviour and Physiology* issued a volume dedicated to different aspects of underwater bioacoustics, with many papers focusing on fishes. Myrberg (1997b) updated earlier reviews with a decade of new data and reaffirmed many of the hypotheses regarding pulse timing as the key informational cue and the behavioral role of overhearing signals by predators. Ladich (1997) focused his review on the role of sound in the agonistic behavior of fishes. Agonistic vocalizations between aggressive fishes are easier to elicit experimentally than are courtship or mating sounds. Ladich noted that during agonistic interactions, male fish are often more vocal than females, with some exceptions. The pattern that male fish are the main sound producers in most species is emerging as a general trend. Bass (1997) reviewed the central and nervous system components regulating the fish vocal/sonic motor system, and Fine (1997) discussed how different hormones influence fish behavior and the production of sounds mediating those behaviors.

The new millennium transitioned from a largely descriptive phase to a more ethological and ecological one. Several publications updated earlier research findings of fish sensory abilities, with an emphasis on communication, especially the behavioral role of sounds. Myrberg (2001) presented a synthesis of the acoustical biology of elasmobranchs. One important conclusion was that several species of sharks were attracted to low frequency pulsed sounds such as typically produced by many fishes during aggression or courtship. Myrberg & Fuiman (2002) reviewed the multiple senses of reef fishes in the context of communication. This was followed next by Collin & Marshall's (2003) edited volume on sensory processing in aquatic environments. This book is notable for the new information it presented on color vision and communication in fishes as well as updated reviews on the mechanisms of hearing and sound production in fishes (Ladich & Bass 2003; Popper et al. 2003a). Bass & McKibben (2003) examined fish acoustic behavior from a neural mechanism perspective. The review by Rosenthal & Lobel (2005) reemphasized that sounds, color, and action patterns all function in concert to transmit elements of communication in fishes. They defined the structure of a pulsed fish sound and illustrated different types of fish mating sounds with sonograms.

Lobel (2001b) drew attention to the possibility of using fish sounds for monitoring fish reproduction by listening for species-specific mating sounds (also Lobel

2002). A symposium was held in 2002 that addressed how science could now apply acoustic technology to fisheries issues based on the fact that a variety of fish species produced distinguishable sounds (Rountree et al. 2003). The theme of this meeting was to review the accumulating evidence that species-specific sounds could be used by passive acoustic detection for monitoring fish behavior. The overall conclusion was that many fish species produce identifiable sounds that can be detected by passive acoustic devices (Rountree et al. 2006).

Two books were recently published that were entirely on fish communication with a significant emphasis on acoustics. Ladich et al. (2006) published a two-volume set entitled *Communication in Fishes*. This book has several chapters on fish acoustic communication including a review of sound generating mechanisms (Ladich & Fine 2006), swimbladder sound mechanisms (Parmentier & Diogo 2006), diversity of fish sounds (Amorim 2006), propagation of fish sounds (Mann 2006), agonistic sounds (Ladich & Myrberg 2006), and reproductive sounds (Myrberg & Lugli 2006). The second book, edited by Webb et al. (2008) was titled *Fish Bioacoustics*. The content is centered on the neurological and sensory aspects of how fishes sense vibrations, water movement, and sounds. The chapter in this book by Bass & Ladich (2008) considers the neural basis for fish communication. They bring forward Myrberg's hypothesis (Myrberg et al. 1978) for the temporal coding of signals by fishes and add new data to support this concept.

Finally, the most recent review of examples of fish sounds and sound producing mechanisms is by Kasumyan (2008). It is detailed and includes many illustrations of different types of sound-producing morphologies and fish sounds. Given this recent publication, we decided not to repeat the same material. The reader is encouraged to read this rich literature, especially the major reviews cited above. These papers do very well in describing the current state of knowledge about fish sound producing mechanisms, hearing, diversity of sounds produced, and how acoustics plays a role in the behavior of fishes.

BEHAVIORAL CONTEXTS FOR SOUND PRODUCTION IN REEF FISHES

Of the more than 179 families of fishes inhabiting coral reefs, and the nearshore tropical marine environment (Choat & Bellwood 1991; Lieske & Myers 1999), 48 families represented by 273 species in 137 genera are currently known as sound producing and hypothesized to use signaling in communication (see Table 10.1). Of these, surprisingly few have had sounds recorded and statistically described in undisturbed, intraspecific social contexts, especially reproduction. Reef fishes produce sounds in the same behavioral contexts as other marine and freshwater fishes (Amorim 2006): disturbance and predator defense (Fish & Mowbray 1970), agonism (e.g., Ladich 1997; Ladich & Myrberg 2006), and reproduction (e.g.,

Myrberg & Lugli 2006). Following below, we review the spectrographic properties of sounds produced by tropical fishes, many associated with reefs, and the behavioral contexts with which they are associated (see Table 10.1) and discuss these aspects of their acoustic biology. Five families, Aploactinidae (Matsubara 1934), Sillaginidae (Walls 1964), Synanceidae (Walls 1964), Tetrarogidae (Walls 1964), and Triglidae (e.g., Uchida 1934; Rauther 1945; Evans 1969, 1970), are known to produce patterned sounds, but both their spectrographic characteristics and details of the behavioral contexts in which they are produced have not been further pursued.

Twenty families of reef fishes produce some kind of agonistic sound. Agonistic encounters included competitive feeding, intraspecific and interspecific chase, territory defense, feeding competition, threat and attack, fights or combat, and interactions with potential predators that involve sound production when either fleeing from or confronting and sometimes attacking the predator. As in other fishes, sounds of tropical marine species play an important role in retreat defense. The pomacentrid, *Stegastes partitus*, has been described as producing a "keep-out" signal (Myrberg 1997a). This species guards a small patch of algae within its territory and uses these sounds to ward off intruders (Myrberg 1972a, b). The acoustic signature of these territorial sounds also communicates an individual's identity to conspecific neighbors (Myrberg & Riggio 1985; Myrberg et al. 1993).

Many of the territorial conflicts and especially male to male combat occur in direct association with the breeding season, although many species that maintain territories all year will continue to vocalize outside the breeding season (Gray & Winn 1961; Miyagawa & Takemura 1986; Ladich 1997). Territory maintenance correlates strongly with the presence of an acoustic signaling system. The selective advantage of the acoustic communication modality as opposed to other categories of signals in aggression is that sounds can enhance signals as they can correlate with individual body size and therefore provide an honest indicator of the likely winner for the outcome of a fight (Ladich 1990; Myrberg et al. 1993). The most recent study to demonstrate this in a common reef fish genus, *Amphiprion*, further supports the common trend for the dominant frequency of male fish agonistic sounds to correlate with body size; smaller males produce sounds with higher dominant frequencies (Colleye et al. 2009). This study has additionally identified pulse duration as correlating with body size. Significantly, this study has also shown that agonistic sounds may be important to both sexes, unlike in reproductive contexts where females are silent or may produce much lower amplitude sounds (Ladich 2007).

Fourteen reef fish families have been demonstrated to produce sounds during the breeding season (see Table 10.1). These reproductive sounds are commonly associated with courtship displays. In some of these species, particularly blenniids (de Jong et al. 2007), gobiids (Malavasi et al. 2003; Amorim & Neves 2007), and pomacentrids (e.g., Myrberg 1972a; Avidor 1974; Lobel & Mann 1995a; Lobel & Kerr 1999), sounds are part of a more elaborate courtship behavior with both acoustic and visual signals functioning to attract females to the territory.

While it is most common for the male to vocalize to attract females, there are examples of males and females exchanging calls during spawning. A case of male and female sound exchanges, or duets, has been observed in a syngnathid (Fish 1953), and a holocentrid (Herald & Dempster 1957). In the hermaphroditic *Hypoplectrus unicolor* (Serranidae), individuals alternate vocalizations while alternating spawning roles (Lobel 1992). Chaetodontid individuals have also been observed to exchange sounds between paired males and females that are hypothesized as mate alert or contact calls (Tricas et al. 2006).

Although the majority of acoustic fish families produce sounds in intraspecific contexts, many also produce disturbance sounds, which may perform a means of communicating interspecifically. Many fishes that produce disturbance sounds also produce sounds in undisturbed behavioral contexts, and therefore, the presence of disturbance sounds is often an indicator that the fish has the capacity to produce sounds for communication (Fish & Mowbray 1970; Lin et al. 2007), hence the purpose for emphasizing the "disturbance sound context" as evidence for potential sound signal production in our review of reef fishes. However, the absence of disturbance sounds in a species does not necessarily indicate total silence, since some acoustic fishes do not produce any kind of sounds when disturbed. In some marine taxa, disturbance sounds are directly associated with defensive weapon displays and are suggestive of acoustic aposematism or an acoustic warning of the fish's ability to damage a predator during a prey attack. Evidence that supports this hypothesis includes the observation that these sounds are produced by structures independent of the defense mechanism itself, hence not incidental noise. In tetraodontids (Sörensen 1894–1895) and diodontids (Burkenroad 1931), sounds occur during defensive displays of body inflation, suggesting a possible signal function in warning potential predators of their ability to defend. Other families erect spines as defensive structures, and these groups do produce sounds directly associated with the weaponized (often venom deployment capable) structures themselves, such as the pectoral fin spines of ariids (Tavolga 1962). Fourteen families have thus far only been reported to produce sounds during disturbance (see Table 10.1). Three of these families—Dactylopteridae (Müller 1857; Sörensen 1894–1895; Fish 1948; Fish & Mowbray 1970), Priacanthidae (Salmon & Winn 1966), and Pempheridae (Takayama et al. 2003)—have specialized acoustic muscles associated with the swimbladder, strongly suggesting that sounds are likely produced in other undisturbed social contexts yet to be observed (Fish & Mowbray 1970).

Fewer than half of the acoustic reef fish families cited in this review have been monitored with hydrophones during reproductive behaviors. Since many recently discovered acoustic fish species are now known to produce sounds inaudible to a casual human observer without the use of an amplified hydro-acoustic monitoring system, we cannot assume any particular taxonomic group is silent simply because we cannot easily hear them produce sounds. Clues to the potential for sound production in monacanthids are reported in a more recent study (Kawase 2005), which observed that males vibrate their dorsal spine rapidly as a common component of their courtship display but the researchers did not use a hydrophone in their study. Early reports on disturbance sounds in this family sometimes describe "creaking" sounds produced by the spines of species in this family. Only acoustic recording concurrent with natural behavior can verify if sounds are indeed generated during these and other species' courtship displays.

During their acoustic monitoring survey of marine fishes, Fish and Mowbray (1970) identified 14 families (Ammodytidae, Antennariidae, Atherinidae, Bothidae, Cyclopteridae, Echeneidae, Lophiidae, Mugilidae, Ogcocephalidae, Pleuronectidae, Soleidae, Stromateidae, Uranoscopidae, and Zoarchidae) as "silent" (i.e., not producing significant patterned sounds). However, even these families were not monitored specifically during reproductive activities. Therefore, we do not assume that they are completely silent, as an absence of evidence in this case does not equal evidence of absence. For example, sounds have been reported from *Uranoscopus scaber*, which is a member of a supposedly "silent" family (Mikhailenko 1973), and the detailed description of the acoustic apparatus of species belonging to numerous families in the order Scorpaeniformes, including some reef species, have only recently been followed up by actual recordings of sounds in undisturbed behavioral contexts (Širović & Demer 2009). It would therefore be difficult to draw broader conclusions about the difference between silent and acoustic reef fish taxa based on our current state of knowledge.

Three additional families (Aulostomidae, Lethrinidae, and Sphyraenidae) were found to produce only noises during swimming but other contexts of behavior have not been acoustically monitored (Fish & Mowbray 1970). Hydrodynamic sounds in general have been recorded in association with stereotypical agonistic and reproductive behaviors in chaetodontids, scarids, and labrids and are described as possible acoustic signals (Lobel 1992; Boyle & Tricas 2006; Tricas et al. 2006). While fishes may produce sounds as a by-product of swimming (Moulton 1960), the breadth of functional significance of swimming sounds is not currently certain. In addition to incidental sounds in fishes, the functional importance of the interception of signals is an area open for study (Myrberg 1981, 1997b). As an example, Steinberg et al. (1965) describe the dramatic sand diving behavior of a labrid (wrasse) in response to its common predator's swimming noises.

SOUND PRODUCTION AND BEHAVIOR
IN POMACENTRIDS (DAMSELFISHES)

The damselfishes (Pomacentridae) are one of the most thoroughly investigated and best understood family of acoustic reef fishes. This is in large part due to their conspicuous presence and abundance in coral reef communities, their demersal and often territorial behavior, and most notably, the sheer volume loudness of their sounds. Scuba divers can easily hear damselfish sounds without the aid of a hydrophone, particularly when a male is aggressively defending his territory. These features of pomacentrid acoustic behavior make them easily accessible to bioacoustic study.

Damselfishes are a diverse and speciose group (e.g., Quenouille et al. 2004; Cooper et al. 2009), and they play a dominant role in coral reef ecosystems worldwide (Bellwood & Hughes 2001). As of this writing, sounds have been recorded from 36 species (Tables 10.1 and 10.2 at end of chapter), representing four of the five subfamilies: Stegastinae, Chrominae, Abudefdufinae, and Pomacentrinae (following the topology and classification of Cooper et al. 2009; the Amphiprioninae, often considered a subfamily, are nested within the Pomacentrinae). The only pomacentrid subfamily from which sounds have not been recorded is the monotypic Lepidozyginae containing *Lepidozygus tapeinosoma*. Given the widespread occurrence of sound production throughout the family, the damselfishes may provide one of the first examples of sounds being a behavioral character uniting the entire family. If so, for any taxa that do not vocalize, this lack of an ability to produce sounds may represent its secondary loss. Damselfish sounds are distinctive among fishes in that they not only have highly repeatable patterned pulses and interpulse intervals in the basic sound display, the "chirp," but they also repeat the "chirp" call type at regular intervals presenting a complex acoustic display with two orders of temporal information, referred to as a call bout (Figure 10.2).

As with many fishes, pomacentrids produce sounds in two main behavioral contexts: aggression and courtship. Aggressive sounds are produced by males defending territories, doing so most fiercely when guarding their demersal eggs. These sounds are of short duration and broadband, and are often produced while resident fish are chasing or biting at conspecific or interspecific intruders (e.g., Mann & Lobel 1998). Agonistic sounds recorded from 26 species show a low mean number of pulses and maximum frequencies often above 1,000 Hz (Parmentier et al. 2006b). The higher-frequency component of the aggressive sound may increase detectability of the call (Fine & Lenhardt 1983; Mann & Lobel 1997; Bass & Clark 2003; Mann 2006) and thereby serve as a warning to other potential intruders.

During courtship displays, many damselfishes combine their repeated, pulsed vocalizations with a prominent visual display, the signal jump (Myrberg 1972a;

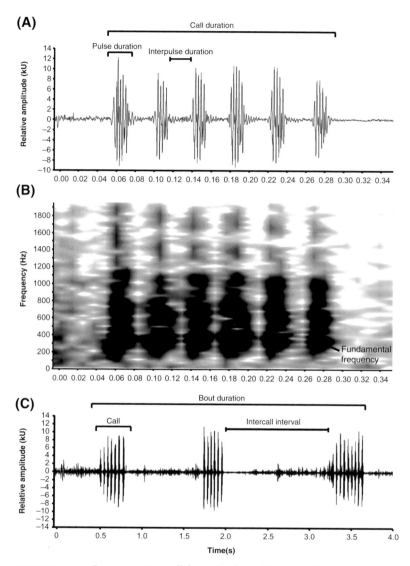

FIGURE 10.2. Representative call from the damselfish *Dascyllus albisella*, recorded from Johnston Atoll. The sounds are represented two ways: (A) the waveform, which shows relative amplitude, or sound pressure, versus time and (B) the spectrogram or sonogram, which shows the frequency content of the call versus time. This example is a single call from *D. albisella* comprised of six pulses. These different views can be used to reveal different temporal and spectral features of an individual sound or series of sounds. (C) Waveform of a calling bout from *D. albisella*, comprised of three calls, shows the broader time duration between calls and pulses.

Avidor 1974; Lobel & Mann 1995b; Mann & Lobel 1998; Parmentier et al. 2009). In the signal jump, fish rise up into the water column and then rapidly swim downwards while vocalizing before returning to the initial position and repeating the pattern. These displaying fish can often be seen and heard from quite a distance. Courtship sounds have been reported from 11 species within the family, but not enough acoustic parameters have been described to enable any broad quantitative analysis of taxon-specific features of courtship sounds.

The specific anatomy involved in sound production in damselfishes has been the subject of speculation for quite some time (Freytag 1968; Myrberg 1972b; Rice & Lobel 2003). Because damselfish sounds exhibit both harmonic and broadband characteristics, it seems likely that the sound-producing mechanism involves muscular and stridulatory elements. Also, given the loud nature of the sound, the swimbladder seems likely to be involved as a resonating device. Thus, the pharyngeal jaw apparatus has been suggested as a possible sound production mechanism (reviewed in Rice & Lobel 2003). Pomacentrids possess well-developed and highly mobile pharyngeal jaws (Galis & Snelderwaard 1997) and it is possible that they have been co-opted for use in sound production (Rice & Lobel 2003).

An alternative hypothesis was proposed from recent work on anemonefish that suggests that the oral jaws are involved in sound production (Parmentier et al. 2007), at least within the genus *Amphiprion*. *Amphiprion clarkii* is able to rapidly close its lower jaw with a novel ligament connecting the ceratohyal and coronoid process of the mandible. The sounds result from the occlusion of the upper and lower jaw teeth, and the jaws themselves are suggested to serve as the source of sound radiation (Parmentier et al. 2007). This work is particularly exciting, since the species-specific sounds produced by *Amphiprion* may correlate with interspecific differences in tooth morphology, suggesting exaptation and perhaps coevolution of feeding and acoustic behaviors (Chen & Mok 1988; Parmentier et al. 2007). Though this acoustic ligament has been suggested to be homologous among pomacentrids (Stiassny & Jensen 1987; Parmentier et al. 2007), its role in sound production across the family remains to be investigated.

FISH SOUNDS: TERMINOLOGY AND CLASSIFICATION

Fish sounds are complex signals and have a diverse combination of different components (Figure 10.2); one or more classes of these different parameters often vary among closely related taxa. Fish sounds have been classified in different ways, usually according to the following criteria: (1) phonetic/onomatopoeia, (2) behavioral context, (3) adaptive or hypothesized function, (4) morphological nature of mechanism, (5) specific acoustic parameters (amplitude or loudness, temporal and frequency), and (6) temporal duration categories (Rountree et al. 2006). Human

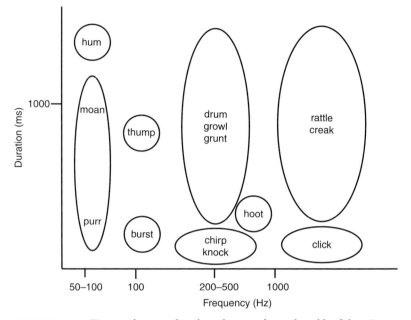

FIGURE 10.3. Terminology used to describe sounds produced by fishes. See Table 10.1 for a listing of fishes and acoustic characteristics on which this figure is based.

appellations for fish sounds— onomatopoeia, or the use of words that imitate the associated sound—commonly capture the temporal and frequency components of a sound (see Figure 10.3). They are widely used for all acoustic animals and are analogous to common names for describing species in addition to a genus and species designation. As observed by Fish & Mowbray (1970), certain general sound terms relate to the types of sound producing mechanisms fish employ. Creaks or clicks are typically higher and broadband frequency stridulation sounds, while the lower frequency sounds (<1,000 Hz) referred to by the greatest diversity of sound terms are typically produced by swimbladder-associated muscles (Figure 10.3). However, onomatopoeia does not provide an absolute, systematic way to differentiate between all sounds. For example, the usefulness of this approach to categorize fish sounds breaks down when onomatopoeia is described in a foreign language or when patterns become more complex than simple time and frequency characteristics. Fish sounds are also often referred to interchangeably as sounds, calls, or signals. These terms are not necessarily always interchangeable and it is likely that further criteria for distinguishing them will follow as we further examine the role of sounds in communication versus noise production.

There have been a number of suggestions for a rigorous and consistent consensus terminology for fish sounds and their components (Winn 1964; Fine et al. 1977; Kihslinger & Klimley 2002)(see Figure 10.2). Fine et al. (1977) provided a classification scheme for fishes that focused on acoustic parameters, specifically temporal patterns of sounds. Such a framework provides a statistically testable system, while a naming system based only on inferred or expected function may result in frequent name changes with new studies. It is important to note that if we can specify what acoustic parameters are significant to the fish, then naming sounds according to only those acoustic parameters that are meaningful will add power to this type of classification scheme.

Marshall (1962) observed that fish sounds produced by different mechanisms sound different to the human observer. He also summarized phonetic descriptions of sounds produced by fish with different mechanisms and classified them into two major categories: stridulation and swimbladder sounds. It is useful when describing a new fish sound to classify it according to the sound production mechanism (SPM), because different types of acoustic mechanisms often are acoustically constrained; that is, they produce sounds that differ in frequency or temporal pattern (e.g., Walls 1964; Myrberg 1972a; Demski et al. 1973). Identifying structures of a likely acoustic mechanism in a museum preparation can provide information about its likely signal characteristics, many fishes are known only from their preserved materials. Inversely, having spectrographic information about a sound signal can provide a clue to identifying a yet unknown acoustic mechanism in a newly discovered acoustic fish species (Rice & Lobel 2002, 2003).

Behaviors are also useful for classifying fish sounds. Naming fish sounds on the basis of their behavioral context or the specific motor patterns with which they occur may involve the use of some of the following terms: male courtship dip display, spawning, nest brooding, agonistic, or schooling sounds. When sounds are additionally described on the basis of their hypothesized function, terms such as appeasement, dominance, mate attraction, fright, or alarm are appended to other descriptions of the sound. The behavioral context description is most valuable as it is purely descriptive and does not imply any function, which can only be determined through experimentation. Naming a fish sound by its presumed function should be carefully qualified as a hypothesis. The use of a strict context or physical activity for naming a fish sound is the better choice as it provides accurate information about the association of a particular behavior with sound(s), without making assumptions about sound function. The advantage of using behavioral context is that it is consistent. A sound produced during courtship alone and named as such will provide a reliable means for categorizing that sound. The disadvantage of using context as a naming device is that a particular sound may be used in more than one context. In such a case, this system does not provide an exclusive naming system. In naming a fish sound, it is therefore useful to ongoing research in the field to describe as many known aspects of the sound as possible, but to treat all such terms as working hypotheses subject to further study.

SPECTROGRAPHIC PROPERTIES
OF REEF FISH SOUNDS

Frequency

Frequency ranges for most fish sounds are broadly overlapping but fall into two general categories: those predominantly below 1,000 Hz in dominant frequency with narrower band frequency overall; and those with broad band frequency ranging up to 8 kHz depending on the mechanism involved (Demski et al. 1973; Ladich & Fine 2006). The tropical reef fish sound spectrum is currently known to extend from below 100 Hz up to 8,000 Hz and is the same as found in many other well-known sound producing fishes. Other frequency indicators that may be more informative regarding biologically significant and evolutionarily selected traits, such as dominant frequency that could indicate age or size (Myrberg et al. 1993; Lobel & Mann 1995b), were not consistently reported for many fish species. The relationship between the spectral properties of the sound and the physical condition of the signaler remains a productive avenue for future research, in essence allowing us to address what properties of the signal reveal about the sender.

Duration

The duration of the basic sound unit of an acoustic display is one measure of temporal patterns that can indicate communicative content in the sounds of fishes in general (McKibben & Bass 1998). Durations for the basic sound unit of reef fishes ranged from 10 ms to just over 6 seconds across acoustic reef fish families (see Table 10.2), similar to the range for fishes found in all other aquatic habitats. The longest acoustic displays known are associated with courtship and spawning among fishes in general (Amorim 2006), where these sounds serve as an advertisement signal to attract females. Sounds with the greatest number of pulses that also had regularly spaced interpulse intervals were most commonly associated with reproductive context calling. Reef fish families with basic sound units of the longest duration, greater than one second, included balistids, batrachoidids, carangids, chaetodontids, holocentrids, monacanthids, pomacentrids, sciaenids, and tetraodontids. A sound display incorporating a series of several different call types by a chorusing tropical sciaenid male, *Johnius macrorhynus*, lasted for 143 minutes (Lin et al. 2007). In some temperate batrachoidids, the basic sound unit of a male reproductive call of an individual fish may last for seconds to minutes, and the call displays continue for up to one hour (Ibara et al. 1983; Brantley & Bass 1994; Rice & Bass 2009).

SPECTRAL PATTERNS:
PULSED VERSUS TONAL SIGNALS

The sounds of many acoustic fishes for which only spectrograms are available are sometimes difficult to classify as either pulsed or non-pulsed, since the waveforms are not always expanded to reveal the presence or absence of interpulse intervals. We provide a conservative classification of sound types for each marine species (Table 10.1), restricting our designation of pulsed, tonal or other call characteristics to those specifically described as such by the authors whose publications we had access to. Many species of sound producing fishes have not had their sounds spectrographically analyzed.

The sounds produced by most fishes in disturbance and escape contexts appear to be short in duration (<500 ms in duration) and have irregular interpulse (i.e., variable interval)(Winn 1972) durations and could be considered simple single pulse displays, although their selective advantage in these contexts remains elusive beyond proposed hypotheses. These sounds could simply be distress noises or a release of sounds produced during other contexts, or they may play a role in a prey species communicating its ability to escape or defend itself. Predator defense weapons, such as the abduction and adduction of venomous pectoral spines, can produce stridulation sounds. In marine catfishes, these sounds are broadband in frequency range and occur in pairs, each consisting of a series of pulses with interpulse intervals that are produced in sound bouts of irregular intervals and thus could be a form of acoustic aposematism.

The basic sound described for many tropical marine fishes in agonistic and reproductive contexts commonly consists of a single pulse or multiple pulses. Comparing the duration of the agonistic and reproductive sounds for our acoustic fish sample we found that the longest duration sounds were produced in association with reproductive contexts, while agonistic sounds were shorter in duration. Multiple pulses consist of regular repeating patterns of interpulse intervals, or off times, between pulses. Basic sound units may be repeated in bouts, which are highly patterned or regular in some damselfishes (*Stegastes*). Sound displays consisting of patterned pulses with stereotypical interpulse off-time durations have been statistically described for members of several families (i.e., Carapidae, Chaetodontidae, Gobiidae, Holocentridae, Pempheridae, Pomacanthidae, Pomacentridae, Sciaenidae, Serranidae, and Terapontidae) and are most commonly associated with male courtship displays. Other families with regularly spaced pulsed waveforms that are agonistic or disturbance sounds, but which have not been statistically described, include Ariidae, Balistidae, Diodontidae, Haemulidae, Ostraciidae, and Priacanthidae.

The most uncommon spectrographic sound type pattern among reef fishes (and all fishes in general) is tonal, that is, sounds that exhibit pure tone frequencies. Of the 48 social sound producing tropical families, only batrachoidids and ostraciids have so far been reported to produce purely tonal sounds. Among tropical batrachoidids, two species in the genus *Opsanus*, and *Batrachomoeus trispinosus*, are known to produce a tonal component to the male courtship display (e.g., Amorim 2006; Rice & Bass 2009). In the ostraciid family *Ostracion meleagris*, males produce tonal sounds during spawning (Lobel 1996). Tonal sounds are produced by the fastest rates of muscle contraction and are therefore extremely physiologically demanding, often requiring specialized muscles to create these sounds (e.g., Rome et al. 1996; Rome 2006). They are temporally indicative of the highest repetition rate for a call, pulses are repeated so rapidly that they are not separated by pulse intervals, and could convey a message of male vigor, and therefore be a male fitness indicator, to females. Tonal sounds are equally rare for other fish families from temperate marine and freshwaters. In these taxa, they are also found as one of multiple call types associated with a male courtship display. In temperate batrachoidids, females are not known to produce tonal sounds, and the sounds they do produce in agonistic contexts are typically single pulse sounds with irregular intervals between successive pulses.

FUNCTIONAL SIGNIFICANCE OF SOUNDS: MALE FITNESS AND FEMALE MATE CHOICE

The hypothesized functional significance of tropical fish sounds is the same as that for other vocal animals (Table 10.3): mate localization; individual discrimination (age, sex, aggressive strength); an indicator of mate fitness (size and vigor or rate of display); species discrimination; and gamete release synchronization.

There are often striking sexual differences in fish acoustic displays indicating the possibility that many fishes can identify the sex of a caller by their sound characteristics. In some species, females lack the major structures involved in sound production, produce quieter sounds or are silent due to the atrophy of their swimbladder drumming muscles in the breeding season, suggesting other physiological priorities over sound communication (Nguyen et al. 2008). There are differences in sounds between sexes in carapid reef fishes. Carapid females, *Carapus boraborensis*, produce longer pulses (Lagardère et al. 2005). In temperate sciaenids, sex differences in sound production, including fundamental frequency, interpulse interval, and number of pulses per call, have been found in several species (Takemura et al. 1978; Ueng et al. 2007).

The evolutionary significance of reproductive sound signaling in fishes, differences in the sounds of species and the ability of individuals to discriminate was first clearly demonstrated in marine fishes for damselfishes. Male display in damselfishes is considered an advertisement signal (Kenyon 1994) and its potential role as an indicator of male fitness and female choice has been proposed (Myrberg et al. 1993). Individual recognition is accomplished by male pomacentrids who can distinguish their neighbors on nearby territories (Myrberg & Riggio 1985). Within marine fish genera, species recognition (i.e., the ability of an individual fish to recognize when another is or is not a member of their own species) that is based on sound production has only been experimentally demonstrated in damselfishes (Myrberg et al. 1986). This was shown by choice experiments in four *Stegastes* species where females preferred male sounds whose pulse numbers matched their own species, thereby allowing for the distinction of conspecifics from heterospecifics. Recent work in acoustic neurophysiology of a damselfish demonstrated that auditory neurons are indeed sensitive to the pulsed temporal components of their own sounds (Maruska & Tricas 2009). Courtship sounds of male *Stegastes* species differ in unit rate, pulse number, frequency, pulse duration, and pulse interval (Myrberg et al. 1978). The role of acoustic parameters other than pulse number for species recognition remains to be tested within this family. Species differences in acoustic male courtship displays underscore its possible significance in reproductive isolation since damselfish females prefer the pulse number totals of their own species. Dialects, or the geographic variation in the male calling display of damselfishes, further support the hypothesis that fish species may diverge on the basis of reproductive-context call characteristics (Parmentier et al. 2005). Congeners of other tropical marine families differ in the acoustic properties of their sounds (Table 10.4). Other spectrographic parameters that vary among reproductive courtship displays of species in the same genus, all of which are also sympatric, include pulse repetition rate in tropical sciaenids (Lin et al. 2007), pulse number, duration, and interpulse interval in pomacentrids (Maruska et al. 2007), number of sound pulses per display in batrachoidids (Fish & Mowbray 1959), and amplitude and duration in gobiids (Stadler 2002). While temporal call differences have been behaviorally demonstrated to provide information to fishes in various communication contexts among damselfishes, other call parameters such as amplitude or frequency may also provide acoustic clues to listeners of other acoustic fish families. Male fitness discrimination by choosey females is hypothesized based on the honest signaling potential of the pomacentrid acoustic system, where a relationship between body size, swimbladder volume, and call dominant frequency exists (Myrberg et al. 1993). Increased body size directly correlating with lower dominant call frequency for male reproductive acoustic display has also been determined for other temperate marine fishes, in particular gobies (Malavasi et al. 2003). In contrast to frequency characteristics, female plainfin midshipmen are attracted to and prefer playbacks of tonal overpulsed male acoustic display components (McKibben & Bass 1998), suggesting that an acoustic feature other than size may be the basis for female choice in this species. This other feature could be male vigor, evidenced by a faster calling rate for tonal versus pulsed sounds. However, further work to determine if and how females choose mates on the basis of these varied male acoustic characteristics is needed.

In fishes with external fertilization, the importance of spawning synchronization cues could be very important. The scarid (*Scarusiseri,* formerly *iserti*) forms spawning aggregations and produces hydrodynamic sounds when groups of individuals rush up in the water column and release gametes (Lobel 1992). These sounds are broadband and adventitious to the swimming movements. Research from several acoustic fish groups, and reef families in particular, have recorded patterned sounds associated with spawning (Lobel 1992). It is presumed that it is mainly for reproduction when groups of individuals are calling together in one specific locality. However, not all fish species producing sounds in choruses have yet been identified as specifically engaging in reproductive behaviors. While closer detailed observations of individual behaviors in these groups are essential, they have been difficult to obtain in the field.

CHORUSING SOUNDS
AND BREEDING AGGREGATIONS

From early hydrophone monitoring, coral reefs were found to be very noisy soundscapes. The sounds noted by the earliest reef fish acoustics researchers were typically highly audible and widely propagating. Reefs can be noisy due to the chorusing behavior of breeding reef fishes (e.g., Cato 1976, 1978; McCauley & Cato 2000). Numerous marine fishes form spawning aggregations (see Knudsen et al. 1948a; Takemura et al. 1978) that occur over specific diel patterns (Breder 1968; Cato 1976, 1978; McCauley & Cato 2000). Most of these species have peaks in their acoustic chorus activities nocturnally with some activity peaks under low light conditions, dusk or dawn. Pomacentrids are notably active predominantly during the day under lighted conditions. Many of these reef sounds appear to be primarily the result of the reproductive activities of pomacentrid, sciaenid, serranid, and batrachoidid species, while the crepuscular calling peaks of holocentrids are due solely to agonistic sounds (Figure 10.4).

SONIC INTERACTIONS

Predators seem to place major selection constraints on fish sound signaling (Nottestad 1998; Luczkovich et al. 2000; Wahlberg & Westerberg 2003; Remage-Healey et al. 2006). Although this work has been demonstrated only in temperate and subtropical marine fishes to date, these species all belong to families that include tropical reef species. Predators elicit a reaction from a temperate toadfish, whose family has representatives in tropical habitats as well. These temperate batrachoidids reduce their calling rate in response to playbacks of dolphin sounds,

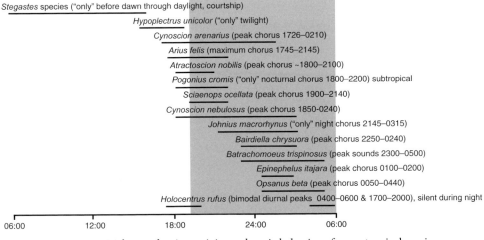

Stegastes species ("only" before dawn through daylight, courtship)
Hypoplectrus unicolor ("only" twilight)
Cynoscion arenarius (peak chorus 1726–0210)
Arius felis (maximum chorus 1745–2145)
Atractoscion nobilis (peak chorus ~1800–2100)
Pogonius cromis ("only" nocturnal chorus 1800–2200) subtropical
Sciaenops ocellata (peak chorus 1900–2140)
Cynoscion nebulosus (peak chorus 1850-0240)
Johnius macrorhynus ("only" night chorus 2145–0315)
Bairdiella chrysuora (peak chorus 2250–0240)
Batrachomoeus trispinosus (peak sounds 2300–0500)
Epinephelus itajara (peak chorus 0100–0200)
Opsanus beta (peak chorus 0050–0440)
Holocentrus rufus (bimodal diurnal peaks 0400–0600 & 1700–2000), silent during night

06:00   12:00   18:00   24:00   06:00

FIGURE 10.4. Diel reproductive activity and sonic behavior of some tropical marine species show three different temporal patterns of acoustic chorusing: diurnal, nocturnal, and crepuscular. Gray shaded area represents night.

indicating that calling in sonic fishes bears risks from intercepting predators (Remage-Healey et al. 2006). It has been proposed that predatory dolphins intercept fish sounds when hunting clupeids, batrachoidids, and sciaenids (Luczkovich et al. 2000; Gannon et al. 2005). On the other hand, an increase in acoustic activity has been reported for the tropical squirrelfish, *Holocentrus adscensionis*, which has been observed confronting predators with pulsed sound bouts and engaging in mobbing behavior as a predator deterrent (Winn et al. 1964).

The timing of chorusing behavior in predominantly nocturnal breeding aggregations of reef fishes may be greatly influenced by the threat to adults of predator interception of the sounds, as well as the threat to freshly fertilized eggs released into the open water in a concentrated area. Calling and spawning at night is hypothesized as a diurnal predator evasion strategy for both adults and spawned eggs. Research on sciaenids supports the notion that cetacean predators can intercept sciaenid sounds (Gannon et al. 2005). Experimental work has demonstrated that concentrations of eggs are under increased predation pressure, indicating that night spawning would result in greater dispersion of eggs by the time visual predators become active the next day (Holt et al. 1985).

These observations raise questions about the evolutionary role of predators in influencing the acoustic displays in fishes. If acoustic calling poses a sufficient risk, then a reduction in the amplitude of acoustic displays is likely. Such adverse selection could even lead to the elimination of sound by previously acoustic species. Evolutionary secondary loss of acoustic activity is a likely scenario given that the homology of neural development of the acoustic motor nucleus indicates an ancestral ability for acoustic communication in vertebrates (Bass et al. 2008).

CONCLUSIONS AND FUTURE DIRECTIONS

The diversity found among coral reef fishes makes them a fascinating study group for scientists in many fields; the field of sound production in reef fishes represents the intersection of these different interests and perspectives. The fact that reef fishes live in warm, clear waters is conducive to underwater study by scuba divers. Reef fish bioacoustics is frequently overlooked in studies of fish behavior, but it offers the potential for major advances in the study of animal communication. Tropical coral reefs around the world are home to more than 4,000 species of fishes (Lieske & Myers 1999; Bellwood & Hughes 2001), and in over 100 years of study, fewer than 10% of taxa have been reported as producing sounds. Given the widespread taxonomic occurrence of sounds in fishes (e.g., Amorim 2006; Bass & Ladich 2008; Fay 2009), and the evidence that all vertebrates share similar brain mechanisms responsible for producing these sounds (Bass et al. 2008), it is clear that bioacoustic signaling is a modality that can no longer be ignored or overlooked in behavioral studies of any fish. For far too long, studies have reported on the courtship or territorial behavior of reef fish groups, with neither mention nor apparent awareness of the possibility that the focal taxon may be using sounds to communicate. Scientific studies of either communication or intraspecific interaction in other (non-fish) animals would be deemed at best incomplete and at worst unacceptable if potential vocalizations were not accounted for. Yet this has been the case for the majority of research on fishes (imagine studying courtship in passerine birds without reporting on the role of song). We raise this issue to serve as encouragement to researchers of coral reef fishes to include the study of sounds as a vital component of behavioral research.

The majority of publications on fish sounds have primarily focused on their temporal occurrence or acoustical characteristics, but the specific behavioral function of such sounds is still relatively understudied. As evidenced by the large and diverse body of work on the bioacoustics of damselfishes described here, understanding how sounds function in behavior requires knowing the diversity of the vocal repertoire, the behavioral response that particular sounds elicit, the hearing range of the receiver, and the acoustic features of the signal to which conspecifics pay attention. These topics have been addressed in only a few fishes (e.g., damselfishes, gobies, and toadfishes). The opportunity to study these components in other taxa is wide open. As Karl von Frisch wrote in 1938, after realizing the co-occurrence of fish sound production and their hearing ability: "There may be much to discover in the future about the language of fishes" (von Frisch 1938, p. 11). His words still hold true today.

ACKNOWLEDGMENTS

Our research on the sounds of coral reef fishes was supported by grants from the Army Research Office (DAAG55-98-1-0304, DAAD19-02-1-0218), Office of Naval Research (N00014-19-J1519, N00014-92-J-1969 and N00014-95-1-1324), the DoD Legacy Resource Management Program (DAMD-17-93-J-3052, DADA87-00-H-0021, DACA87-01-H-0013, W912D4-06-2-0017) and NOAA Seagrant (NA90-AA-D-SG535, NA86-AA-D-SG090, NOAA/NURC-FDU 89-09-NA88A-H-URD20).

REFERENCES

Aalbers SA, Drawbridge MA (2008). White seabass spawning behavior and sound production. Transactions of the American Fisheries Society 137: 542–550.

Akamatsu T, Okumura T, Novarini N, Yan HY (2002). Empirical refinements applicable to the recording of fish sounds in small tanks. Journal of the Acoustic Society of America 112: 3073–3082.

Albrecht H (1981). Aspects of sound communication in some Caribbean reef fishes (*Eupomacentrus* Spec, Pisces, Pomacentridae). Bijdragen tot de Dierkunde 51: 70–80.

Albrecht H (1984). Harmonics in courtship sounds of four Caribbean reef fish species of the genus *Eupomacentrus* (Pomacentridae). Bijdragen tot de Dierkunde 54: 169–177.

Allen GR (1972). *The Anemonefishes: Their Classification and Biology*. TFH Publications, Neptune City, New Jersey.

Amorim MCP (1996a). Acoustic communication in triglids and other fishes. M.Phil. thesis, University of Leicester, Leicester, UK.

Amorim MCP (1996b). Sound production in the blue-green damselfish, *Chromis viridis* (Cuvier, 1830) (Pomacentridae). Bioacoustics 6: 265–272.

Amorim MCP (2006). Diversity of sound production in fish. In: Ladich F, Collin SP, Moller P, Kapoor BG (eds.), *Communication in Fishes*, vol. 1, *Acoustic and Chemical Communication*. Science Publishers, Enfield, New Hampshire, pp. 71–105.

Amorim MCP, Neves ASM (2007). Acoustic signalling during courtship in the painted goby, *Pomatoschistus pictus*. Journal of the Marine Biological Association of the United Kingdom 87: 1017–1023.

Aristotle (1910). *Historia animalium* (translated by DA Thompson). The Internet Classics Archive http://classics.mit.edu//Aristotle/history_anim.html (accessed Feb. 15, 2010).

Avidor A (1974). The signal jump and its associated sound in fish of the genus *Dascyllus* from the Gulf of Eilat. Master's thesis, Tel Aviv University.

Bass AH (1997). Comparative neurobiology of vocal behaviour in teleost fishes. Marine and Freshwater Behaviour and Physiology 29: 47–63.

Bass AH, Clark CW (2003). The physical acoustics of underwater sound communication. In: Simmons AM, Fay RR, Popper AN (eds.), *Acoustic Communication*. Springer, New York, pp. 15–64.

Bass AH, Ladich F (2008). Vocal-acoustic communication: from neurons to behavior. In: Webb JF, Fay RR, Popper AN (eds.), *Fish Bioacoustics*. Springer, New York, pp. 253–278.

Bass AH, McKibben JR (2003). Neural mechanisms and behaviors for acoustic communication in teleost fish. Progress in Neurobiology 69: 1–26.

Bass AH, Gilland EH, Baker R (2008). Evolutionary origins for social vocalization in a vertebrate hindbrain-spinal compartment. Science 321: 417–421.

Bellwood DR, Hughes TP (2001). Regional-scale assembly rules and biodiversity of coral reefs. Science 292: 1532–1535.

Bernoulli AL (1910). Zur frage des hörvermögens der fishce. Pflugers Archiv für die Gesamte Physiologie des Menschen und der Tiere 134: 633–644.

Bierbaum G (1914). Examinations on the structure of the auditory organs of deep sea fish. Zeitschrift für Wissenschaftliche Zoologie 111: 281–380.

Board PA (1956). The feeding mechanism of the fish *Sparisoma cretense* (Linné). Proceedings of the Zoological Society of London 127: 59–77.

Boyle KS, Tricas TC (2006). Sound communication by the forceps fish, *Forcipiger flavissimus* (Chaetodontidae). Journal of the Acoustical Society of America 120: 3104.

Brantley RK, Bass AH (1994). Alternative male spawning tactics and acoustic signals in the plainfin midshipman fish *Porichthys notatus* Girard (Teleostei, Batrachoididae). Ethology 96: 213–232.

Braun CB, Grande T (2008). Evolution of peripheral mechanisms for the enhancement of sound reception. In: Webb JF, Fay RR, Popper AN (eds.), *Fish Bioacoustics*. Springer, New York, pp. 99–144.

Breder CM, Clark E (1947). A contribution to the visceral anatomy, development, and relationships of the Plectognathi. Bulletin of the American Museum of Natural History 88: 293–319.

Breder CM Jr (1968). Seasonal and diurnal occurrences of fish sounds in a small Florida bay. Bulletin of the American Museum of Natural History 138: 327–378.

Bridge TW (1904). Fishes. In: Farmer SF, Shipley AE (eds.), *Cambridge Natural History*. MacMillan, London, pp. 139–537.

Bright CM (1972). Bio-acoustic studies on reef organisms. In: Collette BB, Earle SA (eds.), Results of the tektite program: ecology of coral reef fishes. Science Bulletin 14, Natural History Museum of Los Angeles County, pp. 45–69.

Bright TJ, Sartori JD (1972). Sound production by the reef fishes *Holocentrus coruscus*, *Holocentrus rufus* and *Myripristis jacobus*, Family Holocentridae. Hydro-Lab Journal 1: 11–20.

Burgess WE (1989). *An Atlas of Freshwater and Marine Catfishes: A Preliminary Survey of the Siluriformes*. TFH Publications, Neptune City, New Jersey.

Burke TE, Bright TJ (1972). Sound production and color changes in the dusky damselfish. Hydro-Lab Journal 1: 21–29.

Burkenroad MD (1930). Sound production in the Haemulidae. Copeia 1930: 17–18.

Burkenroad MD (1931). Notes on the sound-producing marine fishes of Louisiana. Copeia 1931: 20–28.

Caldwell DK, Caldwell MC (1967). Underwater sounds associated with aggressive behavior in defense of territory by the pinfish, *Lagodon rhomboides*. Bulletin of the Southern California Academy of Sciences 66: 69–75.

Carlson BA, Bass AH (2000). Sonic/vocal motor pathways in squirrelfish (Teleostei, Holocentridae). Brain, Behavior and Evolution 56: 14–28.

Cato DH (1976). Ambient sea noise in waters near Australia. Journal of the Acoustical Society of America 60: 320–328.

Cato DH (1978). Marine biological choruses observed in tropical waters near Australia. Journal of the Acoustical Society of America 64: 736–743.

Charif RA, Waack AM, Strickman LM (2008). *Raven Pro 1.4 User's Manual*. Cornell Laboratory of Ornithology, Ithaca, New York.

Chen KC, Mok HK (1988). Sound production in the anemonefishes, *Amphiprion clarkii* and *A. frenatus* (Pomacentridae), in captivity. Japanese Journal of Ichthyology 35: 90–97.

Choat JH, Bellwood DR (1991). Reef fishes: their history and evolution. In: Sale PF (ed.), *The Ecology of Fishes on Coral Reefs*. Academic Press, San Diego, pp. 39–66.

Collette BB, Earle SA, eds. (1972). Results of the Tektite program ecology of coral reef fishes. Science Bulletin 14, Natural History Museum of Los Angeles County.

Colleye O, Frederich B, Vandewalle P, Casadevall M, Parmentier E (2009). Agonistic sounds in the skunk clownfish *Amphiprion akallopisos*: size-related variation in acoustic features. Journal of Fish Biology 75: 908–916.

Collin SP, Marshall NJ (2003). *Sensory Processing in Aquatic Environments*. Springer-Verlag, New York.

Colson DJ, Patek SN, Brainerd EL, Lewis SM (1998). Sound production during feeding in *Hippocampus* seahorses (Syngnathidae). Environmental Biology of Fishes 51: 221–229.

Connaughton MA, Fine ML, Taylor MH (2002). Weakfish sonic muscle: influence of size, temperature and season. Journal of Experimental Biology 205: 2183–2188.

Cooper WJ, Smith LL, Westneat MW (2009). Exploring the radiation of a diverse reef fish family: phylogenetics of the damselfishes (Pomacentridae), with new classifications based on molecular analyses of all genera. Molecular Phylogenetics and Evolution 52: 1–16.

Courtenay WR Jr, McKittrick FA (1970). Sound-producing mechanisms in carapid fishes, with notes on phylogenetic implications. Marine Biology 7: 131–137.

Cousteau JY, Dumas F (1953). *The Silent World*. Harper & Brothers, New York.

Cummings WC, Brahy BD, Herrnkind WF (1964). The occurence of underwater sounds of biological origin off the west coast of Bimini, Bahamas. In: Tavolga WN (ed.), *Marine Bio-acoustics*. Pergamon Press, New York, pp. 27–43.

Cummings WC, Brahy BD, Spire JY (1966). Sound production, schooling, and feeding habits of the margate, *Haemulon album* Cuvier, off North Bimini, Bahamas. Bulletin of Marine Science 16: 626–240.

de Jong K, Bouton N, Slabbekoorn H (2007). Azorean rock-pool blennies produce size-dependent calls in a courtship context. Animal Behaviour 74: 1285–1292.

Demski LS, Gerald JW, Popper AN (1973). Central and peripheral mechanisms of teleost sound production. American Zoologist 13: 1141–1167.

Dobrin MB (1947). Measurements of underwater noise produced by marine life. Science 105: 19–23.

Dobrin MB, Loomis WE (1943). Acoustic measurements at the John G. Shedd Aquarium, Chicago, Illinois. Naval Ordnance Laboratory Memorandum 3416: 1–17.

Dufossé L (1858a). De l'ichthyopsophie, ou des différents phénomènes physiologiques nommés voix des poissons (deuxième partie). Comptes Rendus Hebdomadaires des Séances de l'Académie des Sciences 46: 916.

Dufossé L (1858b). Des différents phénomènes physiologiques nommés voix des poissons. Comptes Rendus Hebdomadaires des Séances de l'Académie des Sciences 46: 352–356.

Dufossé L (1862). Sur les différents phénomènes physiologiques nommés voix des poissons, ou sur l'ichthyopsophose (troisième partie). Comptes Rendus Hebdomadaires des Séances de l'Académie des Sciences 54: 393–395.

Dufossé L (1874a). Recherches sur les bruits et les sons expressifs que font entendre les poissons d'Europe et sur les organes producteurs de ces phénomènes acoustiques ainsi que sur, les appareils de l'audtion de plusieurs de ces animaux. Première partie. Annales des Sciences Naturelles Cinquième Série: Zoologie et Paléontologie 19: 1–53.

Dufossé L (1874b). Recherches sur les bruits et les sons expressifs que font entendre les Poissons d'Europe et sur les organes producteurs de ces phenomenes acoustiques ainsi que sur les appareils de l'audtion de plusieurs de ces animaux. Annales des Sciences Naturelles Cinquième Série: Zoologie et Paléontologie 20: 1–134.

Emery AR (1973). Comparative ecology and functional osteology of fourteen species of damselfish (Pisces: Pomacentridae) at Alligator Reef, Florida Keys. Bulletin of Marine Science 23: 649–770.

Evans RR (1969). Phylogenetic significance of sound producing mechanisms of Western Atlantic fishes of the family Triglidae and Peristediidae. Ph.D. dissertation, Department of Zoology, Boston University, Ann Arbor, Michigan.

Evans RR (1970). Phylogenetic significance of teleost sound producing mechanisms. Journal of the Colorado-Wyoming Academy of Science 7: 9–10.

Fay RR (2009). Fish bioacoustics. In: Havelock D, Kuwano S, Vorländer M (eds.), Handbook of Signal Processing in Acoustics. Springer, New York, pp. 1851–1860

Fay RR, Popper AN (2000). Evolution of hearing in vertebrates: the inner ears and processing. Hearing Research 149: 1–10.

Fine ML (1997). Endocrinology of sound production in fishes. Marine and Freshwater Behavior and Physiology 29: 23–45

Fine ML, Lenhardt ML (1983). Shallow-water propagation of the toadfish mating call. Comparative Biochemistry and Physiology A 76A: 225–231.

Fine ML, Winn HE, Olla BL (1977). Communication in fishes. In: Sebeok TA (ed.), How Animals Communicate. Indiana University Press, Bloomington, pp. 472–518.

Fish JF, Cummings WC (1972). A 50-dB increase in sustained ambient noise from fish (Cynoscion xanthulus). Journal of the Acoustical Society of America 52: 1266–1270.

Fish MP (1948). Sonic Fishes of the Pacific. Woods Hole Oceanographic Institution, Woods Hole, Massachussetts.

Fish MP (1953). The production of underwater sound by the northern seahorse, Hippocampus hudsonius. Copeia 1953: 98–99.

Fish MP (1954). The character and significance of sound production among fishes of the western North Atlantic. Bulletin of the Bingham Oceanographic Collection 14: 1–109.

Fish MP, Mowbray WH (1959). The production of underwater sound by Opsanus sp., a new toadfish from Bimini, Bahamas. Zoologica 44: 71–79.

Fish MP, Mowbray WH (1970). Sounds of the Western North Atlantic Fishes. Johns Hopkins University Press, Baltimore.

Fish MP, Kelsey AS, Mowbray WH (1952). Studies on the production of underwater sound by North Atlantic coastal fishes. Journal of Marine Research 11: 180–193.

Fourmanoir P, Laboute P (1976). *Poissons des mers Tropicales: Nouvelle Calédonie, Nouvelles Hébrides*. Hachette, Tahiti.

Freytag G (1968). Ergebnisse zur marinen bioakustik. Protokolle zur Fischereitechnik 11: 252–352.

Gainer H, Kusano K, Mathewson RF (1965). Electrophysiological and mechanical properties of squirrelfish sound-producing muscle. Comparative Biochemistry and Physiology 14: 661–671.

Galis F, Snelderwàârd P (1997). A novel biting mechanism in damselfishes (Pomacentridae): the pushing up of the lower pharyngeal jaw by the pectoral girdle. Netherlands Journal of Zoology 47: 405–410.

Gannon DP, Barros NB, Nowacek DP, Read AJ, Waples DM, Wells RS (2005). Prey detection by bottlenose dolphins, *Tursiops truncatus*: an experimental test of the passive listening hypothesis. Animal Behaviour 69: 709–720.

Gilmore RG (2003). Sound production and communication in the spotted seatrout. In: Bortone SA (ed.), *Biology of the Spotted Sea Trout*. CRC Press, Boca Raton, Florida, pp. 177–195.

Graham R (1992). Sounds fishy. Australia's Geographic Magazine 14: 76–83.

Gray GA, Winn HE (1961). Reproductive ecology and sound production of the toadfish, *Opsanus tau*. Ecology 42: 274–282.

Griffin DR (1950). Underwater sounds and the orientation of marine animals, a preliminary survey. Cornell University Project NR 162-429, Contract N.6 ONR. 264 t.o.9. between the Office of Naval Research and Cornell University, Technical Report 3.

Guest WC, Lasswell JL (1978). A note on courtship behavior and sound production of red drum. Copeia 1978: 337–338.

Hardenberg JDF (1934). Ein Töne erzeugender Fisch. Zoologischer Anzeiger 108: 224–227.

Hawkins AD (1986). Underwater sound and fish behavior. In: Pitcher TJ (ed.), *The Behaviour of Teleost Fishes*. Chapman & Hall, London, pp. 129–169.

Hawkins AD, Myrberg AA, Jr. (1983). Hearing and sound communication under water. In: Lewis B (ed.), *Bioacoustics: A Comparative Approach*. Academic Press, London, pp. 347–405.

Herald ES, Dempster RP (1957). Courting activity in the white-lipped squirrelfish. Aquarium Journal 28: 43–44.

Holt GJ, Holt SA, Arnold CR (1985). Diel periodicity of spawning in sciaenids. Marine Ecology Progress Series 27: 1–7.

Holt SA (2002). Intra- and inter-day variability in sound production by red drum (Sciaenidae) at a spawning site. Bioacoustics 12: 227–229.

Holzberg S (1973). Beobachtungen zur oekologie und zum socialverhalten des korallenbarsches *Dascyllus marginatus* Rueppell (Pisces: Pomacentridae). Zeitschrift für Tierpsychologie 33: 492–513.

Horch K, Salmon M (1973). Adaptations to the acoustic environment by the squirrelfishes *Myripristis violaceus* and *M. pralinius*. Marine Behaviour and Physiology 2: 121–139.

Ibara RM, Penny LT, Ebeling AW, van Dykhuizen G, Cailliet G (1983). The mating call of the plainfin midshipman fish, *Porichthys notatus*. In: Noakes DLG, Lindquist DG, Helfman GS, Ward JA (eds.), *Predators and Prey in Fishes*. Dr. W. Junk Publishers, The Hague, pp. 205–212.

James PL, Heck KL (1994). The effects of habitat complexity and light intensity on ambush predation within a simulated seagrass habitat. Journal of Experimental Marine Biology and Ecology 176: 187–200.

Jordan DS, Richardson RE (1907). Fishes from the islands of the Philippine Archipelago. Bulletin of the Bureau of Fisheries 27: 233–295.

Kasumyan AO (2008). Sounds and sound production in fishes. Journal of Ichthyology 48: 981–1030.

Kawase H (2005). Spawning behavior of the pygmy leatherjacket *Brachaluteres jacksonianus* (Monacanthidae) in southeastern Australia. Ichthyological Research 52: 194–197.

Kenyon TN (1994). The significance of sound interception to males of the bicolor damselfish, *Pomacentrus partitus*, during courtship. Environmental Biology of Fishes 40: 391–405.

Kihslinger RL, Klimley AP (2002). Species identity and the temporal characteristics of fish acoustic signals. Journal of Comparative Psychology 116: 210–214.

Kim SH (1977). Sound production and phonotactic behaviour of the yellowtail, *Seriola quinqueradiata* Temminck et Schlegel. Bulletin of the National Fisheries University, Busan (Natural Sciences) 17: 17–25.

Knudsen VO, Alford RS, Emling JW (1948a). Survey of Underwater Sound. Report No. 3: Ambient Noise. Office of Scientific Research and Development, National Defense Research Committee, Washington, DC.

Knudsen VO, Alford RS, Emling JW (1948b). Underwater ambient noise. Journal of Marine Research 7: 410–429.

Koenig O (1957). Erfahrungen mit korallenfischen. Aquarien und Terrarien Zeitschrift 10: 154–156.

Kronengold M, Dann R, Green WC, Loewenstein JM (1964). Description of the system. In: Tavolga WN (ed.), *Marine Bio-acoustics*. Pergamon Press, New York, pp. 11–26.

Kumpf HE (1964). Use of underwater television in bioacoustic research. In: Tavolga WN (ed.), *Marine Bio-acoustics*. Pergamon Press, New York, pp. 45–57.

Ladich F (1990). Vocalization during agonistic behaviour in *Cottus gobio* L. (Cottidae): an acoustic threat display. Ethology 84: 193–201.

Ladich F (1997). Agonistic behaviour and significance of sounds in vocalizing fish. Marine and Freshwater Behaviour and Physiology 29: 87–108.

Ladich F (2000). Acoustic communication and the evolution of hearing in fishes. Philosophical Transactions of the Royal Society of London B 355: 1285–1288.

Ladich F (2007). Females whisper briefly during sex: context- and sex-specific differences in sounds made by croaking gouramis. Animal Behaviour 73: 379–387.

Ladich F, Bass AH (2003). Underwater sound generation and acoustic reception in fishes with some notes on frogs. In: Collin SP, Marshall NJ (eds.), *Sensory Processing in Aquatic Environments*. Springer-Verlag, New York, pp. 173–193.

Ladich F, Fine ML (2006). Sound-generating mechanisms in fishes: a unique diversity in vertebrates. In: Ladich F, Collin SP, Moller P, Kapoor BG (eds.), *Communication in Fishes, Vol. 1. Acoustic and Chemical Communication*. Science Publishers, Enfield, New Hampshire, pp. 3–43.

Ladich F, Myrberg AA Jr. (2006). Agonistic behavior and acoustic communication. In: Ladich F, Collin SP, Moller P, Kapoor BG (eds.), *Communication in Fishes, Vol. 1. Acoustic and Chemical Communication*. Science Publishers, Enfield, New Hampshire, pp. 121–148.

Ladich F, Collin SP, Moller P, Kapoor BG (2006). *Communication in Fishes*. Science Publishers, Inc., Enfield, New Hampshire.

Lafite-Dupont JA (1907). Recherches sur l'audition des poissons. Comptes Rendus Des Séances de la Société de Biologie et de ses Filiales 63: 710–711.

Lagardère JP, Fonteneau G, Mariani A, Morinière P (2003). Les émissions sonores du poisson-clown mouffette *Amphiprion akallopisos*, Bleeker 1853 (Pomacentridae), enregistrées dans l'aquarium de la Rochelle. Annales de la Société des Sciences Naturelles de la Charente-Maritime 9: 281–288.

Lagardère JP, Millot S, Parmentier E (2005). Aspects of sound communication in the pearlfish *Carapus boraborensis* and *Carapus homei* (Carapidae). Journal of Experimental Zoology 303A: 1066–1074.

Lammers MO, Brainard RE, Au WWL, Mooney TA, Wong KB (2008). An ecological acoustic recorder (EAR) for long-term monitoring of biological and anthropogenic sounds on coral reefs and other marine habitats. Journal of the Acoustical Society of America 123(3): 1720–1728.

Lauder GV, Liem KF (1983). The evolution and interrelationships of the actinopterygian fishes. Bulletin of the Museum of Comparative Zoology 150: 95–197.

Leis JM, Lockett MM (2005). Localization of reef sounds by settlement-stage larvae of coral-reef fishes (Pomacentridae). Bulletin of Marine Science 76: 715–724.

Lieske E, Myers R (1999). *Coral Reef Fishes*. Princeton University Press, Princeton, New Jersey.

Limbaugh C (1964). Notes on the life history of two California pomacentrids: Garibaldis, *Hypsypops rubicunda* (Girard), and blacksmiths, *Chromis punctipinnis* (Cooper). Pacific Science 28: 41–50.

Lin YC, Mok HK, Huang BQ (2007). Sound characteristics of big-snout croaker, *Johnius macrorhynus* (Sciaenidae). Journal of the Acoustical Society of America 121: 586–593.

Lobel PS (1978). Diel, lunar, and seasonal periodicity in the reproductive behavior of the pomacanthid fish, *Centropyge potteri*, and some other reef fishes in Hawaii. Pacific Science 32: 193–207.

Lobel PS (1992). Sounds produced by spawning fishes. Environmental Biology of Fishes 33: 351–358.

Lobel PS (1996). Spawning sound of the trunkfish, *Ostracion meleagris* (Ostraciidae). Biological Bulletin 191: 308–309.

Lobel PS (2001a). Acoustic behavior of cichlid fishes. Journal of Aquariculture and Aquatic Sciences 9: 89–108.

Lobel PS (2001b). Fish bioacoustics and behavior: passive acoustic detection and the application of a closed-circuit rebreather for field study. Marine Technology Society Journal 35: 19–28.

Lobel PS (2002). Diversity of fish spawning sounds and the application of acoustic monitoring. Bioacoustics 12: 286–289.

Lobel PS (2003a). Fish courtship and mating sounds: unique signals for acoustic monitoring. In: Rountree RA, Goudey CA, Hawkins AD (eds.), *Listening to Fish: Passive*

Acoustic Applications in Marine Fisheries. Proceedings from an International Workshop in Passive Acoustics. MIT Press, Cambridge, Massachussetts, pp. 54–58

Lobel PS (2003b). Synchronized underwater audio-video recording. In: Rountree RA, Goudey CA, Hawkins AD (eds.), *Listening to Fish: Passive Acoustic Applications in Marine Fisheries. Proceedings from an International Workshop in Passive Acoustics.* MIT Press, Cambridge, Massachussetts, pp. 127–130.

Lobel PS (2005). Scuba bubble noise and fish behavior: a rationale for silent diving technology. Proceedings of the American Academy of Underwater Sciences Symposium, University of Connecticut at Avery Point, Groton, Connecticut, pp. 49–59.

Lobel PS, Kerr LM (1999). Courtship sounds of the Pacific Damselfish, *Abudefduf sordidus* (Pomacentridae). Biological Bulletin 197: 242–244.

Lobel PS, Mann DA (1995a). Courtship and mating sounds of *Dascyllus albisella* (Pomacentridae). Bulletin of Marine Science 57: 705.

Lobel PS, Mann DA (1995b). Spawning sounds of the damselfish, *Dascyllus albisella* (Pomacentridae), and relationship to male size. Bioacoustics 6: 187–198.

Locascio JV, Mann DA (2008). Diel periodicity of fish sound production in Charlotte Harbor, Florida. Transactions of the American Fisheries Society 137: 606–615.

Luczkovich JJ, Sprague MW, Johnsen SE, Pullinger RC (1999). Delimiting spawning areas of weakfish, *Cynoscion regalis* (Family Sciaenidae) in Pamlico Sound, North Carolina using passive hydroacoustic surveys. Bioacoustics 10: 143–160.

Luczkovich JJ, Daniel HJ, Hutchinson M, Jenkins T, Johnson SE, Pullinger RC, Sprague MW (2000). Sounds of sex and death in the sea: bottlenose dolphin whistles silence mating choruses of silver perch. Bioacoustics 10: 323–334.

Luczkovich JJ, Mann DA, Rountree RA (2008a). Passive acoustics as a tool in fisheries science. Transactions of the American Fisheries Society 137: 533–541.

Luczkovich JJ, Pullinger RC, Johnson SE, Sprague MW (2008b). Identifying sciaenid critical spawning habitats by the use of passive acoustics. Transactions of the American Fisheries Society 137: 576–605.

Luh HK, Mok HK (1986). Sound production in the domino damselfish, *Dascyllus trimaculatus* (Pomacentridae) under laboratory conditions. Japanese Journal of Ichthyology 33: 70–74.

Malavasi S, Torricelli P, Lugli M, Pranovi F, Mainardi D (2003). Male courtship sounds in a teleost with alternative reproductive tactics, the grass goby, *Zosterisessor ophiocephalus*. Environmental Biology of Fishes 66: 231–236.

Mann DA (2006). Propagation of fish sounds. In: Ladich F, Collin SP, Moller P, Kapoor BG (eds.), *Communication in Fishes, Vol. 1. Acoustic and Chemical Communication.* Science Publishers, Enfield, New Hampshire, pp. 107–120.

Mann DA, Grothues TM (2009). Short-term upwelling events modulate fish sound production at a mid-Atlantic Ocean observatory. Marine Ecology Progress Series 375: 65–71.

Mann DA, Lobel PS (1995). Passive acoustic detection of spawning sounds produced by the damselfish, *Dascyllus albisella* (Pomacentridae). Bioacoustics 6: 199–213.

Mann DA, Lobel PS (1997). Propogation of damselfish (Pomacentridae) courtship sounds. Journal of the Acoustic Society of America 101: 3783–3791.

Mann DA, Lobel PS (1998). Acoustic behavior of the damselfish *Dascyllus albisella*: behavioral and geographic variation. Environmental Biology of Fishes 51: 421–428.

Mann DA, Lu Z, Popper AN (1997). A clupeid fish can detect ultrasound. Nature 389: 341.

Mann DA, Casper BM, Boyle KS, Tricas TC (2007). On the attraction of larval fishes to reef sounds. Marine Ecology Progress Series 338: 307–310.

Mann DA, Locascio JV, Coleman FC, Koenig CC (2009). Goliath grouper *Epinephelus itajara* sound production and movement patterns on aggregation sites. Endangered Species Research 7: 229–236.

Marage L (1906). Contribution à l'étude de l'audition des poissons. Comptes Rendus Hebdomadaires des Séances de l'Académie des Sciences 143: 852–853.

Marshall NB (1962). The biology of sound-producing fishes. Symposia of the Zoological Society of London 7: 45–60.

Maruska KP, Tricas TT (2009). Encoding properties of auditory neurons in the brain of a soniferous damselfish: response to simple tones and complex conspecific signals. Journal of Comparative Physiology A 195: 1071–1088.

Maruska KP, Boyle KS, Dewan LR, Tricas TC (2007). Sound production and spectral hearing sensitivity in the Hawaiian sergeant damselfish, *Abudefduf abdominalis*. Journal of Experimental Biology 210: 3990–4004.

Matsubara K (1934). Studies on the Scorpaenoid fishes of Japan. I. Descriptions of one new genus and five new species. Journal of the Imperial Fisheries Institute of Tokyo 80: 199–210.

Matsuno Y, Hujieda S, Chung YJ, Yamanaka Y (1994). Underwater sound in the net pen at the culture grounds in the innermost area of Kagoshima Bay. Bulletin of the Japanese Society of Fisheries Oceanography 58: 11–20.

McCauley RD, Cato DH (2000). Patterns of fish calling in a nearshore environment in the Great Barrier Reef. Philosophical Transactions of the Royal Society of London B 355: 1289–1293.

McCauley RD, Fewtrell J, Popper AN (2003). High intensity anthropogenic sound damages fish ears. Journal of the Acoustical Society of America 113: 638–642.

McKibben JR, Bass AH (1998). Behavioral assessment of acoustic parameters relevant to signal recognition and preference in a vocal fish. Journal of the Acoustical Society of America 104: 3520–3533.

Mikhailenko NA (1973). Organ of sound formation and electro generation in the Black Sea stargazer *Uranoscopus scaber* (Uranoscopidae). Zoologicheskii Zhurnal 52: 1353–1359.

Miyagawa M, Takemura A (1986). Acoustical behaviour of the scorpaenoid fish *Sebasticus marmoratus*. Bulletin of the Japanese Society of Scientific Fisheries 52: 411–415.

Möbius K (1889). *Balistes aculeatus*, ein trommelander Fisch. Sitzungsberichte der Preussischen Akademie der Wissenschaften 46: 999–1006.

Moulton JM (1958). The acoustical behavior of some fishes in the Bimini area. Biological Bulletin 114: 357–374.

Moulton JM (1960). Swimming sounds and the schooling of fishes. Biological Bulletin 119: 210–223.

Moulton JM (1962). Marine animal sounds of the Queensland coast. American Zoologist 2: 542.

Moulton JM (1963). Acoustic behaviour of fishes. In: Busnel RG (ed.), *Acoustic Behaviour of Animals*. Elsevier, Amsterdam, pp. 655–685.

Moulton JM (1964a). Acoustic behaviour of fishes. In: Busnel RG (ed.), *Acoustic Behaviour of Animals*. Elsevier, New York, pp. 655–693.

Moulton JM (1964b). Underwater sound: biological aspects. Oceangraphy and Marine Biology Annual Review 2: 425–454.

Moulton JM (1969). The classification of acoustic communicative behavior among teleost fishes. In: Sebeok TA, Ramsay A (eds.), *Approaches to Animal Communication*. Mouton & Co. N.V., The Hague, pp. 146–178.

Moyer JT (1975). Reproductive behavior of the damselfish *Pomacentrus nagasakiensis* at Miyake-jima, Japan. Japanese Journal of Ichthyology 22: 151–163.

Moyer JT (1979). Mating strategies and reproductive behavior of ostraciid fishes at Miyake-jima, Japan. Japanese Journal of Ichthyology 26: 148–160.

Moyer JT, Thresher RE, Collin PL (1983). Courtship, spawning and inferred social organization of American angelfishes (genera *Pomacanthus, Holacanthus* and *Centropyge*; *Pomacanthidae*). Environmental Biology of Fishes 9: 25–39.

Müller J (1857). Über die Fische, welche Töne von sich geben, und die Entstchung dieser Töne. Archiv für Anatomie, Physiologie und wissenschaftliche Medicin: 249–279.

Myrberg AA, Jr. (1972a). Ethology of the bicolor damselfish, *Eupomacentrus partitus* (Pisces: Pomacentridae): a comparison of laboratory and field behaviour. Animal Behaviour Monographs 5: 199–283.

Myrberg AA Jr (1972b). Social dominance and territoriality in bicolor damselfish, *Eupomacentrus partitus* (Poey) (Pisces: Pomacentridae). Behaviour 41: 207–230.

Myrberg AA Jr (1980). Fish bio-acoustics: its relevance to the "not so silent world." Environmental Biology of Fishes 5: 297–304.

Myrberg AA Jr (1981). Sound communication and interception in fishes. In: Tavolga WN, Popper AN, Fay RR (eds.), *Hearing and Sound Communication in Fishes*. Springer-Verlag, New York, pp. 395–426.

Myrberg AA Jr (1996). Fish bioacoustics: serendipity in research. Bioacoustics 7: 143–150.

Myrberg AA Jr. (1997a). Sound production by a coral reef fish (*Pomacentrus partitus*): evidence for a vocal, territorial "keep-out" signal. Bulletin of Marine Science 60: 1017–1025.

Myrberg AA Jr (1997b). Underwater sound: its relevance to behavioral functions among fishes and marine mammals. Marine and Freshwater Behavior and Physiology 29: 3–21.

Myrberg AA Jr (2001). The acoustical biology of elasmobranchs. Environmental Biology of Fishes 60: 31–45.

Myrberg AA Jr, Fuiman LA (2002). The sensory world of coral reef fishes. In: Sale PF (ed.), *Coral Reef Fishes: Dynamics and Diversity in a Complex Ecosystem*. Elsevier Academic Press, San Diego, pp. 123–148.

Myrberg AA Jr, Lugli M (2006). Reproductive behavior and acoustical interactions. In: Ladich F, Collin SP, Moller P, Kapoor BG (eds.), *Communication in Fishes, Vol. 1. Acoustic and Chemical Communication*. Science Publishers, Enfield, New Hampshire, pp. 149–176.

Myrberg AA Jr, Riggio RJ (1985). Acoustically mediated individual recognition by a coral reef fish (*Pomacentrus partitus*). Animal Behaviour 33: 411–416.

Myrberg AA Jr, Spires JY (1972). Sound discrimination by the bicolor damselfish, *Eupomacentrus partitus*. Journal of Experimental Biology 57: 727–735.

Myrberg AA Jr, Stadler JH (2002). The significance of the sounds by male gobies (Gobiidae) to conspecific females: similar findings to a study made long ago. Bioacoustics 12: 255–257.

Myrberg AA Jr, Thresher RE (1974). Interspecific aggression and its relevance to the concept of territoriality in reef fishes. American Zoologist 14: 81–96.

Myrberg AA Jr, Banner A, Richard JD (1969). Shark attraction using a video-acoustic system. Marine Biology 2: 264–376.

Myrberg AA Jr, Spanier E, Ha SJ (1978). Temporal patterning in acoustical communication. In: Reese ES, Lighter FJ (eds.), *Contrasts in Behavior.* Wiley Interscience, New York, pp. 137–179.

Myrberg AA Jr, Mohler M, Catala JD (1986). Sound production by males of a coral reef fish (*Pomacentrus partitus*): its significance to females. Animal Behaviour 34: 913–923.

Myrberg AA Jr, Ha SJ, Shamblott MJ (1993). The sounds of bicolor damselfish (*Pomacentrus partitus*): predictors of body size and a spectral basis for individual recognition and assessment. Journal of the Acoustical Society of America 94: 3067–3070.

Nakazato M, Takemura A (1987). Acoustical behavior of Japanese parrot fish *Oplenathus* [sic] *fasciatus.* Nippon Suisan Gakkaishi 53: 967–973.

Nelson EM (1955). The morphology of the swim bladder and auditory bulla in the Holocentridae. Fieldiana: Zoology 37: 121–130.

Nguyen TK, Lin H, Parmentier E, Fine ML (2008). Seasonal variation in sonic muscles in the fawn cusk-eel Lepophidium profundorum. Biology Letters 4: 707–710.

Nottestad L (1998). Extensive gas bubble release in Norwegian spring-spawning herring (*Clupea harengus*) during predator avoidance. ICES Journal of Marine Science 55: 1133–1140.

Okumura T, Akamatsu T, Yan HY (2002). Analyses of small tank acoustics: empirical and theoretical approaches. Bioacoustics 12: 330–332.

Oliver SJ (2001). Mate choice and sexual selection in domino damselfish, *Dascyllus albisella.* Ph.D. dissertation, Boston University, Boston, Massachussetts.

Onuki A, Somiya H (2007). Innervation of sonic muscles in teleosts: occipital vs. spinal nerves. Brain, Behavior and Evolution 69: 132–141.

Pappe L (1853). *Synopsis of the Edible Fishes at the Cape of Good Hope.* Van de Sandt de Villiers & Tier, Capetown.

Parker GH (1903). The sense of hearing in fishes. American Naturalist 37: 185–204.

Parmentier E, Diogo R (2006). Evolutionary trends of swimbladder sound mechanisms in some teleost fishes. In: Ladich F, Collin SP, Moller P, Kapoor BG (eds.), *Communication in Fishes, Vol. 1. Acoustic and Chemical Communication.* Science Publishers, Enfield, New Hampshire, pp. 45–70.

Parmentier E, Vandewalle P, Lagardère JP (2003). Sound-producing mechanisms and recordings in Carapini species (Teleostei, Pisces). Journal of Comparative Physiology A 189: 283–292.

Parmentier E, Lagardère JP, Vandewalle P, Fine ML (2005). Geographical variation in sound production in the anemonefish *Amphiprion akallopisos.* Proceedings of the Royal Society of London B 272: 1697–1703.

Parmentier E, Fine M, Vandewalle P, Ducamp J-J, Lagardère J-P (2006a). Sound production in two carapids (*Carapus acus* and *C. mourlani*) and through the sea cucumber tegument. Acta Zoologica 87: 113–119.

Parmentier E, Lagardere J-P, Braquegnier J-B, Vandewalle P, Fine ML (2006b). Sound production mechanism in carapid fish: first example with a slow sonic muscle. Journal of Experimental Biology 209: 2952–2960.

Parmentier E, Vandewalle P, Frédérich B, Fine ML (2006c). Sound production in two species of damselfishes (Pomacentridae): *Plectroglyphidodon lacrymatus* and *Dascyllus aruanus*. Journal of Fish Biology 69: 491–503.

Parmentier E, Colleye O, Fine ML, Frédérich B, Vandewalle P, Herrel A (2007). Sound production in the clownfish *Amphiprion clarkii*. Science 316: 1006.

Parmentier E, Lecchini D, Frederich B, Brie C, Mann D (2009). Sound production in four damselfish (*Dascyllus*) species: phyletic relationships? Biological Journal of the Linnean Society 97: 928–940.

Parvulescu A (1967). The acoustics of small tanks. In: Tavolga WN (ed.), *Marine Bioacoustics*. Pergamon Press, Oxford, pp. 7–14.

Pauly D (2004). *Darwin's Fishes: An Encyclopedia of Ichthyology, Ecology and Evolution*. Cambridge University Press, Cambridge.

Picciulin M, Costantini M, Hawkins AD, Ferrero EA (2002). Sound emissions of the Mediterranean damselfish *Chromis chromis* (Pomacentridae). Bioacoustics 12: 236–238.

Popper AN (2003). Effects of anthropogenic noise on fishes. Fisheries 28: 24–31.

Popper AN, Hastings MC (2009). The effects of human-generated sound on fish. Integrative Zoology 4: 43–52.

Popper AN, Salmon M, Parvulescu A (1973). Sound localization by the Hawaiian squirrelfishes *Myripristis berndti* and *M. argyromus*. Animal Behaviour 21: 86–97.

Popper AN, Fay RR, Platt C, Sand O (2003a). Sound detection mechanisms and capabilities of teleost fishes. In: Collin SP, Marshall NJ (eds.), *Sensory Processing in Aquatic Environments*. Springer-Verlag, New York, pp. 3–38.

Popper AN, Fewtrell J, Smith ME, McCauley RD (2003b). Anthropogenic sound: effects on the behavior and physiology of fishes. Marine Technology Society Journal 37: 35–40.

Quenouille B, Bermingham E, Planes S (2004). Molecular systematics of the damselfishes (Teleostei: Pomacentridae): Bayesian phylogenetic analyses of mitochondrial and nuclear DNA sequences. Molecular Phylogenetics and Evolution 31: 66–88.

Radford CA, Jeffs AG, Tindle CT, Cole RG, Montgomery JC (2005). Bubbled waters: the noise generated by underwater breathing apparatus. Marine and Freshwater Behaviour and Physiology 38: 259–267.

Radford C, Jeffs A, Tindle C, Montgomery JC (2008a). Resonating sea urchin skeletons create coastal choruses. Marine Ecology Progress Series 362: 37–43.

Radford CA, Jeffs AG, Tindle CT, Montgomery JC (2008b). Temporal patterns in ambient noise of biological origin from a shallow water temperate reef. Oecologia 156: 921–929.

Ramcharitar J, Gannon DP, Popper AN (2006). Bioacoustics of fishes of the family Sciaenidae (croakers and drums). Transactions of the American Fisheries Society 135: 1409–1431.

Rauther M (1945). Uber die schwimmblase und die zu ihr in beziehung tretenden somatischen muskeln bei den Triglidae und anderen Scleroparei. Zoologische Jahrbucher 69: 159–250.

Remage-Healey L, Nowacek DP, Bass AH (2006). Dolphin foraging sounds suppress calling and elevate stress hormone levels in a prey species, the Gulf toadfish. Journal of Experimental Biology 209: 4444–4451.

Rice AN, Bass AH (2009). Novel vocal repertoire and paired swimbladders of the three-spined toadfish, *Batrachomoeus trispinosus*: insights into the diversity of the Batrachoididae. Journal of Experimental Biology 212: 1377–1391.

Rice AN, Lobel PS (2002). Enzyme activities of pharyngeal jaw musculature in the cichlid *Tramitichromis intermedius*: implications for sound production in cichlid fishes. Journal of Experimental Biology 205: 3519–3523.

Rice AN, Lobel PS (2003). The pharyngeal jaw apparatus of the Cichlidae and Pomacentridae (Teleostei: Labroidei): function in feeding and sound production. Reviews in Fish Biology and Fisheries 13: 433–444.

Richard JD (1968). Fish attraction with pulsed low frequency sound. Journal of the Fisheries Research Board of Canada 25: 1441–1452.

Ripley JL, Foran CM (2007). Influence of estuarine hypoxia on feeding and sound production by two sympatric pipefish species (Syngnathidae). Marine Environmental Research 63: 350–367.

Rome LC (2006). Design and function of superfast muscles: new insights into the physiology of skeletal muscle. Annual Review of Physiology 68: 193–221.

Rome LC, Syme DA, Hollingworth S, Lindstedt SL, Baylor SM (1996). The whistle and the rattle: the design of sound producing muscles. Proceedings of the National Academy of Sciences of the United States of America 93: 8095–8100.

Rosenthal GG, Lobel PS (2005). Communication. In: Sloman KA, Wilson RW, Balshine S (eds.), *Behaviour and Physiology of Fish*. Fish Physiology series, vol. 24. Academic Press, San Diego, pp. 39–78.

Rountree RA, Goudey CA, Hawkins AD, Luczkovich JJ, Mann DA (2003). Listening to fish: passive acoustic applications in marine fisheries. Massachusetts Institute of Technology, Sea Grant Digital Oceans, Cambridge. MITSG 0301 http://web.mit.edu/seagrant/digitalocean/listening.pdf.

Rountree RA, Gilmore RG, Goudey CA, Hawkins AD, Luczkovich JJ, Mann DA (2006). Listening to fish: applications of passive acoustics to fisheries science. Fisheries 31: 433–446.

Sadovy de Mitcheson Y, Liu M (2008). Functional hermaphroditism in teleosts. Fish and Fisheries 9: 1–43.

Salmon M (1967). Acoustical behavior of the menpachi, *Myripristis berndti*, in Hawaii. Pacific Science 21: 364–381.

Salmon M, Winn HE (1966). Sound production by priacanthid fishes. Copeia 1966: 869–872.

Salmon M, Winn HE, Sorgente N (1968). Sound production and associated behavior in triggerfishes. Pacific Science 22: 11–20.

Santangelo N, Bass AH (2006). New insights into neuropeptide modulation of aggression: field studies of arginine vasotocin in a territorial tropical damselfish. Proceedings of the Royal Society of London B 273: 3085–3092.

Santiago JA, Castro JJ (1997). Acoustic behaviour of *Abudefduf luridus*. Journal of Fish Biology 51: 952–959.

Sartori JD, Bright TJ (1973). Hydrophonic study of the feeding activities of certain Bahamian parrot fishes, family Scaridae. Hydro-Lab Journal 2: 25–56.

Schneider H (1961). Neuere ergebnisse de lautforschung bei fischen. Naturwissenschaften 48: 512–518.

Schneider H (1964a). Bioakustische untersuchungen an anemonenfischen der gattung Amphiprion (Pisces). Zeitschrift für Morphologie und Ökologie der Tiere 53: 453–474.

Schneider H (1964b). Physiologische und morphologie untersuchungen zur bioakustik der tigerfische (Pisces, Theraponidae). Zeitschrift für Vergleichende Physiologie 47: 493–558.

Schneider H, Hasler AD (1960). Laute und lauterzeugung bein suesswassertrommler Aplodinotus grunniens Rafinesque (Sciaenidae, Pisces). Zeitschrift für Vergleichende Physiologie 43: 499–517.

Schwarz AL (1985). The behavior of fishes in their acoustic environment. Environmental Biology of Fishes 13: 3–15.

Sebeok TA (1977). How Animals Communicate. Indiana University Press, Bloomington

Shishkova EV (1958a). Concerning the reactions of fish to sounds and the spectrum of trawler noise. Rybnoye Khozyaystvo 34: 33–39.

Shishkova EV (1958b). Recordings and analysis of noise made by fish. Trudy Vsesoiuznyi Nauchno-Isseledovatel'skii Institut Morskogo Rybnogo Khoziaistva i Okeanografii 36: 280–294.

Simpson SD, Meekan MG, McCauley RD, Jeffs A (2004). Attraction of settlement-stage coral reef fishes to reef noise. Marine Ecology Progress Series 276: 263–268.

Simpson SD, Meekan M, Montgomery J, McCauley R, Jeffs A (2005). Homeward sound. Science 308: 221.

Širović A, Demer DA (2009). Sounds of captive rockfishes. Copeia 2009: 502–509.

Sörensen W (1894–1895). Are the extrinsic muscles of the air-bladder in some Siluroidae and the "elastic spring"' apparatus of others subordinate to the voluntary production of sounds? What is, according to our present knowledge, the function of the weberian ossicles? A contribution to the biology of fishes. Journal of Anatomy and Physiology 29: 109–139; 205–229; 399–423; 518–552.

Spanier E (1970). Analysis of sounds and associated behavior of domino damselfish Dascyllus trimaculatus (Rueppell, 1828). (Pomacentridae). Master's thesis, Tel-Aviv University.

Spanier E (1979). Aspects of species recognition by sound in four species of damselfishes, genus Eupomacentrus (Pisces: Pomacentridae). Zeitschrift für Tierpsychologie 51: 301–316.

Stadler JH (2002). Evidence for hydrodynamic mechanism of sound production by courting males of the notchtongue goby Bathygobius curacao (Metzelaar). Bioacoustics 13: 145–152.

Steinberg JC, Cummings WC, Brahy BD, Spires JYM (1965). Further bioacoustic studies off the west coast of North Bimini, Bahamas. Bulletin of Marine Science 15: 942–963.

Stiassny MLJ, Jensen JS (1987). Labroid intrarelationships revisited: morphological complexity, key innovations, and the study of comparative diversity. Bulletin of the Museum of Comparative Zoology 151: 269–319.

Takayama M, Onuki A, Yosino T, Yoshimoto M, Ito H, Kohbara J, Somiya H (2003). Sound characteristics and the sound producing system in silver sweeper, *Pempheris schwenkii* (Perciformes: Pempheridae). Journal of the Marine Biological Association of the United Kingdom 83: 1317–1320.

Takemura A (1983). Studies on the underwater sound. VIII. Acoustical behaviour of clownfishes (*Amphirion* spp.). Bulletin of the Faculty of Fisheries Nagasaki University 54: 21–27.

Takemura A, Takita T, Mizue K (1978). Studies on the underwater sound. VII. Underwater calls of the Japanese marine drum fishes (Sciaenidae). Bulletin of the Japanese Society of Scientific Fisheries 44 (2): 121–125.

Takemura A, Nishida N, Kobayashi Y (1988). The attraction effect of natural feeding sound in fish. Nagasaki Daigaku Suisangakauba (Sasebo Japan) Kenkyu Hokoku 63: 1–4.

Tavolga WN (1958a). The significance of underater sounds produced by males of the gobiid fish, *Bathygobius soporator*. Physiological Zoology 31: 259–271.

Tavolga WN (1958b). Underwater sounds produced by two species of toadfish, *Opsanus tau* and *Opsanus beta*. Bulletin of Marine Science of the Gulf and Caribbean 8: 278–284.

Tavolga WN (1960). Sound production and underwater communication in fishes. In: Lanyon WE, Tavolga W (eds.), *Animal Sounds and Communication*. American Institute of Biological Sciences, Washington, DC, pp. 93–136.

Tavolga WN (1962). Mechanisms of sound production in the Ariid catfishes *Galeichthys* and *Bagre*. Bulletin of the American Museum of Natural History 124: 1–30

Tavolga WN (1964). *Marine Bio-acoustics*. Pergamon Press, New York.

Tavolga WN (1968). Fishes. In: Sebeok TA (ed.), *Animal Communication: Technique of Study and Results of Research*. Indiana University Press, Bloomington, pp. 271–288.

Tavolga WN (1976). *Sound Reception in Fishes*. Dowden, Hutchinson & Ross, Stroudsburg, Pennsylvania.

Tavolga WN (1977). *Sound Production in Fishes*. Dowden, Hutchinson & Ross, Stroudsburg, Pennsylvania.

Tavolga WN, Popper AN, Fay RR (1981). *Hearing and Sound Communication in Fishes*. Springer-Verlag, New York, pp. 608.

Taylor M, Mansueti RJ (1960). Sounds produced by very young crevalle jack, *Caranx hippos*, from the Maryland seaside. Chesapeake Science 1: 115–116.

Thorson RF, Fine ML (2002a). Acoustic competition in the gulf toadfish *Opsanus beta*: acoustic tagging. Journal of the Acoustical Society of America 111: 2302–2307.

Thorson RF, Fine ML (2002b). Crepuscular changes in emission rate and parameters of this boatwhistle advertisement call of the gulf toadfish, *Opsanus beta*. Environmental Biology of Fishes 63: 321–331.

Thresher RE (1982). Courtship and spawning in the emperor angelfish *Pomacanthus imperator*, with comments on reproduction by other pomacanthid fishes. Marine Biology 70: 149–156.

Tricas TC, Boyle K (2009). Sound production and hearing in coral reef butterflyfishes. Journal of the Acoustical Society of America 125: 2487.

Tricas TC, Kajiura SM, Kosaki RK (2006). Acoustic communication in territorial butterflyfish: test of the sound production hypothesis. Journal of Experimental Biology 209: 4994–5004.

Uchida K (1934). Sound producing fishes of Japan. Report of Japan Science Association 9: 369–375.

Ueng J-P, Huang B-Q, Mok H-K (2007). Sexual differences in the spawning sounds of the Japanese croaker, *Argyrosomus japonicus* (Sciaenidae). Zoological Studies 46: 103–110.

Vasconcelos RO, Amorim MCP, Ladich F (2007). Effects of ship noise on the detectability of communication signals in the Lusitanian toadfish. Journal of Experimental Biology 210: 2104–2112.

Verwey J (1930). Coral reef studies. I. The symbiosis between damselfishes and sea anemones in Batavia Bay. Treubia 12: 305–355.

von Frisch K (1936). About the sense of hearing in fish. Biological Reviews of the Cambridge Philosophical Society 11: 210–246.

von Frisch K (1938). The sense of hearing in fish. Nature 141: 8–11.

Wahlberg M, Westerberg H (2003). Sounds produced by herring (*Clupea harengus*) bubble release. Aquatic Living Resources 16: 271–275

Walls PD (1964). The anatomy of the sound producing apparatus of some Australian teleosts. Honors thesis, Bowdoin College, Brunswick, Maine.

Warner LH (1932). The sensitivity of fishes to sound and to other mechanical stimulation. Quarterly Review of Biology 7: 326–339.

Webb JF, Fay RR, Popper AN (2008). *Fish Bioacoustics*. Springer, New York.

Whitley GP (1957). A kennel of frogfishes. Australian Museum Magazine 12: 139–142

Winn HE (1964). The biological significance of fish sounds. In: Tavolga WN (ed.), *Marine Bio-acoustics*. Pergamon Press, New York, pp. 213–231.

Winn HE (1972). Acoustic discrimination by the toadfish with comments on signal systems. In: Winn HE, Olla BL (eds.), *Behavior of Marine Animals: Current Perspectives in Research*. Vol. 2, Vertebrates. Plenum Press, New York, pp. 361–385.

Winn HE, Marshall JD (1963). Sound-producing organ of the squirrelfish, *Holocentrus rufus*. Physiological Zoology 36: 34–44.

Winn HE, Marshall JD, Hazlett B (1964). Behavior, diel activities, and stimuli that elicit sound production, and reaction to sounds in the longspine squirrelfish. Copeia 1964: 413–425.

Yokoyama K, Kamei Y, Toda M, Hirano K, Iwatsuki Y (1994). Reproductive behavior, eggs, and larvae of a caesionine fish, *Pterocaesio digramma*, observed in an aquarium. Japanese Journal of Ichthyology 41: 261–274.

TABLE 10.1. Catalog of reef fish sounds and behaviors.

Species are included for which any information or mention (even anecdotal) of the occurrence of sound production in coral reef fishes is reported.

Family/species	Sound pattern	No. sound types	No. pulses	Frequency range (hz)	Sound duration (ms)	Behavioral contexts	References
ACANTHURIDAE (4 species)							
Acanthurus bahianus	–	–	–	150–4,700	100	chase conspecific, hydrodynamic sound	Steinberg et al. 1965
A. bahianus	–	2	–	–	–	disturbance	Fish & Mowbray 1970
A. chirurgus	–	–	–	–	–	disturbance, escape	Fish & Mowbray 1970
A. coeruleus	–	3	–	–	–	disturbance, escape	Fish & Mowbray 1970
Paracanthurus hepatus	–	–	–	–	–	disturbance*	Fish 1948
APLOACTINIDAE (1 species)							
Paraploactis trachyderma	–	–	–	–	–	unknown	Walls 1964
ARIIDAE (2 species)							
Bagre marinus	short pulse series	3	–	–	~200	chorus of individuals	Tavolga 1960
B. marinus	descending frequency	–	–	400–850	110–200	spontaneous in social groups	Tavolga 1960
B. marinus	descending frequency	–	–	350–275	420–550	spontaneous in social groups	Tavolga 1960
Arius felis	–	3	–	low < 100	10	disturbance	Tavolga 1962
A. felis	–	–	–	low 300–450	45	disturbance	Tavolga 1962
A. felis	–	–	–	high 2000	30–50	disturbance	Tavolga 1962
A. felis	–	2	–	–	–	disturbance*	Burkenroad 1931
A. felis	–	3	–	~100–500	~75–180	swimming social groups	Tavolga 1960
A. felis	broadband frequency	–	–	harmonics to 4,000	30–50	disturbance: fishing line caught	Tavolga 1960

(continued)

TABLE 10.1. (continued)

Family/species	Sound pattern	No. sound types	No. pulses	Frequency range (hz)	Sound duration (ms)	Behavioral contexts	References
Arius felis	–	–	–	–	–	nocturnal chorus, schooling individuals	Tavolga 1960
A. felis	–	–	–	–	–	chorus of seasonal aggregations	Breder 1968
A. felis	–	–	–	–	–	disturbance	Fish & Mowbray 1970
AULOSTOMIDAE (1 species)							
Aulostomus maculatus	harmonic intervals	–	–	–	–	lunge at prey, hydrodynamic swimming noise	Bright 1972
BALISTIDAE (14 species)							
Balistes sp.	–	–	–	–	–	sonic morphology	Fish 1948
B. sp.	–	–	–	–	–	sonic morphology	Bridge 1904; Fish 1948
B. sp.	–	–	–	–	–	sounds reported	Fish 1948
B. (= Sufflamen) bursa	–	–	–	75–9,600	–	disturbance	Salmon et al. 1968
B. capriscus	–	–	–	–	–	interspecific agonism, disturbance	Fish 1948
B. capriscus	–	3	–	–	–	intraspecific agonism, escape	Fish & Mowbray 1970
B. capriscus	–	2	–	–	–	escape	Fish 1954
B. capistratus	–	–	–	75–9,600	–	disturbance	Salmon et al. 1968
Balistapus undulatus	–	–	–	–	–	sonic morphology	Fish 1948
B. vetula	–	–	–	–	–	sonic morphology	Möbius 1889; Sörensen 1894–1895

B. vetula	–	–	0–5,800	30	disturbance in air	Moulton 1958
B. vetula	–	–	600–2,900	150	free swimming during feeding in fish pen	Moulton 1958
B. vetula	multipulsed	≤11	50–2,000	1,410–2,400	attack territory intruders	Steinberg et al. 1965
B. vetula	–	–	1,000–2,350	100	feeding	Steinberg et al. 1965
B. vetula	–	–	–	–	feeding	Tavolga 1968
B. vetula	–	2	–	–	chase intraspecific	Salmon et al. 1968
B. vetula	–	–	–	–	disturbance, feeding	Fish & Mowbray 1970
B. vetula	–	–	75–9,600	99 ±19	disturbance	Salmon et al. 1968
B. vetula	–	–	–	–	chase conspecific	Salmon et al. 1968
Canthidermis sufflamen	–	4	–	–	competitive feeding, disturbance, spine raising	Fish & Mowbray 1970
Melichthys buniva	–	–	75–9,600	–	disturbance	Salmon et al. 1968
M. niger	–	–	–	–	sound reported	Uchida 1934
M. niger	–	–	–	–	feeding	Moulton 1958
M. niger	multipulsed	3	100–350	100–300	disturbance defense "trigger spine raised"	Fish & Mowbray 1970
M. niger	–	–	–	–	territory chase intraspecific, disturbance	Salmon et al. 1968
M. piceus (=*niger*)	–	–	0–8,000	60–100	disturbance in air	Moulton 1958
M. piceus (=*niger*)	–	–	0–8,000	20–40	disturbance in air	Moulton 1958
M. piceus (=*niger*)	–	–	75–9,600	212 ±92	disturbance	Salmon et al. 1968
M. vidua	pulse groups	–	–	–	disturbance	Salmon et al. 1968
Odonus niger	–	–	–	–	sonic morphology	Schneider 1961
Rhinecanthus aculeatus	–	2	–	–	sonic morphology	Fish 1948
R. aculeatus	–	–	–	–	sonic morphology	Möbius 1889
R. aculeatus	–	–	–	–	sonic morphology	Sörensen 1894–1895
R. rectangulus	–	–	75–9,600	180 (mean)	disturbance	Salmon et al. 1968
R. rectangulus	–	–	150–1,200	100–200	disturbance	Salmon et al. 1968
Sufflamen bursa	–	–	–	–	chase, intraspecific agonism, disturbance	Salmon et al. 1968
S. frenatum	–	–	–	–	disturbance	Salmon et al. 1968

(continued)

TABLE 10.1. (continued)

Family/species	Sound pattern	No. sound types	No. pulses	Frequency range (hz)	Sound duration (ms)	Behavioral contexts	References
BATRACHOIDIDAE (4 species)							
Batrachomoeus trispinosus	long sound	4	–	–	285–6,077	undisturbed in aquarium	Rice & Bass 2009
B. trispinosus	short sound	–	–	–	276 (mean)	undisturbed in aquarium	Rice & Bass 2009
B. trispinosus	series of short sounds	–	–	–	28 per series (mean)	undisturbed in aquarium	Rice & Bass 2009
B. trispinosus	acoustic beats	–	–	–	147 (mean)	disturbance, undisturbed in aquarium	Rice & Bass 2009
Halophryne diemensis	–	–	–	low 85	1,000–3,000	disturbance	Whitley 1957
H. diemensis	–	–	–	–	–	sounds reported	Walls 1964
Opsanus beta	single pulses in series	2	–	100–1,800	30–70	reproductive	Tavolga 1958b
O. beta	tonal	–	–	200–2,100	160–410	reproductive	Fish & Mowbray 1959
O. beta	multi-note: pulsed & tonal	–	–	~200–900	214–1,409 (mean)	male-male agonism, male display on nest	Thorson & Fine 2002a
O. beta	single & double pulsed	–	–	–	–	reproductive	Thorson & Fine 2002b
O. beta	single-note tonal	–	–	~200–900	201–506 (mean)	male-male agonism, male display on nest	Tavolga 1958b
O. phobetron	multiple pulse series	–	–	~50–800	–	reproductive	Fish & Mowbray 1959, 1970
O. phobetron	pulsed & tonal	–	–	–	–	disturbance	Fish & Mowbray 1970

BLENNIIDAE (1 species) Hypsoblennius hentz	–	–	–	150–1,000	150–350	male courtship	Tavolga 1960
CAESIONIDAE (1 species) Pterocaesio digramma	–	–	–	–	–	pre-reproductive swimming aggregation	Yokoyama et al. 1994
CAPROIDAE (1 species) Capros aper	–	–	–	–	–	disturbance	Fish 1948
CARAPIDAE (7 species) Carapus acus	pulse bursts	–	–	1.7–22k	600–7,600 (sequence)	interaction inside holothurian host	Parmentier et al. 2003
C. acus	12–13 peaks/sound	–	–	250–1600	35 (mean)	silent inside sea star host, solitary spontaneous swim outside host	Parmentier et al. 2006a, c
C. mourlani	11–20 peaks/sound	–	–	570–100	16–30	compete for entrance into sea star	Parmentier et al. 2006c
C. boraborensis	10–28 sounds/sequence	3	–	55–800	83 (mean)	agonism,* social context inside sea cucumber	Lagardère et al. 2005
C. boraborensis	16–83 sounds/sequence	–	–	55–800	136 (mean)	agonism,* social context inside sea cucumber	Lagardère et al. 2005
C. boraborensis	11–30 pulses/sound	–	–	80–800	25–30 seconds	interaction inside holothurian host	Parmentier et al. 2003
C. homei	10 sounds / sequence	–	–	90–4,450	218 (mean)	agonism,* social context inside sea cucumber	Lagardère et al. 2005
C. homei	–	–	–	90–>10,000	262 (mean)	interaction inside holothurian host	Parmentier et al. 2003
Encheliophis gracilis	single beat	–	–	<600	362 (mean)	interaction inside holothurian host	Parmentier et al. 2003
E. gracilis	sequence	–	–	<600	<1,000	interaction inside holothurian host	Parmentier et al. 2003

(continued)

TABLE 10.1. (continued)

Family/species	Sound pattern	No. sound types	No. pulses	Frequency range (hz)	Sound duration (ms)	Behavioral contexts	References
Onuxodon margaritiferae	–	–	–	–	–	sonic morphology	Courtenay & McKittrick 1970
O. parvibrachium	–	–	–	–	–	sonic morphology	Courtenay & McKittrick 1970
CARANGIDAE (24 species)							
Alectis ciliaris	–	–	–	–	–	disturbance, escape	Fish 1954
A. crinitis	–	2	–	~400–1,500	~30–80	disturbance, escape	Fish & Mowbray 1970
Carangoides bartholomaei	–	2	–	–	–	electric stimulation only	Fish & Mowbray 1970
C. equula	–	–	–	–	–	disturbance	Uchida 1934
C. ruber	–	–	–	–	–	swimming	Moulton 1960
C. ruber	–	–	–	10–500	700	swimming	Cummings et al. 1964
C. ruber	–	–	–	–	–	disturbance	Fish & Mowbray 1970
Caranx crysos	–	–	–	–	–	disturbance	Moulton 1960
C. crysos	–	–	–	–	–	chase prey as a group, schooling group	Steinberg et al. 1965
C. crysos	–	2	–	~100–600	~100–120	disturbance, escape	Fish & Mowbray 1970
C. hippos	–	–	–	–	–	feeding	Fish 1948
C. hippos	–	–	–	–	–	disturbance	Burkenroad 1931
C. hippos	–	–	–	0–8,000	60	disturbance	Moulton 1958
C. hippos	–	–	–	–	–	disturbance, spontaneous solitary	Taylor & Mansueti 1960

Species						Behavior	Reference
C. hippos	–	–		~100–2,000	~60–80	disturbance	Fish & Mowbray 1970
C. latus	–	–		–	–	disturbance	Fish 1954
C. latus	–	–		–	–	swimming	Moulton 1960
C. latus	–	–	regular pattern	~100–1,200	~100	disturbance, escape, feeding	Fish & Mowbray 1970
Chloroscombrus chrysurus	–	–		–	–	disturbance	Fish 1954
C. chrysurus	–	–	regular pattern	~150–1,500	~400–600	disturbance	Fish & Mowbray 1970
Elagatis bipinnulata	–	–		–	–	escape response to sharks	Fish & Mowbray 1970
Oligoplites saurus	–	–		~100–600	~100–200	escape	Fish & Mowbray 1970
Selar crumenophthalmus	–	–	regular pattern	~400–1,600	~60–100	disturbance hook & line	Fish & Mowbray 1970
Selene brevoortii	–	–		–	–	disturbance	Fish 1948
S. declavifrons	–	–		–	–	spontaneous, continuous	Fish 1948
S. vomer	–	–		–	–	disturbance	Fish 1948, 1954
S. vomer	–	–		–	–	disturbance	Fish & Mowbray 1970
S. setipinnis	–	–		–	–	disturbance	Fish 1948, 1954
S. setipinnis	–	–		~100–1,500	~600–1,300	disturbance	Fish & Mowbray 1970
Seriola sp.	–	–		–	–	disturbance "speared"	Steinberg et al. 1965
Seriola dumerili	–	–		–	–	feeding	Matsuno et al. 1994
S. dumerili	–	–		~60–240	~100–200	feeding competition	Fish & Mowbray 1970
S. quinqueradiata	–	–		–	–	feeding	Kim 1977
S. zonata	–	–		–	–	disturbance, escape	Fish 1954
S. zonata	–	–		~50–180	~300–800	disturbance	Fish & Mowbray 1970
Trachinotus falcatus	–	–		–	–	escape, flee predator	Fish & Mowbray 1970
T. ovatus	–	–		–	–	disturbance	Fish 1954
T. ovatus	–	–		–	–	escape, feeding	Fish & Mowbray 1970
T. paitensis	–	–		–	–	dorsal fin display	Knudsen et al. 1948b

(*continued*)

TABLE 10.1. *(continued)*

Family/species	Sound pattern	No. sound types	No. pulses	Frequency range (hz)	Sound duration (ms)	Behavioral contexts	References
Trachinotus paitensis	–	–	–	–	–	sound recorded	Fish 1948
T. paitensis	–	–	–	–	–	swimming	Moulton 1960
Trachurus japonicus	–	–	–	–	–	feeding	Uchida 1934
T. trachurus	–	–	–	–	–	schooling	Shishkova 1958a, b
CHAETODONTIDAE (7 species)							
Chaetodon multicinctus	hydrodynamic	6	1	40–98	154 (mean)	resident aggressive	Tricas et al. 2006
C. multicinctus	single pulse	–	1	2449–5,190	10 (mean)	resident aggressive	Tricas et al. 2006
C. multicinctus	multipulsed	–	6	346–593	300 (mean)	resident aggressive	Tricas et al. 2006
C. multicinctus	single pulse	–	1	179–260	17 (mean)	resident & bottled intruder	Tricas et al. 2006
C. multicinctus	single pulse	–	1	113–137	115 (mean)	resident aggressive	Tricas et al. 2006
C. multicinctus	multipulsed	–	17	132–228	5,700 (mean)	agonism, mated pairs "alert call"*	Tricas et al. 2006
C. auriga	–	–	–	–	–	pairing behavior	Tricas & Boyle 2009
C. ocellatus	–	–	–	–	–	electric stimulation only	Fish & Mowbray 1970
C. ornatissimus	–	–	–	–	–	pairing behavior	Tricas & Boyle 2009
C. striatus	–	–	–	–	–	sonic morphology	Sörensen 1894–1895
C. striatus	–	–	–	–	–	electric stimulation only	Fish & Mowbray 1970
C. ulietensis	–	–	–	–	–	spontaneous in loose school	Lobel pers. comm.
Forcipiger flavissimus	single pulse	3	–	318 (peak)	21	head movement	Boyle & Tricas 2006
F. flavissimus	low-frequency pulse	–	–	41 (peak)	9	hydrodynamic fin movement	Boyle & Tricas 2006

	broadband frequency					
F. flavissimus	–	–	7,924 (peak)	3	hydrodynamic tail slap	Boyle & Tricas 2006
F. flavissimus	–	–	–	–	pairing behavior	Tricas & Boyle 2009
DACTYLOPTERIDAE (2 species)						
Dactyloptena orientalis	–	–	–	–	sonic morphology	Müller 1857
D. orientalis	–	–	–	–	sonic morphology	Bridge 1904; Fish 1948
D. orientalis	–	–	–	–	sonic morphology	Sörensen 1894–1895
D. orientalis	–	–	–	–	sonic morphology	Jordan & Richardson 1907; Fish 1948
Dactylopterus volitans	–	–	~100–800	~80–120	disturbance	Fish & Mowbray 1970
DIODONTIDAE (7 species)						
Chilomycterus atringa	multipulsed	–	–	–	disturbance defense inflation	Fish & Mowbray 1970
C. schoepfii	–	–	–	–	sonic morphology	Sörensen 1894–1895
C. schoepfii	–	–	–	–	feeding competition, disturbance defense inflation	Fish 1954
C. schoepfii	–	–	~100–1,500	~60–80	disturbance defense inflation	Fish & Mowbray 1970
C. spinosus	–	–	–	–	disturbance defense inflation	Burkenroad 1931
Dicotolichthys punctulatus	–	–	–	–	defense, feeding	Graham 1992
D. hystrix	–	–	–	–	disturbance defense inflation & non-air sound	Sörensen 1894–1895

(continued)

TABLE 10.1. *(continued)*

Family/species	Sound pattern	No. sound types	No. pulses	Frequency range (hz)	Sound duration (ms)	Behavioral contexts	References
Dicotolichthys hystrix	–	–	–	–	–	disturbance defense inflation	Uchida 1934; Fish 1948
D. hystrix	–	–	–	0–8,000	90	disturbance defense inflation, feeding	Moulton 1958
D. hystrix	–	–	–	~50–5,000	~30–50	disturbance defense inflation, feeding	Fish & Mowbray 1970
Diodon holocanthus	–	–	–	–	–	disturbance defense inflation, feeding	Uchida 1934; Fish 1948
Tragulichthys jaculiferus	–	–	–	–	–	disturbance defense inflation	Walls 1964
ENOPLOSIDAE (1 species)							
Enoplosus armatus	–	–	–	–	–	teeth grinding during normal activities	Graham 1992
EPHIPPIDAE (2 species)							
Chaetodipterus faber	–	2	–	–	–	disturbance	Burkenroad 1931
C. faber	–	–	–	220	150	in aquarium	Knudsen et al. 1948b
C. faber	–	2	–	75–150	–	free swimming in aquarium	Fish et al. 1952
C. faber	–	2	–	–	–	competitive feeding, escape	Fish & Mowbray 1970
C. ocellatus	–	2	–	–	–	electrical stimulation only	Fish & Mowbray 1970
GERREIDAE (2 species)							
Eucinostomus gula	–	–	–	–	–	disturbance in net	Fish & Mowbray 1970
Gerres cinereus	–	–	–	–	–	disturbance	Fish & Mowbray 1970

GOBIIDAE (2 species)

Bathygobius curacao	–	2	–	–	–	courting male	Stadler 2002
B. curacao	broadband	–	1	100–1,500	50	courting male	Myrberg & Stadler 2002
B. curacao	multipulsed	–	≤19	100–200 peak	35 per pulse	courting male	Myrberg & Stadler 2002
B. soporator	nonharmonic	–	–	100–500	150–350	courting male	Tavolga 1958a

HAEMULIDAE (14 species)

Anisotremus surinamensis	–	1	–	–	–	disturbance	Fish & Mowbray 1970
A. virginicus	multipulsed	1	–	~100–600	60–100	disturbance	Fish & Mowbray 1970
Haemulon album		–	–	–	–	feeding competition, disturbance, schools spontaneous	Fish & Mowbray 1970
H. album	burst feeding stridulation	–	–	50–1,600	–	feeding	Cummings et al. 1964; Kumpf 1964; Cummings et al. 1966
H. album	plankton feeding noise	–	–	20–700	–	feeding	Cummings et al. 1964; Kumpf 1964; Cummings et al. 1966
H. album	hydrodynamic swimming noise	–	–	10–500	–	swimming	Cummings et al. 1964; Kumpf 1964; Cummings et al. 1966
H. aurolineatum	multipulsed	–	–	~100–1,500	50–100	disturbance	Fish & Mowbray 1970
H. carbonarium	multipulsed	–	–	~100–1,000	30–80	disturbance	Fish & Mowbray 1970
H. flavolineatum	multipulsed	–	–			disturbance "mild annoyance" & netted	Fish & Mowbray 1970

(continued)

TABLE 10.1. (continued)

Family/species	Sound pattern	No. sound types	No. pulses	Frequency range (hz)	Sound duration (ms)	Behavioral contexts	References
Haemulon flavolineatum	–	1	–	–	–	disturbance, feeding	Moulton 1958
H. macrostomum	multipulsed	–	–	~100–600	~50–150	disturbance, escape	Fish & Mowbray 1970
H. melanurum	–	–	–	–	–	swimming	Cummings et al. 1964
H. melanurum	multipulsed	–	–	~50–900	~50–100	feeding competition, disturbance, school escape	Fish & Mowbray 1970
H. parra	multipulsed	–	–	–	–	disturbance	Fish & Mowbray 1970
H. plumierii	–	–	–	–	–	disturbance	Burkenroad 1930
H. plumierii	multipulsed	–	–	~100–1,200	~60–100	disturbance	Graham 1992
H. sciurus	–	–	–	–	–	disturbance	Burkenroad 1930
H. sciurus	rasp pulses	–	–	0–8,000	20–100	disturbance	Moulton 1958
H. sciurus	multipulsed	–	–	~100–1,100	80–150	disturbance, escape	Fish & Mowbray 1970
H. striatum	–	–	–	–	–	disturbance, escape	Fish & Mowbray 1970
Plectorhincus sp.	–	–	–	–	–	sounds reported	Fish 1948
Pomadasys maculata (=mauclatus)	–	–	–	–	–	sounds reported and recorded	Walls 1964
HEMIRAMPHIDAE (1 species)							
Hyporhamphus unifasciatus	–	–	–	–	–	disturbance	Burkenroad 1931
HOLOCENTRIDAE (12 species)							
Holocentrus adscensionis	–	–	–	–	–	sonic morphology	Fish 1948
H. adscensionis	–	–	–	–	–	sonic morphology	Myrberg & Stadler 2002
H. adscensionis	–	2	–	0–4,000	40–100	hydrophone attacked, disturbance, escape	Moulton 1958

H. adscensionis	–	–	–	–	–	sonic morphology	Winn & Marshall 1963
H. adscensionis	–	2	–	–	–	mobbing, crevice territory defense	Winn et al. 1964
H. adscensionis	multipulsed	–	–	–	–	agonism intraspecific, disturbance, escape	Fish & Mowbray 1970
H. adscensionis	–	3	–	–	–	–	Bright 1972
H. rufus	–	1	–	–	–	–	Winn & Marshall 1963
H. rufus	multipulsed		2–5	85–4,500	22–72	disturbance	Winn & Marshall 1963
H. rufus	–	2	–	–	–	chase, territorial, intra- & interspecific, disturbance, chorus	Winn et al. 1964
H. rufus	–	–	–	–	–	sonic morphology	Gainer et al. 1965
H. rufus	–	–	–	–	–	chase, territorial, intra- & interspecific	Salmon 1967
H. rufus	multipulsed		2–10	500	–	sonic morphology	Bright & Sartori 1972
H. rufus	–	–	–	–	–	sonic morphology	Carlson & Bass 2000
Myripristis amaena	–	–	–	–	–	chasing, nonterritorial schools, chorus	Nelson 1955
M. amaena	–	4	–	–	–		Salmon 1967
M. amaena	–	–	–	–	–	predator alarm calls	Popper et al. 1973
M. berndti	–	–	–	–	>3,000	agonism, congeners school over reefs	Salmon 1967
M. berndti	–	–	–	–	–	at predator (moray), disturbance	Salmon 1967

(continued)

TABLE 10.1. (continued)

Family/species	Sound pattern	No. sound types	No. pulses	Frequency range (hz)	Sound duration (ms)	Behavioral contexts	References
Myripristis berndti	–	–	–	–	10	chase small fish	Salmon 1967
M. berndti	–	–	–	–	–	disturbed by predator	Salmon 1967
M. berndti	multipulsed	–	7–10	75–4,800	80–118	disturbance	Salmon 1967
M. berndti	–	–	–	–	–	predator alarm calls	Popper et al. 1973
M. jacobus	–	–	–	–	–	sounds reported	Winn 1964
M. jacobus	multipulsed	–	3–5	–	–	disturbance	Fish & Mowbray 1970
M. jacobus	multipulsed	–	1–2	150–500	80–100	chase, intra- & interspecific, disturbance, chorus	Bright & Sartori 1972
M. jacobus	hydrodynamic	–	–	–	–	territorial defense	Bright 1972
M. pralinia	–	4	–	–	–	disturbance	Horch & Salmon 1973
M. violacea	–	–	–	–	–	agonism: circle, tail beat, chase, disturbance	Horch & Salmon 1973
M. violacea	multipulsed	–	many	–	–	agonism, disturbance: response to diver	Horch & Salmon 1973
M. violacea	3–7 groups	–	–	–	–	–	Horch & Salmon 1973
M. violacea	up to 10 groups	–	–	–	300–2,000	–	Horch & Salmon 1973
Neoniphon sammara	–	–	–	–	–	sonic morphology	Carlson & Bass 2000
Sargocentron cornutum	–	–	–	–	–	sonic morphology	Carlson & Bass 2000
S. coruscus	–	2	–	450–1,500	–	agonism, intra- & interspecific, chorus	Bright & Sartori 1972
S. coruscus	–	–	–	–	–	disturbance	Bright 1972
S. coruscus	–	3	–	–	–	agonism	Winn 1964
S. seychellense	–	–	–	–	–	sonic morphology	Carlson & Bass 2000
S. xantherythrum	–	–	2–9	–	–	courtship duet	Herald & Dempster 1957

					sounds reported	
KYPHOSIDAE (2 species)						
Atypichthys strigatus	–	–	–	–	sounds reported	Graham 1992
Kyphosus sectator	~50–250	~20	3	–	disturbance, spontaneous, two fish in tank	Fish & Mowbray 1970
LABRIDAE (4 species)						
Choerodon venustus	–	–	–	pulse series	swimming noise	Moulton 1964b
Halichoeres bivittatus	50–2,400	200 (series 3,000)	–	–	feeding	Steinberg et al. 1965
H. bivittatus	~100–300	~100–200	–	–	electrical stimulation only	Fish & Mowbray 1970
H. radiatus	–	–	–	–	feeding noises	Fish & Mowbray 1970
Lachnolaimus maximus	~100–200	~100	–	–	disturbance "body twist"	Fish & Mowbray 1970
LETHRINIDAE (1 species)						
Gymnocranius audleyi	none	none	–	–	swimming noises only	Moulton 1964b
LUTJANIDAE (7 species)						
Lutjanus analis	–	–	–	–	electrical stimulation only	Fish & Mowbray 1970
L. apodus	–	–	–	–	disturbance	Fish & Mowbray 1970
L. apodus	~100–400	~100–150	–	–	escape	Fish & Mowbray 1970
L. griseus	~50–100	500	–	–	disturbance = in nets	Fish & Mowbray 1970
L. griseus	~100–300	~100–200	3	–	disturbance = in nets	Fish & Mowbray 1970
L. jocu	–	~500	–	–	escape	Fish & Mowbray 1970
L. synagris	~50–200	~100–190	–	–	disturbance	Fish & Mowbray 1970
Ocyurus chrysurus	~80–300	~200–300	–	–	competitive feeding	Fish & Mowbray 1970
O. chrysurus	–	–	–	–	escape	Fish & Mowbray 1970
Rhomboplites aurorubens	–	–	2	–	electrical stimulation only	Fish & Mowbray 1970
MONACANTHIDAE (8 species)						
Aluterus scripta	–	–	–	–	disturbance	Fish & Mowbray 1970
A. shoepfi	~200–1,000	~80–1,200	–	–	competitive feeding, spontaneous	Fish & Mowbray 1970

(continued)

TABLE 10.1. *(continued)*

Family/species	Sound pattern	No. sound types	No. pulses	Frequency range (hz)	Sound duration (ms)	Behavioral contexts	References
Aluterus shoepfi	–	–	–	~200–500	~100	disturbance	Fish & Mowbray 1970
A. shoepfi	–	1	–	–	–	feeding competition, escape	Fish 1954
A. shoepfi	–	–	–	50–4,800	–	feeding	Fish et al. 1952
Cantherhines pardalus	–	–	–	–	–	sonic morphology	Sörensen 1894–1895
C. pullus	–	–	–	–	–	defensive trigger raised	Fish & Mowbray 1970
Monacanthus filicauda	–	–	–	–	–	sounds recorded	Walls 1964
Paramonacathus oblongus	–	–	–	–	–	sounds recorded	Walls 1964
Stephanolepis cirrhifer	–	–	–	–	–	disturbance, swimming, feeding	Fish 1948
S. hispidus	–	–	–	–	–	disturbance	Burkenroad 1931
S. hispidus	–	2	–	–	–	disturbance, defense, spine raising	Fish & Mowbray 1970
S. hispidus	–	–	–	–	–	feeding noise	Fish & Mowbray 1970
S. hispidus	–	2	–	–	–	feeding competition, disturbance, escape	Fish 1954
MONOCENTRIDAE (1 species)							
Monocentris japonica	–	–	–	100–600	–	disturbance	Onuki & Somiya 2007
MULLIDAE (2 species)							
Mulloidichthys martinicus	–	–	–	~50–450	~100	disturbance	Fish & Mowbray 1970
M. martinicus	–	–	–	–	–	feeding noises very low intensity	Bright 1972

Species							Reference
Pseudopeneus maculatus	–	–	–	–	–	feeding noises	Bright 1972
P. maculatus	–	–	–	~50–200	~100	disturbance	Fish & Mowbray 1970
OPHICHTHYIDAE							
Ophichthys spp.	–	–	–	–	–	disturbance & threat: "opercular vibrations"	Carlson & Bass 2000
OPLEGNATHIDAE (1 species)							
Oplegnathus fasciatus	–	–	–	160	13–20	territorial, chase (reproductive season), chorus	Nakazato & Takemura 1987
O. fasciatus	–	–	–	–	–	sound reported	Uchida 1934
OSTRACIIDAE (9 species)							
Acanthostracion quadricornis	multipulsed	–	–	1,000–3,400	80 and 140	chorus: multiple calls from sea grass beds most common	Steinberg et al. 1965
A. quadricornis	–	–	–	~50–350	~100	disturbance	Fish & Mowbray 1970
Lactophrys bicaudalis	–	–	–	–	–	feeding noises	Fish & Mowbray 1970
L. triqueter	–	–	–	–	–	disturbance	Fish & Mowbray 1970
L. triqueter	–	–	–	–	–	chewing food	Fish & Mowbray 1970
L. triqueter	–	–	–	–	–	swimming	Bright 1972
L. trigonus	–	–	–	–	–	disturbance*	Fish 1948
L. trigonus	–	–	–	~100–200	~300–700	disturbance, swimming noise	Fish & Mowbray 1970
Lactoria fornasini	–	–	–	–	–	spawning chants	Moyer 1979
Ostracion immaculatus	–	–	–	–	–		Uchida 1934
O. meleagris	tonal, frequency modulation	–	–	100–300	300	interspecific w/damselfish	Lobel 1996
O. meleagris	tonal & broadband	–	–	–	–	agonism	Lobel 1998

(continued)

TABLE 10.1. (*continued*)

Family/species	Sound pattern	No. sound types	No. pulses	Frequency range (hz)	Sound duration (ms)	Behavioral contexts	References
Ostracion meleagris	broadband	–	–	<4,000	9.9–10.6	combat: two males head-butting	Lobel 1996
O. meleagris	–	–	–	<1,000	130–209	male interrupt mating, disturbance	Lobel 1996
O. meleagris	tonal	–	–	<500	–	spawning sound	Lobel 1996
O. meleagris	low vibration	–	–	–	–	–	Pappe 1853
O. cubicus	–	–	–	–	–	swimming noise	Moulton 1964b
O. cubicus	–	–	–	–	–	sonic morphology	Sörensen 1894–1895
O. cubicus	–	–	–	–	–	sounds recorded	Walls 1964
Tetrosomus reipublicae	low frequency	–	–	–	–	sound reported	Graham 1992
PEMPHERIDAE (2 species)							
Pempheris japonica	–	–	–	–	–	disturbance	Uchida 1934; Fish 1948
P. japonica	–	–	–	–	–	sounds reported	Uchida 1934; Fish 1948
P. schwenkii	multipulsed	–	2–7	100–300	56 (mean)	disturbance: while schooling near diver	Takayama et al. 2003
P. schwenkii	–	–	–	100–300	60,75, 75	disturbance	Onuki & Somiya 2007
POLYNEMIDAE (1 species)							
Polydactylus virginicus	–	–	–	~100–500	~10–20	escape	Fish & Mowbray 1970
PLOTOSIDAE (1 species)							
Plotosus lineatus	–	–	–	–	–	sounds reported	Uchida 1934; Fish 1948
P. lineatus	–	–	–	–	–	disturbance	Burgess 1989

Species						
POMACANTHIDAE (7 species)						
Centropyge potteri	–	2	–	–	sounds reported	Lobel 1978
Holacanthus ciliaris	–	–	100–200	~100	electric stimulation only "body twist"	Fish & Mowbray 1970
H. ciliaris	–	–	100–400	~100	electric stimulation only	Fish & Mowbray 1970
H. ciliaris	–	–	<500	60–100	free in sea tank	Moulton 1958
H. isabelita	–	–	~50–300	~200	disturbance	Fish & Mowbray 1970
H. tricolor	–	–	~100–400	~100	escape	Fish & Mowbray 1970
Pomacanthus arcuatus	–	–	–	–	field monitoring	Fish & Mowbray 1970
P. arcuatus	–	–	<500	40–200	pair near hydrophone, feeding, startled, free in sea tank	Moulton 1958
P. arcuatus	–	–	–	–	interspecific "butting" by pair	Moulton 1969
P. arcuatus	–	–	–	200	pair call to each other	Moulton 1969
P. arcuatus	–	–	~100–300	~200	escape	Fish & Mowbray 1970
P. arcuatus	–	–	–	–	courtship	Moyer et al. 1983
P. paru	–	–	~100–400	~100	escape, feeding agonistic	Fish & Mowbray 1970
P. imperator	–	–	–	–	male chase female after spawn	Fourmanoir & Laboute 1976
P. imperator	–	–	–	–		Thresher 1982
P. imperator	multipulsed	3–12	101–3,778	30.8–128.3 per phrase	heterospecific chase, approach, lateral display	Amorim 1996a
POMACENTRIDAE (36 species)						
Abudefduf abdominalis	multipulsed	11 (mean)	–	1,793 (mean)	male courtship	Maruska et al. 2007
A. abdominalis	multipulsed	1–2	90–380	161 (mean)	aggressive, hydrodynamic quick body moves	Maruska et al. 2007
A. abdominalis	–	3–13	–	1,013 (mean)	aggressive	Maruska et al. 2007
A. abdominalis	–	6 (mean)	–	1,425 (mean)	male nest prep.	Maruska et al. 2007

(*continued*)

TABLE 10.1. *(continued)*

Family/species	Sound pattern	No. sound types	No. pulses	Frequency range (hz)	Sound duration (ms)	Behavioral contexts	References
Abudefduf abdominalis	–	–	6 (mean)	–	949 (mean)	male looping female	Maruska et al. 2007
A. abdominalis	–	–	1 (mean)	–	1,133 (mean)	male mouth-pushing in nest	Maruska et al. 2007
A. luridus	–	–	2	–	64–82	agonism	Santiago & Castro 1997
A. luridus	pulse series	–	–	–	700–1,000	agonism	Santiago & Castro 1997
A. luridus	multipulsed	–	15–20	–	15–20	agonism	Santiago & Castro 1997
A. saxatilis	–	1	–	–	–	feeding	Fish & Mowbray 1970
A. saxatilis	–	2	–	–	–	escape	Fish & Mowbray 1970
A. sordidus	multipulsed	–	5 (mean)	–	620 (mean)	agonism	Lobel & Kerr 1999
Amphiprion akallopisos (Madag.)	pulsed chirp	4	8 (mean)	665 (mean)	89 (train mean)	territory defense	Parmentier et al. 2005
A. akallopisos (Madag.)	single pulse	–	1	1,097 (mean peak)	8 (mean)	territory defense	Parmentier et al. 2005
A. akallopisos (Madag.)	single pulse	–	1	896 (mean peak)	13.3 (mean)	territory defense	Parmentier et al. 2005
A. akallopisos (Madag.)	single pulse	–	1	724 (mean peak)	12.8 (mean)	territory defense	Parmentier et al. 2005
A. akallopisos (Indon.)	single pulse	3	1	1,088 (mean peak)	7.4 (mean)	territory defense	Parmentier et al. 2005
A. akallopisos (Indon.)	single pulse	–	1	655 (mean peak)	11.7 (mean)	territory defense	Parmentier et al. 2005

A. akallopisos (Indon.)	single pulse	–	1	572 (mean peak)	12.7 (mean)	territory defense	Parmentier et al. 2005
A. akallopisos	multipulsed	–	7–11	–	–	agonism	Lagardère et al. 2003
A. akallopisos	–	–	–	–	–	sounds reported	Verwey 1930
A. (=Premnas) biaculeatus	–	–	–	.	–	sounds reported	Takemura 1983
A. bincinctus	–	–	–	–	–	sounds reported	Schneider 1964a
A. bincinctus	–	–	–	–	–	sounds reported	Chen & Mok 1988
A. chrysopterus	–	1	–	–	35–45	agonism: threat attack, submission	Onuki & Somiya 2007
A. chrysopterus	–	–	–	–	–	sounds reported	Fish & Mowbray 1970
A. clarkii	multipulsed train	–	1–8	450–800	26 (mean)	conspecific approaches territory	Parmentier et al. 2007
A. clarkii	–	3	–	500	45–60	fight	Burgess 1989
A. clarkii	–	–	–	500	250–400	agonism "shake"	Schneider 1964a
A. clarkii	–	–	–	600	25–30	threat	Schneider 1964a
A. clarkii	single pulse	–	1 pulse	–	–	agonism	Takemura 1983
A. clarkii	multipulsed	–	pulse series	–	–	agonism	Takemura 1983
A. clarkii	multipulsed	–	1–17	1,000–1,500	–	agonism	Chen & Mok 1988
A. clarkii	multipulsed	–	1–2	1,000–2,000	–	agonism	Chen & Mok 1988
A. clarkii	single pulse	–	1 pulse	–	–	agonism	Takemura 1983
A. frenatus	–	–	–	5,000–6,000	–	agonism	Takemura 1983
A. frenatus	–	–	–	200–600	–	agonism	Takemura 1983
A. frenatus	multipulsed	2	–	–	–	sounds reported	Fish 1948
A. frenatus	single pulse	–	1	–	56	agonism	Takemura 1983
A. frenatus	–	–	–	–	45–6	"fighting sound"	Schneider 1964a
A. frenatus	multipulsed	–	1–2	–	50	agonism	Chen & Mok 1988
A. frenatus	multipulsed	–	1–7	–	50	agonism	Chen & Mok 1988
A. melanopus	–	–	–	up to 8,000	–	agonism	Takemura 1983
A. melanopus	–	–	–	250–500	–	agonism	Takemura 1983

(*continued*)

TABLE 10.1. *(continued)*

Family/species	Sound pattern	No. sound types	No. pulses	Frequency range (hz)	Sound duration (ms)	Behavioral contexts	References
Amphiprion melanopus	single pulse	–	1	–	64	agonism	Takemura 1983
A. ocellaris	–	–	–	1.25–2.8 k	–	agonism	Takemura 1983
A. ocellaris	–	–	–	<2,500	–	agonism	Takemura 1983
A. ocellaris	multipulsed series	–	–	>3,000	–	agonism	Takemura 1983
A. ocellaris	single pulse	–	1	–	–	agonism	Takemura 1983
A. percula	–	–	–	–	–	threat, fight, shake, feeding	Schneider 1964a
A. perideraion	–	–	–	–	100	threat attack, submission	Allen 1972
A. perideraion	–	–	–	–	–	intraspecific agonism	Chen & Mok 1988
A. polymnus	–	–	–	–	–	sounds reported	Verwey 1930
A. polymnus	–	–	–	–	–	threat, fight, attack, feeding	Schneider 1964a
A. polymnus	–	–	–	–	–	agonism	Takemura 1983
A. polymnus	multipulsed	–	4–5	2.5–5 k	–	agonism	Takemura 1983
A. polymnus	single pulse	–	1	4,000	–	agonism	Takemura 1983
A. polymnus	single pulse	–	1	>8,000	–	agonism	Takemura 1983
A. sandaracinos	single pulse	–	1	200–3,500	–	agonism	Takemura 1983
A. sandaracinos	multipulsed	–	<3	200–3,500	–	agonism	Takemura 1983
A. sandaracinos	–	–	–	–	–	agonism	Takemura 1983
Chromis viridis	multipulsed	–	1–5	500–2,000	4.9–20.8	conspecific agonism	Amorim 1996b
C. chromis	single pulse	–	1	max. 340–1420	–	agonism, courtship	Picciulin et al. 2002
Chrysiptera leucopoma	–	–	–	–	–	sounds reported	Graham 1992
Dascyllus albisella	–	2	–	–	–	sounds reported	Fish 1948
D. albisella	multipulsed	–	1–14	–	4–25 per pulse	courtship	Oliver 2001

Species	Sound type					Context	Reference
D. albisella	multipulsed	–	6 (mean)	–	262	courtship	Lobel & Mann 1995a
D. albisella	multipulsed	–	3 (mean)	–	127	mating	Lobel & Mann 1995a
D. albisella	multipulsed	–	1–2	–	17	agonism	Mann & Lobel 1998
D. albisella	multipulsed	–	3–11	–	16	agonism	Mann & Lobel 1998
D. albisella	multipulsed	–	5 (mean)	–	–	courtship	Mann & Lobel 1998
D. albisella	multipulsed	–	–	–	–	agonism only, female sound	Mann & Lobel 1998
D. albisella	multipulsed	–	–	–	–	male courtship	Mann & Lobel 1997
D. albisella	multipulsed	–	3–10	–	–	courtship	Schneider 1964a
D. albisella	multipulsed	–	–	401±4	45.6±0.4	courtship	Parmentier et al. 2009
D. aruanus	–	–	–	–	–	sounds reported	Avidor 1974
D. aruanus	–	–	–	–	–	sounds reported	Graham 1992
D. aruanus	single pulse	–	1	>4,000	6.4	agonism	Parmentier et al. 2009
D. aruanus	multipulsed	–	12–42	–	0.8	agonism	Parmentier et al. 2006a
D. aruanus	multipulsed	–	3–9	474 (max mean)	<400	courtship	Parmentier et al. 2006a
D. aruanus	–	–	–	466±4	30±0.3	courtship	Parmentier et al. 2009
D. carneus	–	–	–	–	–	sounds reported	Koenig 1957
D. flavicaudus	multipulsed	–	3–10	348±2	48±0.4	courtship	Parmentier et al. 2009
D. marginatus	–	–	–	–	–	signal jump courtship	Avidor 1974
D. marginatus	–	–	–	–	–	reproductive*	Holzberg 1973
D. trimaculatus	–	–	–	–	–	sounds reported	Fish 1948
D. trimaculatus	–	–	–	–	–	sounds reported	Spanier 1970
D. trimaculatus	single pulse	–	1	up to 10000	13–55	agonism only, none	Luh & Mok 1986
D. trimaculatus	multipulsed	–	3–6	up to 10000	13–55	agonism only, none	Luh & Mok 1986
D. trimaculatus	multipulsed	–	3–6	up to 10000	13–55	agonism only, none	Luh & Mok 1986
D. trimaculatus	multipulsed	–	3–11	465±5	49±1	courtship	Parmentier et al. 2009
Hypsypops rubicundus	–	–	–	7,400	11	sounds reported	Knudsen et al. 1948b
H. rubicundus	–	–	–	75–100	–	adult group in an aquarium	Dobrin 1947
H. rubicundus	–	–	–	–	–	reported sounds	Fish 1948

(continued)

TABLE 10.1. (continued)

Family/species	Sound pattern	No. sound types	No. pulses	Frequency range (hz)	Sound duration (ms)	Behavioral contexts	References
Hypsypops rubicundus	–	–	–	–	–	reported sounds	Limbaugh 1964
H. rubicundus	–	3	–	–	–	competitive feeding	Fish & Mowbray 1970
Microspathodon chrysurus	–	–	–	–	–	disturbance	Emery 1973
Plectroglyphidodon lacrymatus	multipulsed	–	2–5	100–1,000	56	agonism	Parmentier et al. 2006a
Pomacentrus nagasakiensis	–	–	–	–	–	enticement	Moyer 1975
Premnas biaculeatus	–	–	–	–	–	territorial, intraspecific agonism	Takemura 1983
Stegastes dorsopunicans (=adustus)	–	3	–	–	–	territorial defense	Burke & Bright 1972
S. dorsopunicans (=adustus)	multipulsed	–	3–9	–	–	male courtship	Spanier 1979
S. dorsopunicans (=adustus)	multipulsed	–	–	–	–	male courtship	Albrecht 1981
S. dorsopunicans (=adustus)	–	–	–	–	–	male courtship	Albrecht 1984
S. fuscus	–	–	–	–	–	male courtship display	Dobrin & Loomis 1943
S. fuscus	–	–	–	–	–	sounds reported	Myrberg 1972a
S. leucostictus	–	–	–	0–1,500	20	male pursue/chase others	Moulton 1958
S. leucostictus	–	–	–	–	–	chase, competitive feeding	Fish & Mowbray 1970
S. leucostictus	multipulsed	–	2–6	–	–	male courtship	Spanier 1979
S. leucostictus	multipulsed	–	–	–	–	male courtship	Albrecht 1981
S. leucostictus	–	–	–	–	–	male courtship	Albrecht 1984
S. leucostictus	14-chirp series	–	–	50–2,000	150	chase	Steinberg et al. 1965
S. leucostictus	multipulsed	–	4	–	7–10	agonism	Myrberg & Spires 1972

S. partitus	—	—	—	—	—	male courtship	Myrberg & Spires 1972
S. partitus	single pulse	—	1	—	10–40	agonism	Myrberg 1972a
S. partitus	multipulsed	—	3–6	—	8–15	agonism	Myrberg & Spires 1972
S. partitus	multipulsed	—	3	<50–2,000	10–20 (120–240 sequence)	colony of individuals	Myrberg 1972a
S. partitus	multipulsed	—	4–6	<50–2,000	10–20 (160–500 sequence)	colony of individuals	Myrberg 1972a
S. partitus	—	—	variable	<50–1,300	10–20	colony of individuals	Myrberg 1972a
S. partitus	multipulsed	—	8–12	350–1,000	20–30	colony of individuals	Myrberg 1972a
S. partitus	single pulse	—	1	<50–2,000	10–30	colony of individuals	Myrberg 1972a
S. partitus	—	—	variable	<50–10,000	20–30	colony of individuals	Myrberg 1972a
S. partitus	—	—	—	—	—	male courtship	Graham 1992
S. partitus	multipulsed	—	1–4	—	—	male courtship	Spanier 1979
S. partitus	multipulsed	—	—	—	—	male courtship	Albrecht 1981
S. partitus	—	—	—	—	—	male courtship	Albrecht 1984
S. partitus	—	—	—	—	—	territorial display, chasing, male courtship	Myrberg & Riggio 1985
S. partitus	multipulsed	—	3	—	—	male courtship	Myrberg & Riggio 1985
S. partitus	—	—	—	—	—	male courtship	Myrberg et al. 1986
S. partitus	—	—	—	—	—	male courtship	Myrberg et al. 1993
S. partitus	—	2	—	—	—	courtship, pre-mating	Kenyon 1994
S. partitus	—	—	—	—	—	territory "keep-out" signal	Myrberg 1997a
S. planifrons	multipulsed	—	—	—	—	male courtship	Albrecht 1981
S. planifrons	—	—	—	—	—	male courtship	Albrecht 1984
S. planifrons	multipulsed	—	2–6	—	—	male courtship	Spanier 1979

(continued)

TABLE 10.1. *(continued)*

Family/species	Sound pattern	No. sound types	No. pulses	Frequency range (hz)	Sound duration (ms)	Behavioral contexts	References
Stegastes planifrons	multipulsed	–	4	–	7–12	agonism	Myrberg & Spires 1972
S. variabilis	–	several	–	–	–	male courtship dip	Myrberg & Spires 1972
PRIACANTHIDAE (3 species)							
Heteropriacanthus cruentatus	multipulsed	–	–	75–1,200	76–839	chased; 6–10 in caves, disturbance	Salmon & Winn 1966
Priacanthus meeki	multipulsed	–	–	–	64–318	chased; 6–10 in caves, disturbance	Salmon & Winn 1966
P. macracanthus	–	–	–	–	–	unknown	Moulton 1962
P. macracanthus	–	–	–	–	–	sonic morphology	Walls 1964
SCARIDAE (11 species)							
Scarus coeruleus	–	–	–	–	~100	escape	Fish & Mowbray 1970
S. coelestinus	–	–	–	–	–	escape when prodded	Fish & Mowbray 1970
S. guacamaia	–	–	–	~100–600	~100–200	disturbance	Fish & Mowbray 1970
S. iseri	–	–	–	–	–	escape held by tail	Fish & Mowbray 1970
S. iseri	hydrodynamic	–	–	30–1,200	750	mating sound (= noise), schools spawning noise	Lobel 1992
S. vetula	–	–	–	–	–	escape fast turn in field pen, feeding noise	Fish & Mowbray 1970
Sparisoma aurofrenatum	–	–	–	–	–	disturbance, feeding noise	Fish & Mowbray 1970
S. cretense	–	–	–	–	–	feeding noises	Board 1956

Species	Sound type		No. pulses		Behavior	Reference
S. chrysopterum	-	-	-	-	escape	Fish & Mowbray 1970
S. radians	-	-	-	-	spontaneous, solitary individual	Fish & Mowbray 1970
S. rubripinne	-	-	-	-	disturbance, feeding noises	Fish & Mowbray 1970
Sparisoma viride	-	-	-	-	escape, feeding	Fish & Mowbray 1970
SCIAENIDAE (17 species)						
Atractoscion nobilis	sound series	-	-	7–55 sec	spawning chants	Aalbers & Drawbridge 2008
A. nobilis	multipulsed	-	-	473 (mean)	courtship	Aalbers & Drawbridge 2008
A. nobilis	pulsed series	-	-	697–1450 (mean)	courtship	Aalbers & Drawbridge 2008
A.nobilis	multipulsed	-	-	230–360	spawning	Aalbers & Drawbridge 2008
A.nobilis	non-pulsed	-	-	200–310	spawning	Aalbers & Drawbridge 2008
A.nobilis	non-pulsed	13–450	-	-	spawning , burst swimming	Aalbers & Drawbridge 2008
A. nobilis	sound series	-	-	5–45 sec	–	Aalbers & Drawbridge 2008
Bairdiella chrysoura	-	-	-	-	reproductive aggregations	Locascio & Mann 2008
Cynoscion arenarius	multipulsed	~200–500	2–12	235±93 (mean ± SD)	chorusing	Locascio & Mann 2008
C. nebulosus	multipulsed	100–1,150	2	140–210	courtship, spawning	Gilmore 2003
C. nebulosus	multipulsed	100–1,150	3–5	140–450	courtship, spawning	Gilmore 2003
C. nebulosus	single pulse	90–1,150	1	175–367	courtship, spawning	Gilmore 2003
C. nebulosus	multipulsed	30–1,300	>16	822.5	courtship, spawning	Gilmore 2003
C. nebulosus	-	-	-	-	reproductive aggregations	Locascio & Mann 2008

(continued)

TABLE 10.1. (continued)

Family/species	Sound pattern	No. sound types	No. pulses	Frequency range (hz)	Sound duration (ms)	Behavioral contexts	References
Cynoscion regalis	multipulsed	–	6–10	–	–	disturbance	Connaughton et al. 2002
C. xanthulus	multipulsed	–	4–13	400–1,200	~50	reproductive aggregations	Fish & Cummings 1972
Equetus lanceolatus	–	–	–	–	–	sonic morphology	Schneider & Hasler 1960
Johnius australis	–	–	–	–	–	sounds recorded	Walls 1964
J. belengerii	–	–	–	–	–	disturbance	Lin et al. 2007
J. macrorhynus	multipulsed	–	8–28	–	111–281 (143 min display)	disturbance, field recordings	Lin et al. 2007
J. macrorhynus	multipulsed	–	2	–	46 (mean)	disturbance, field recordings	Lin et al. 2007
J. tingi	–	–	–	–	–	disturbance	Lin et al. 2007
Odontoscion dentex	–	–	–	–	–	escape	Fish & Mowbray 1970
Otolithes ruber	–	–	–	–	–	disturbance	Lin et al. 2007
Pareques acuminatus	–	2	–	–	–	disturbance	Fish & Mowbray 1970
Pennahia macrocephalus	–	–	–	–	–	disturbance	Lin et al. 2007
P. pawak	–	–	–	–	–	disturbance	Lin et al. 2007
Sciaenops ocellata	–	–	–	–	–	male courtship display, chorus	Guest & Lasswell 1978
S. ocellata	–	–	–	240–1,000	–	courtship, spawning	Holt 2002
S. ocellata	–	2	–	–	–	disturbance	Fish & Mowbray 1970

SCORPAENIDAE (2 species)						
Scorpaena plumieri	–	–	–	–	heterospecific in cave territory	Fish & Mowbray 1970
Scorpaenopsis gibbosa	–	–	–	–	sounds reported	Fish & Mowbray 1970
SEBASTIDAE (1 species)						
Sebastiscus marmoratus	–	–	85–125	100–160	agonism	Miyagawa & Takemura 1986
S. marmoratus	–	multiple	–	–	agonism	Miyagawa & Takemura 1986
SERRANIDAE (18 species)						
Alphestes afer	–	2	–	–	disturbance, escape	Fish & Mowbray 1970
Cephalopholis fulva	–	2	–	–	disturbance, escape	Fish & Mowbray 1970
C. cruentata	–	–	–	–	chase heterospecific	Bright 1972
C. cruentata	–	2	–	–	electrical stimulation only	Fish & Mowbray 1970
Diplectrum formosum	–	2	–	–	disturbance, escape	Fish & Mowbray 1970
Epinephelus adscensionis	–	2	–	–	feeding competition, disturbance, escape	Fish & Mowbray 1970
E. adscensionis	–	2	<900	40–100	retreat from hydrophone, disturbance, swimming accompanying feeding	Moulton 1958
E. drummondhayi	–	–	–	–	feeding competition, flee predator, escape	Fish & Mowbray 1970
E. guttatus	–	–	–	–	feeding competition interspecific, disturbance	Fish & Mowbray 1970
E. itajara	single pulse	1	–	132 (mean)	spawning aggregations	Mann et al. 2008
E. itajara	–	–	–	–	disturbance, flee predator	Fish & Mowbray 1970
E. morio	–	–	–	–	feeding competition	Fish & Mowbray 1970
E. nigritus	–	2	–	–	disturbance, escape	Fish & Mowbray 1970
E. striatus	–	3	–	–	feeding competition, feeding	Fish et al. 1952

(continued)

TABLE 10.1. (continued)

Family/species	Sound pattern	No. sound types	No. pulses	Frequency range (hz)	Sound duration (ms)	Behavioral contexts	References
Epinephelus striatus	–	1	–	0–2,000	100–200	retreat from hydrophone, quickened swimming accompanying feeding	Moulton 1958
E. striatus	sound series	–	–	–	–	unknown	Moulton 1958
E. striatus	–	–	–	–	–	disturbance	Moulton 1958
E. striatus	–	–	–	–	–	feeding competition	Fish & Mowbray 1970
E. striatus	–	–	–	–	–	escape (fleeing predator)	Bright 1972
E. striatus	–	–	–	low <600	50–125	agonism, disturbance	Moulton 1958
Hypoplectrus unicolor	multipulsed	3	–	500	200–1,500	courtship	Lobel 1992
H. unicolor	frequency modulated	–	–	600–200	150	mating sound	Lobel 1992
H. unicolor	broadband	–	–	350–1,650	1,250	mating (gamete release)	Lobel 1992
Mycteroperca bonaci	multipulsed	–	–	–	–	disturbance	–
M. bonaci	multipulsed	–	4–5	100–400	~150	territorial:* ariids swim by cave, disturbance	Tavolga 1960
M. bonaci	multipulsed	–	4–6	~40–100	~500	disturbance, escape, spontaneous	Fish & Mowbray 1970
M. microlepis	–	–	–	–	–	electrical stimulation only	Fish & Mowbray 1970
M. venenosa	–	–	–	–	–	disturbance, feeding	Fish & Mowbray 1970
Rypticus bistrispinus	–	–	–	–	–	electrical stimulation only	Fish & Mowbray 1970
R. saponaceus	–	–	–	–	–	electrical stimulation only	Fish & Mowbray 1970
Serranus tigrinus	–	–	–	–	–	electrical stimulation only	Fish & Mowbray 1970

SILLAGINIDAE (1 species)

| Sillago maculata | – | – | – | – | – | sounds recorded | Walls 1964 |

Family / Species					Behavior	Reference
SPARIDAE (8 species)						
Archosargus sp.	—	—	5,800	26	feeding noise*	Knudsen et al. 1948a
A. probatocephalus	—	—	—	—	feeding noise	Fish & Mowbray 1970
A. rhomboidalis	—	—	—	—	escape	Fish & Mowbray 1970
Calamus bajonad	—	—	~50–400	~100	escape	Fish & Mowbray 1970
C. calamus	—	—	~100–400	~50–100	escape	Fish & Mowbray 1970
C. calamus	—	2	—	—	disturbance in net	Fish & Mowbray 1970
C. penna	—	—	—	—	escape	Fish & Mowbray 1970
Diplodus argenteus	—	—	—	—	disturbance, escape	Fish & Mowbray 1970
Lagodon rhomboides	—	—	2,000–3,000	50	chase intraspecific	Caldwell & Caldwell 1967
SPHYRAENIDAE (1 species)						
Sphyraena barracuda	—	2	—	—	swimming	Fish & Mowbray 1970
SYNANCEIDAE (1 species)						
Erosa erosa	—	—	—	—	sound recorded	Walls 1964
SYNGNATHIDAE (4 species)						
Hippocampus erectus	—	—	—	—	male-female duet breeding	Dufossé 1874
H. erectus	—	—	—	—	feeding noise	Colson et al. 1998
H. erectus	—	—	—	—	feeding noise	James & Heck 1994
H. erectus	—	—	<50–4,800	—	spontaneous new surroundings	Fish 1953
H. erectus	single pulse	—	<50–4,800	—	unknown	Fish 1954
H. erectus	pulse series	2–5	50–4,800	—	spontaneous, solitary individual	Fish 1954
H. erectus	pulse series	2–5	<50–1,600	—	unknown	Fish et al. 1952
H. erectus	—	—	~100–1,200	~200	male-female duet, explore new surroundings "orientation"	Fish & Mowbray 1970

(continued)

TABLE 10.1. (continued)

Family/species	Sound pattern	No. sound types	No. pulses	Frequency range (hz)	Sound duration (ms)	Behavioral contexts	References
Hippocampus zostera	–	–	–	2,650–3,430 peak	–	feeding noise	Colson et al. 1998
Syngnathus floridae	–	–	–	–	–	feeding chorus	Ripley & Foran 2007
S. acus	–	–	–	–	–	disturbance*	Burkenroad 1931
TERAPONTIDAE (4 species)							
Pelates quadrilineatus	–	–	–	–	–	sound reported	Moulton 1964b
P. quadrilineatus	–	–	–	–	–	sound reported	Graham 1992
Therapon (=Terapon) jarbua	multipulsed	–	3–30	800	10, 60–150	agonism	Schneider 1964b
T. (=Terapon) jarbua	sound series	–	–	–	5,000–10,000 sec	agonism	Schneider 1964b
T. puta	–	–	–	–	–	sound recorded	Walls 1964
T. theraps	–	–	–	–	–	chase	Hardenberg 1934
T. theraps	multipulsed	–	9–12	–	79–105	free-ranging in ocean	McCauley & Cato 2000
T. theraps	–	–	–	–	7	fight	Schneider 1964b

TETRAODONTIDAE (6 species)

Amblyrhynchotes honckenii	–	–	disturbance defense inflation	Pappe 1853
Canthigaster rivulata	–	–	sound reported	Uchida 1934
Lagocephalus sceleratus	–	–	sonic morphology	Moulton 1964b
Sphoeroides nephelus	–	–	defense inflation (during & after)	Burkenroad 1931
S. spengleri	–	–	defense inflation (during & after)	Fish 1948
S. spengleri	0–8,000	–	defense inflation	Fish & Mowbray 1970
S. spengleri	–	70	disturbance	Moulton 1958
S. testudineus	~40–6,000	~80–1,100	defense inflation	Fish & Mowbray 1970

TETRAROGIDAE (2 species)

Centropogon australis	–	–	sounds recorded	Walls 1964
C. marmoratus	–	–	sounds recorded	Walls 1964

TRIGLIDAE (1 species)

Lepidotrigla argus	–	–	sounds recorded	Walls 1964

* Based on authors' interpretation of the published data.

TABLE 10.2 Interspecific and intraspecific acoustic properties of tropical fish sounds

Family	No. species recorded	Frequency range (hz)	Sound duration (ms)	Behavioral contexts
Acanthuridae	1	150–4,700	100	agonism
Ariidae	2	100–4,000	10–550	disturbance, schooling chorus
Balistidae	7	50–9,600	20–2,400	disturbance, defense, feeding competition, territoriality
Batrachoididae	4	50–2,100	30–6,077	disturbance, agonism, reproduction
Blenniidae	1	150–1,000	150–350	reproduction
Carangidae	11	10–8,000	30–1,300	disturbance, escape, feeding competition
Carapidae	5	55–4,450	16–362	agonism hypothesized
Chaetodontidae	2	40–5,190	3–5,700	agonism, reproduction
Dactylopteridae	2	~100–800	~80–120	disturbance
Diodontidae	2	50–8,000	30–90	disturbance defense display
Ephippidae	1	75–220	150	disturbance, spontaneous in aquarium
Gobiidae	2	100–1,500	35–350	reproduction
Haemulidae	8	~50–8,000	20–200	disturbance, escape, feeding competition, schooling
Holocentridae	8	75–4,800	10–3,000	disturbance agonism, reproduction
Kyphosidae	1	50–250	20	"spontaneous" in social group
Labridae	2	~100–300 [50–2,400]	~100–200 [200]	escape, [feeding noise]
Lutjanidae	10	~50–400	~100–500	escape, feeding competition
Monacanthidae	1	200–1,000	~80–1,200	disturbance, feeding competition,
Monocentridae	1	100–600	unknown	disturbance
Mullidae	2	~50–450	~100	escape
Oplegnathidae	1	160	13–20	agonism, reproductive chorus
Ostraciidae	3	~50–4,000	10–700	disturbance, agonism, spawning, chorus
Pempheridae	1	100–300	56–75	disturbance, agonism
Polynemidae	1	~100–500	~10–20	escape
Pomacanthidae	6	100–3,800	31–200	escape, agonism, reproduction
Pomacentridae	23	200–7,400	7–1,793	agonism, reproduction
Priacanthidae	2	75–1,200	64–839	disturbance, agonism
Scaridae	3	30–1200	~100–750	disturbance, escape, spawning
Sciaenidae	7	13–1,300	5–1,450	disturbance, escape, reproductive chorus aggregation
Sebastidae	1	85–125	100–160	agonism
Serranidae	5	~40–2,000	40–200	disturbance, agonism, chorus spawning

TABLE 10.2 *(continued)*

Family	No. species recorded	Frequency range (hz)	Basic sound unit duration (ms)	Behavioral contexts
Sparidae	4	~50–5,800	26–100	escape, agonism
Syngnathidae	2	100–4,800 [2.6–3.4k]	~200 [5–20]	reproduction "duet," solitary [feeding noise]
Terapontidae	2	200–1,500	7–105	agonism, spontaneous in ocean
Tetraodontidae	2	40–8,000	70–1,100	disturbance, defense display

Approximations "~" are from Fish & Mowbray (1970) spectrograms when no other data was available. Square brackets "[ ]" indicate feeding noise data. "Disturbance" in this table refers only to artificially manipulated or restrained specimens. Only contexts with recorded data are noted.

TABLE 10.3 Possible information included in fish acoustic signals, in order of increasing complexity of signal interpretation

Message	Acoustic clue
Mate location	Sound occurrence
Readiness to spawn Synchronization of gamete release (= Mating or spawning sound)	Sound occurrence
Vigor/aggressiveness	Duration of call and/or call repetition rate
Individual size	Dominant frequency
Species identity	Variation in pulse repetition rate in a call, number of pulses in a call, variation in pulse amplitude, call duration, plus color patterns and behavior
Individual identity	Combination of all above clues, plus other features of behavior

TABLE 10.4 Coral reef fish congeners whose sounds have been recorded and for which comparisons of temporal patterns and frequency can be made

Family/ genus	No. species	Species	Behavioral contexts	Species sounds differed in . . .
Balistidae (Salmon et al. 1968)				
Melichthys	3	*buniva, niger, vidua*	agonism & disturbance	waveform
Balistes	3	*bursa, capistratus, vetula*	agonism & disturbance	waveform
Batrachoididae (Fish & Mowbray 1959)				
Opsanus	2	*beta, phobetron*	reproductive	sound pulse number per display
Carapidae (Courtenay & McKittrick 1970; Lagardère et al. 2005; Parmentier et al. 2006b)				
Carapus	4	*acus, boraborensis, mourlani, homei*	agonism?	pulse duration, interpulse interval
Gobiidae (Stadler 2002)				
Bathygobius	2	*curacao, soporator*	male courtship display	amplitude, duration
Holocentridae (Winn et al. 1964; Bright & Sartori 1972; Popper et al. 1973)				
Holocentrus	2	*adscensionis, rufus*	disturbance	sounds similar
Myripristis	3	*brendti, jacobus, violacea*	agonism	sounds similar
Pomacanthidae (Thresher 1982; Moyer et al. 1983; Amorim 1996a)				
Pomacanthus	2	*arcuatus, imperator*	agonism, pair interactions	duration
Pomacentridae (e.g., Parmentier et al. 2006a; Maruska et al. 2007; Parmentier et al. 2009)				
Abudefduf	3	*abdominalis, luridus, sordidus*	male courtship	pulse number, duration
Amphiprion	8	*alkallopisos, chrysopterus, clarkii, frenatus, melanopus, ocellaris, polymnus, sandaracinos*	agonism	sound similar
Chromis	2	*viridis, chromis*	agonism, male courtship	frequency
Dascyllus	4	*albisella, aruanus, flavicaudus, trimaculatus*	agonism, male courtship	pulse number & interval, duration

(*continued*)

TABLE 10.4 *(continued)*

Family/genus	No. species	Species	Behavioral contexts	Species sounds differed in . . .
Stegastes	4	*dorsopunicans, leucostictus, partitus, planifrons*	male courtship display	pulse number & interval, duration
Sciaenidae (Ramcharitar et al. 2006; Lin et al. 2007; Locascio & Mann 2008)				
Cynoscion	2	*arenarius, nebulosus*	courtship, spawning	pulse repetition rate
Johnius	3	*belangeri, macrorhynus, tingi*	disturbance	pulse repetition rate

AFTERWORD

At its inception, this volume was envisaged as an opportunity to shine a spotlight on a number of topics associated with reproductive biology among marine fishes. Many of the topics have either received scant coverage in the past or were ready for a comprehensive and updated treatment. Accordingly, contributions to the volume were chosen to fill in previous gaps in our knowledge, to present new approaches to existing problems, and to generate new testable models for questions related to the evolution of reproductive and sexual biology in marine fishes.

As a topic, the biology of vertebrate reproduction encompasses a wide variety of subdisciplines ranging from the study of molecules, cells, tissues, and organ systems to that of individuals, populations, metapopulations, and species. Among a considerable number of marine fish taxa, an additional (and unusual, for vertebrates) feature is that phenotypic sexuality is not always dictated by genotype. In these cases, sexual function involves the strategic apportionment of ova and sperm production within a single individual, and the nature of expressed sexual patterns reflects the combined influences of social system, reproductive mode, and reproductive advantage. Sexuality, instead of being fixed, is labile, and sexual patterns are characterized by functional hermaphroditism. Not surprisingly, labile sexual patterns found among marine fish taxa are inextricably intertwined with their reproductive biology. When viewed from several different perspectives as provided by a number of chapters within this volume, it becomes clear that understanding the interrelationships between reproduction and sexuality is central to understanding the significant evolutionary and adaptive success of a number of functionally hermaphroditic marine fish taxa.

This volume also brings a new focus to some relatively small-sized and inconspicuous marine fish taxa. There has always been a fascination with the classical archetype of a coral reef fish comprising an exotic body shape and brilliant poster coloration. This is reflected in the many early studies of reproductive biology that focused on some of the more conspicuous fish species occupying accessible marine environments. Recently however, there has been increasing recognition that a significant portion of biodiversity, biomass, trophic cycling, and energy flow within complex marine fish communities is linked to small, and typically cryptic, fishes. These fishes—which are usually either underrepresented in, or entirely excluded from, fish surveys—constitute a hidden underworld, the reproductive biology of which remains predominantly unknown. Accordingly, this volume highlights a number of aspects of the reproductive biology and sexuality found among some of the smallest and least conspicuous of marine fish taxa, the blennioids and gobiids.

A recurring theme provides a subtext for many of the included chapters: What is the present state of our ocean environments, and how will marine fishes fare with possible changes to the ocean environments of tomorrow? A number of chapters are particularly timely in this regard. Increasing fishing pressure is raising concerns regarding the sustainability of commercially important species at present harvesting rates. Anthropogenic influences on marine environments are increasingly becoming associated with physiological and behavioral abnormalities among marine community inhabitants. The steadily increasing wealth of data documenting substantial changes in saltwater environments in terms of pH, salinity, CO_2 levels, and temperatures—the latter at all depths from sea surface to bottom water—signal significant ongoing changes in our oceans. And yet, with a few exceptions, we know very little regarding the biology, and particularly the reproductive biology, of most marine fishes.

As our awareness of marine environments has grown over the last several decades, it is evident that our preconceptions as to the constancy and apparent invulnerability of these aquatic regions have been misplaced. Marine environments are changing, and the distribution and abundance of marine organisms are undergoing changes as well. Our ability to make effective decisions regarding local and global issues that bear directly on the health and sustainability of marine environments and their resident communities depends directly on our ability to make informed choices. And that ability rests on having a sound understanding of the biological interrelationships within and among marine habitats and marine communities.

Collectively, the chapters included here offer new insights into patterns and processes of reproduction and sexuality among marine fishes. The contents, in providing the latest information available in the covered topics, hold the promise of suggesting new approaches to existing problems and point the way for future research. It is hoped that the efforts of the contributing authors to produce this volume will not only advance our knowledge of reproductive biology among marine fishes, but will also move us a step closer to the optimistic goal of ensuring a future for all marine fishes.

Kathleen S. Cole, editor

INDEX

Citations ending with a "t" indicate a table; those with an "f" indicate a figure.